CONCRETE

PLANET

ROBERT COURLAND

CONCRETE PLANET

The Strange and Fascinating Story
of the World's Most Common
Man-Made Material

Prometheus Books

Essex, Connecticut

Prometheus Books

An imprint of Globe Pequot, the trade division of
The Rowman & Littlefield Publishing Group, Inc.
4501 Forbes Blvd., Ste. 200
Lanham, MD 20706
www.rowman.com

Distributed by NATIONAL BOOK NETWORK

British Library Cataloguing in Publication Information available

Library of Congress Cataloging-in-Publication Data

Courland, Robert, 1952–
Concrete planet : the strange and fascinating story of the world's most common man-made
    material / by Robert Courland.
p. cm.
Includes bibliographical references and index.
1. Concrete—History. I. Title.
TA439.C588 2011
620.1'36—dc23

                                                                                        2011028696

ISBN 978–1–63388–816–6 (pbk : alk. paper)
ISBN 978–1–63388–869–2 (ebook)

∞™ The paper used in this publication meets the minimum requirements of American
National Standard for Information Sciences—Permanence of Paper for Printed Library
Materials, ANSI/NISO Z39.48-1992

Contents

List of Illustrations

ACKNOWLEDGMENTS

A couple years ago, a close friend asked me what kind of book I was working on. I told him, "It's the story of concrete." He gave me a strange look and then a sly smile and said, "Right! Seriously, what is your book *really* about?"

I kept thinking about my friend's reaction after I began submitting my book proposal to literary agents, most of whom seemed to have simply scanned the cover sheet before sending a rejection letter or e-mail. For this reason, I am especially grateful to Barbara Braun of the Barbara Braun Literary Agency for taking the time to read the proposal and for deciding that the book was interesting enough for her to represent me. Representing a book on this theme was hardly an easy task for Barbara (good agents earn every penny of their commissions), but she did find someone who was willing to take the time to read the proposal and the sample chapters and see the book's potential. This someone was Linda Regan, executive editor at Prometheus Books. Both Barbara and Linda have guided me through the process of researching and writing *Concrete Planet*, providing invaluable advice and solid direction. I am supremely grateful to both of them.

I would also like to take this opportunity to thank Bob Nason, whose hospitality, large library, and long experience (his structural damage assessment of the 1906 earthquake and fire for the United States Geological Survey proved invaluable) contributed greatly to this book. My friend, Randolph Langenbach, an eminent architect and author who has also studied the structural damage caused to both masonry and reinforced concrete buildings in earthquakes around the world, reviewed an early draft of the book and made a number of very useful suggestions that helped to improve the clarity of the text.

Another person to whom I owe thanks is fellow writer Dennis Smith. We were having a discussion over lunch one day when I mentioned the incredible material I had unearthed about problems related to reinforced concrete. I also expressed qualms about doing a book about the subject, since I was neither a civil engineer nor an architect. "That's why you need to write the book," he said. "You don't have a career to lose." I am glad I took his advice.

The most pleasant aspect of writing *Concrete Planet* was the tremendous assistance I received from many generous individuals around the world who took the time to send me material and images, much of which were unavailable from libraries, photo archives, or Internet sources. Among the people who I would like to thank in this regard are Dr. Patty Jo Watson, Dr. Richard Anderson, Dr. Mehmet Özdoğan, Kate Tarasenko, Emilio Labrador, Matthias Kabel, Christopher Newberry, Martin Olsson, and many others. I am also grateful to Gladys and Richard Hansen for allowing me access to the archives of the Museum of the City of San Francisco, which provided some very obscure but very important material. The museum, which runs one of the best historical websites in the United States, has suffered from the recent economic upheavals. Those wishing to help this fine institution by making a donation should go to http://www.sfmuseum.org.

I would also like to thank Dr. Michael Kapphahn for helping me track down important historical papers and articles in Germany about the last years of William Aspdin. The Library of Congress, as always, has been extraordinarily helpful. It is regrettable that so many other institutions, many of which are financially well-endowed, do not possess the same spirit of open cooperation.

I would like to also thank Julia Sandison for the images of the Hockley Viaduct. Her small volunteer group, Friends of the Hockley Viaduct, is fighting hard to save the viaduct—perhaps the most beautiful and architecturally significant surviving concrete structure of the nineteenth century—from vandals and neglect. English Heritage declined to include the viaduct in its list of historic buildings because it is not a "pure" concrete structure. By the same criteria, Rome's concrete Pantheon and Aelian Bridge would also be excluded from preservation status. To assist the Friends of the Hockley Viaduct in their noble efforts to save this remarkable edifice, please contact Ms. Sandison at Julia@ntcom.co.uk.

Though he is no longer with us, I would like to thank the late Major A. C. Francis, the British historian of modern concrete's early days. Without the results of his dogged research, much of chapters 5 and 6 would have been very different, and certainly less accurate.

Finally, this book would have been impossible to write without the continual encouragement, research assistance, and translation work provided by my wife, Josephine. Words are an inadequate medium in which to express my deep appreciation and love for her.

FOREWORD

I never imagined that inherent in the word *concrete*, a term I often use, there could be so much history, discovery, engineering and construction science, and profound, unrelenting, and seemingly irresolvable controversy. Indeed, the controversy can be very well described as a life and death misunderstanding and misrepresentation of the viability of concrete, both in threatening and mortal circumstances, such as earthquakes and fires, and in normal circumstances, where reinforced concrete is compromised by the crippling effects of steel corrosion and the simple wear, tear, and weakening that comes with age.

It has long been believed that when our ancestors were still wearing animal skins and communicating largely by pointing to things, someone started a sustained and very hot fire on a bed of flat rock. That person, man or woman, noticed that there was a white ash, clumped and brittle at the edges of the fire. Someone stepped on it and complained that the bottom of his foot was in pain. Someone else poured water over the foot to cool it, but instead the water added significantly to the pain. If scatology had been invented at the time, the air would have been filled with it. It was further noticed that whenever pieces of this material that fell from the bottom of the foot were mixed with water, they magically hardened (in a process we call hydration) into something that the Romans would later call *caementis*, and we call today "concrete."

It is a good story, as it goes, and indeed a scenario as related above could have happened, but as you will learn in these pages, the fire would need to have reached kiln-level heat to then transform the stone into something called calcium oxide, which, when then mixed with water and the carbon dioxide in air, would have resulted in a very hard and sustainable material. It is in the word *sustainable* where the importance of *Concrete Planet* shines.

In today's world we are constructing buildings in which people live, work, do commerce, and are entertained—structures that may last a hundred years. And, we are building memorials to commemorate certain people, like the 9/11 Memorial at the Pentagon in Washington, DC. These buildings may last for a hundred years because they are designed to last a century. And the architects and engineers who build them are proud of the fact that they will last a century. Imagine what the designers of the Roman Pantheon would have said if people told them their building would last just a century, or what the Medici family might have said if Michelangelo told them that Lorenzo's tomb would last for ten whole decades.

Though knowing the shortcomings of steel and concrete, architects continue to create buildings in our modern era by utilizing reinforced concrete, or concrete poured over steel rods that add to the forms' strength. The problem is that, over time, these forms will deteriorate because of natural attacks—not on the concrete, but in the interior steel. Cracks that occur in a structure may be repaired, but not before air, moisture, and many other possible chemicals seep into the form to cause rust. The rust on the steel expands the rod's diameter, destroying the surrounding concrete and debilitating the structure. This was a fact that many engineers and architects until recently refused to admit, or if they did admit it, they seemed to think it did not matter. That it *does* matter is one of the themes of this extraordinary book.

The World Trade Center buildings, before they were constructed and even after the first attack of 1993, were never subjected to a fire-load analysis. And they fell because of the stress on and weakening of the steel caused by the fires. But when it comes to concrete, what architects and engineers do not accept, because of their own self-imposed guidelines, can lead to disaster. Bridges may fall; buildings may collapse. Natural disasters will come and test the level of viable strength. This type of blindness can lead to destruction beyond the devastations that result from the natural disasters, themselves. Just think of the ninety-two children and three nuns killed in Chicago when fire destroyed Our Lady of the Angels school in 1958; or the coal slag mountain that slid down to destroy another school in Aberfan, Wales, in 1966, killing 141 people, including 116 schoolchildren; or the earthquake that occurred in Szechuan in 2008, killing tens of thousands of people, among which were over five thousand children and teens who perished in shoddily built schools. All of these hundreds of children could have survived if some thoughtful person had anticipated the disaster and pre-planned a mitigating protocol.

Great tragedies are circumnavigated every day, sometimes by fire-fighters or police officers who are trained to see unusual circumstances and act quickly to avert a building collapse or the locking of an exit door or a shooting or a killing. There is no controversy in these acts. But controversy always exists when someone calls into question accepted practices within a profession, especially if those practices impact on public safety. Nonetheless, Robert Courland anticipates the dangers of continuing to build as we have built for the last hundred years and more, and he illuminates these problems in a thoughtful and persuasive manner.

An important controversy found within these pages is the idea that concrete has a far longer lifespan when not reinforced by steel rods (called rebar), and that alternative materials for producing rebar should allow the building of structures with a thousand-year lifespan instead of a single century, facts that the author writes about here convincingly.

Humans might know that the universe is theorized to be 14 billion years old or that the Milky Way was formed 8 billion years ago, but the way we feel about ourselves in relation to a 4.5 billion-year-old earth is not much different from the way indigenous people studying a night sky might have felt about themselves anywhere on earth ten thousand years ago. The subject of what can possibly happen on earth is simply too mind-boggling for most of us to handle if we are to continue to be an optimistic race. A Canary Island mountain can fall into the sea off the coast of Africa and create a tsunami that would wash over most buildings in New York City a few hours later. A dam at the gateway to the San Fernando Valley could be torn by an earthquake in California and kill more than a half million people. The fault under New York City could slip and create such destruction that it would take decades to rebuild. (Aside from the politics of construction—it took New York City's leadership more than ten years to rebuild four square city blocks after 9/11, while it took the Chinese less than three years to rebuild an entire region of 2 million people after the Szechuan earthquake of 2008 killed more than seventy thousand.)

We know that natural cataclysms will occur and that we should plan against them. We should make our buildings stronger, of course. Can we say that they have been made as strong as they can be? This is the question we should be thinking about in a more serious manner. There are facts found in the pages of this book that we should not let pass into an obscure scientific history, for remembering them will undoubtedly help ensure a safer future for all on our planet.

We must remember first that earthquakes seldom kill people. It is the environment of falling structures that kills. *Concrete Planet* suggests ways in which we can make this environment safer. But there is so much more to Courland's book than the safety of buildings. For instance, does anyone think that the bridges being built today are as beautiful as those of the Victorian age? Just a simple aesthetic question, but the answer, at least partly, has to do with changing the use and vitality of concrete. Have you ever thought, as I had not, what concrete had to do with the success of the Roman expansion? Courland has.

I first met Robert Courland when I was determined to write a book, which came to be called *San Francisco Is Burning*, about the 1906 earthquake. Realizing that I needed to find someone who knows more than just about anyone about earthquakes, in general, and San Francisco's earthquake, in particular, I was advised by several librarians to seek out the author of this book. It was very good advice. A day did not pass during the seven consecutive months I worked with Courland that I did not learn new and fascinating things about the world we live in and the tempestuous earth beneath our feet. One of the first things I learned about the destruction in San Francisco that killed about three thousand people and collapsed or burned over twenty-eight thousand buildings was that some buildings survived collapse while buildings on either side were left flattened to a heap of stone. The answers to such enigmas are to be found in the pages of this book.

I keep remembering insights, as I hope you will, from this fascinating book, that help me recognize the issues of building construction on every street I walk—that many if not most buildings and bridges pose significant design, municipal planning, and expense problems for which there are no easy answers. And where there are solutions, every solution is harder than it sounds. Despite the importance and complexity of the information presented, *Concrete Planet* is, above all, a truly entertaining read.

—Dennis Smith, New York City

The Most Common Man-made Substance on Earth

We use it for our buildings, roads, and infrastructure. We walk, drive, and ride on it. Many of us live or work within its walls, and, after our deaths, a few of us will be buried within vaults constructed of it. The equivalent of forty tons of this material exists for every person on the planet, and an additional one ton per person is added with every passing year. Should human beings suddenly disappear from the face of the earth, the last century of our existence will be clearly discernable one hundred million years in the future by a unique, rust-colored layer of sediment found all over the planet. Like the K-T boundary that marks the end of the Age of Dinosaurs, this particular stratum will denote the reign of *Homo sapiens*, for it will mostly consist of crushed and recrystallized concrete, tinged reddish brown by the oxidation of its now-vanished steel reinforcement bars. In comparison, our bones will be a relatively rare find.

Despite its ubiquitous presence, most of us know very little about it.

As with all commonplace things, we take concrete for granted; our eyes pass over it, while our minds usually register only the end product: a building, a bridge, a something else. Most people with an adequate education have at least a vague understanding about how our automobile engines and electric lamps operate, but few of us know what makes concrete work. We have images of a wet, mud-like substance being poured through a trough and into a mold, eventually setting to form some off-white, solid object. That's about it. Most of us would be surprised to learn that concrete is formed by one of the most amazing chemical processes known to science. Most would also be surprised to discover that the story of concrete includes some of the most famous and colorful people in history. It is a tale full of mystery, intrigue, power politics, and quixotic dreams.

A civil engineer once asked me to imagine a world without concrete. It was plain that it was a question he often enjoyed asking, and one that no doubt caused the person being so queried to hesitate a moment in order to visualize such a place. By luck or happenstance, the answer quickly came to me. I said, "Easy. The world would resemble the nineteenth century." Without concrete, a larger share of today's population would be in the building trades, either as carpenters or masons, and they would be building to a smaller scale. Steel-frame skyscrapers would exist, but they probably would not be as tall, and their interior and curtain walls would be made of brick. Mountains of granite would need to be quarried for stone veneer to cover the brick surfaces of major buildings to give them a more elegant appearance. Dams would inevitably be of earth-fill construction or huge masonry affairs utilizing large blocks of stone held in place more by their awesome weight than by lime mortar.

There would still be beautiful buildings in this parallel world, but the more rapturous forms of architecture, such as Frank Lloyd Wright's Guggenheim Museum in New York, or Jørn Utzon's stunning Sydney Opera House in Australia, would not exist. Concrete can be molded to almost any form, but the same cannot be said of masonry or steel-frame construction. Electricity would still bring light and propel subways, and internal-combustion engines would still power automobiles, but road surfaces would be smooth only after the latest layer of asphalt had been poured and cured. Potholes would certainly be more common. Environmental issues would remain, as the carbon dioxide generated by the kilning of billions of bricks would probably be equivalent to the vast volume of greenhouse gases currently being produced by the cement industry (estimated to equal that produced by twenty-two million automobiles in the United States alone). Despite the other technical achievements of this alternative world, which may even include microprocessors and space travel, to our eyes it would still appear somewhat old-fashioned and backward, all because concrete would be missing from the visual environment.

In a way, the story of concrete is also the story of civilization: its roots reach back to prehistoric times and even predate agriculture and the wheel. The Romans were the first people to realize concrete's potential, and they used it to erect dazzling buildings that stand to this day. After the fall of their empire, the formula for making concrete vanished for over a thousand years, and its use did not begin to accelerate until the late nineteenth century. By the early twentieth century, concrete had captured the imagination of many visionary architects, while others in the profession remained suspicious of the new material, feeling it had not yet proven its long-term structural worthiness. The skeptics were soon brushed aside, and known deficiencies of concrete were ignored or covered up by the concrete industry. By the second half of the twentieth century, we were using concrete not only for most buildings and infrastructure but also for such counterintuitive applications as Frisbees® and ships.

As we initially rushed to use concrete to build almost everything, we were still ignorant of the limitations of this novel substance. As we shall see, the Romans knew more about certain aspects of concrete than we did. For this reason, almost all the concrete structures you see today are doomed to a limited life span. The concrete Roman Senate House and Pantheon still stand after almost two millennia, but hardly any of the concrete structures that now exist are capable of enduring two centuries, and many will begin disintegrating after fifty years. In short, we have built a disposable world using a short-lived material, the manufacture of which generates millions of tons of greenhouse gases. Most of the concrete structures built at the beginning of the twentieth century have begun falling apart, and most will be, or already have been, demolished. If a modern concrete building has enough historic value for us to want to preserve it, the restoration costs are staggering—often many times the original construction costs in inflation-adjusted currency.

This is *not* a technical book and is, in some ways, an *anti*-technical book. Really, it is the *human* story of concrete, with emphasis on the people who discovered—and rediscovered—this building material, and who also pioneered novel ways of using it. The book also explains how concrete profoundly impacts society. In order to make *Concrete Planet* accessible to as many readers as possible, I have avoided chemical equations, choosing instead to describe basic processes in ways accessible to almost anyone.

My two principal aims in writing this book were to enlighten and to entertain. I hope that I have accomplished those ends, but that is for the reader to decide.

Chapter 1

ORIGINS

Israeli geologists caused a minor stir in the 1960s when they announced the discovery of a concrete compound in a twelve-million-year-old rock formation in the Negev desert.[1] This news raised eyebrows, particularly among engineering historians, archaeologists, and paleontologists, for the discovery predated not only the earliest known use of concrete but also the earliest known hominids. Twelve million years ago, our ancestors were hardly more advanced than today's lemurs. Who made the concrete?

Predictably, the facts are less sensational than the headlines. The pseudo-concrete was created when geologic forces gradually brought a limestone outcrop into contact with oil shale and, with water as a catalyst, produced a natural cement compound. This compound would not be considered concrete by chemists or engineers; rather, it could more accurately be described as "bad asphalt." For this reason, I think it best to confine our examination of concrete's origins to those early human societies whose approach to its invention, while sometimes not quite scientific, was more methodical and successful than Nature's random products.

Because lime—sometimes called "quicklime"—is the essential ingredient of concrete, it is perhaps best to begin our story with the discovery of that remarkable substance. Lime is derived from the principal component of limestone: calcium carbonate. Limestone is created from the physical remains of countless generations of corals and shellfish that eventually formed a thick sedimentary layer. This layer was eventually crushed and crystallized by powerful geologic forces, resulting in a whitish rock. Pick up a sun-bleached seashell lying on the beach, and you are, in effect, holding a pure form of limestone, for the shell consists almost entirely of calcium carbonate. Limestone in its abundance provides silent testimony to the massive volume of life that thrived in the oceans for hundreds of millions of years before human beings came into existence. And those human beings would eventually discover that limestone contained hidden properties that, to their primitive eyes, seemed nothing short of miraculous. Even for us today, long inured to technical wonders, these properties still seem a little eerie and preternatural.

We now know that lime was discovered sometime toward the end of the Paleolithic Age, approximately twelve thousand years ago. The Paleolithic (Old Stone) Age reaches back almost three million years, and technically begins with the first stone tools. Consequently, it encompasses both the hominids and anatomically modern humans who used stone tools. For *Homo sapiens*, the Paleolithic Age begins with our species, approximately two hundred thousand years ago.

Until the invention of carbon-14 analysis and other sophisticated dating technologies, it was often difficult for archaeologists to differentiate the age of objects found at various sites where our Paleolithic ancestors once lived. Were the objects one hundred thousand years old? Sixty thousand years old? There was sometimes not enough variation among the artifacts to easily chart a firm evolutionary path in toolmaking. For us, technological breakthroughs happen constantly within a single lifetime; for our hunter-gatherer ancestors, they took place perhaps once every hundred generations or more. For most of our existence on this planet, we were pretty slow on the uptake when it came to technology.

Then something happened between fifty and sixty thousand years ago. Stone tools gradually improved, and the diversity of artifacts increased. Around this time, the earliest bow was invented, and the sophistication of spear and axes improved.[2] The first artwork arose; people began painting on cave walls and carving primitive figures in bone. One result of these small but immeasurably important developments was that when our ancestors first entered Europe forty thousand years ago, they possessed a decisive technological edge over the Neanderthals they encountered there. If modern humans had immigrated to Europe twenty thousand years earlier, this might be a completely different book.

After this technological uptick, a period of stasis returned that lasted millennia. The Neolithic (New Stone) Age began around 10,000 BCE. Compared to the slow pace of change during the Paleolithic Age, the Neolithic Age witnessed an explosive transformation in human societies and technologies. Ceramic figurines appeared, followed by pottery. Sheep were domesticated, and cloth weaving appeared soon after. Intertribal communities arose with shared languages and belief systems, and the larger labor pool led to the first great building projects. During this time, agriculture was gaining importance, increasingly supplementing the diet of hunter-gatherers with sowed grains and legumes. The modest leisure time afforded by food surpluses allowed for the discovery and development of new crafts. Some villages grew to become towns, and some of these towns would become the first cities.

Complex human societies had begun to emerge, and with them, what we may broadly call "civilization" had also arrived.

Archaeologists believed for a long time that the dramatic changes that took place during the Neolithic Age were primarily due to the discovery of agriculture and animal domestication. Only recently have we come to realize that the story is much more complicated than that. Technological and societal revolutions began centuries before agriculture arose.[3] The old demarcations between the Neolithic and Paleolithic Ages have recently been pushed back and become blurred.[4] These societal changes may have begun with the discovery of that remarkable substance: lime.

The most popular theory of lime's discovery is the "campfire on limestone" scenario.[5] It runs something like this: At some point in the distant past, a group of hunter-gatherers pick a convenient spot for foraging and settle in for a few weeks or months. As a safety precaution, they locate their fire pit in a small depression on a stony outcrop well isolated from any dry brush. On this occasion, they build their fire in a limestone declivity. After some days or weeks have passed, they notice that the stone near the flames becomes desiccated and breaks off into clumps that easily collapse into a white powder. This is lime. After the fire has been extinguished and the pit becomes cool, someone picks up one of the clumps and crushes it in his hand. He feels a slightly painful irritation caused by the caustic powder.

After first shaking his hand violently to remove the powder, he then pours some water on the affected skin to remove the nuisance. This normal response to an irritant is, on this occasion, not a very wise move. The water provokes a powerful reaction with the powder, and what had merely caused irritation a moment earlier now produces a chemical burn. Frenzied with pain, he continues to pour water on his palm and fingers. Fortunately, the more water he pours, the greater the dilution of the reactive compound, and soon the pain is lessened. Red, blistery patches—the scarring caused by second-degree burns—remain on his hand.

Of course, the victim goes to a doctor. It does not matter that the physician in question is a witch doctor, for the tribe's medical practitioner is not only the font of countless spells accumulated from an oral tradition that dates back from time immemorial, but he is also the chief pharmacist, possessing a repertoire of cures or palliatives that comprise hundreds of plant and animal parts. Perhaps less than a quarter of the ingredients in his medical arsenal represent true cures or analgesics, and the rest are placebos. However, between the panaceas and a few real herbal remedies, the shaman probably enjoys a high success rate—most people survive their illnesses naturally—which reinforces his people's faith in him.

After he dresses his patient's wound and gives him the Paleolithic equivalent of two aspirins, the shaman decides to investigate the curious powder that caused the problem. He knows that people confronted with a painful experience are not the best witnesses, and so he probably doubts the story about water causing the powder to burn. After all, water is used to *put out* fires, right? Bathing in it on a hot day cools the body and drinking it slakes one's thirst. It soothes pain but does not cause it.

The shaman goes to the fire pit and uses a stick to scrape some of the powdery rock into a small basket or clay bowl. Perhaps he first sticks the tip of his left pinkie into the powder and discovers that, yep, it is a mild irritant. Then he adds a little water to the powder. What happens next no doubt amazes him: the mixture begins to generate a flameless heat. His patient was not deranged after all; this is some serious stuff. As he continues to watch the bubbling concoction, he observes that after a few minutes the foaming dies down and the heat diminishes. If the witch doctor is patient, he will

notice that the substance soon becomes solid. He taps the material with a
stick and confirms that it has become very hard. A rock has been created!

Today we know what the shaman did not: that the heat of the fire transformed the calcium carbonate of the limestone. All the carbon dioxide and
water within the rock (yes, all rocks contain a small degree of water) will
have evaporated, leaving behind calcium oxide: the caustic powder we call
lime or quicklime. When water is added to the powder, a violent chemical
reaction takes place: heat is generated—up to 150°C (over 300°F)—and
calcium hydroxide is formed. This new compound craves the carbon
dioxide it lost during the baking process and so pulls it from the atmosphere
like some alien creature dying of asphyxiation. Calcium carbonate begins to
form within the mixture, and after a short time, it hardens and becomes, in
effect, limestone once again. This artificially created calcium carbonate is
very white and very hard. One might view it as the purest form of concrete.

Until very recently, the importance of this discovery was not fully
appreciated. Archaeologists reasoned that a humble substance created by a
simple campfire could not have had that much effect on the course of
human technological development. This perspective was almost universally
held until the 1990s, when new excavations and a closer look at the empir-
ical evidence would challenge that assumption.[6] Indeed, it is possible that
the revolutions that led to the Neolithic Age may have begun with the
invention of concrete, and that lime's discovery was far stranger and more
interesting than anyone had previously supposed.

A RADICAL NEW VIEW OF THE LATE STONE AGE

In 1963, members of a joint American-Turkish archaeological survey team
from the University of Istanbul and the Oriental Institute of the University
of Chicago combed the landscape of southeastern Turkey, looking for sites
to catalog for possible future excavation. In the small province of Şanlıurfa,
not far from the Syrian border, they found an exceptional number of potential sites for exploration. Toward the end of their survey in the province, the
archaeologists came to a large hill with a rounded top that the Turks called

Göbekli Tepe, which means "potbellied mount." The hill is a little over 300 m (*ca.* 1,000 ft) high and lies at the base of the Taurus Mountains. Nothing about the hill sets it apart from the others, except that its particular position provides superb views in all directions, the most spectacular being the mountains to the north and the Harran Plain to the south. However, the hill's immediate vicinity holds a particular and powerful interest for historians and biblical scholars. Turkey has a very rich past, and the southeastern region of the country, known since ancient times as Anatolia, has an even richer heritage. The tiny Anatolian province of Şanlıurfa is especially drenched in prehistory, history, myth, and—incorporating varying mixtures of all three—tales central to Judeo-Christian and Islamic traditions.

Over the millennia, this diminutive region has been conquered and ruled by the Sumerians, Akkadians, Babylonians, Assyrians, Hittites, Hurris, Aramaeans, Medes, Persians, Macedonians, Romans, Parthians, Armenians, Arabs, Kurds, Crusaders, and, finally, the Turks. Near Göbekli Tepe is the ancient city of Harran, for which the plain is named. According to local Islamic tradition, Harran was the first spot where Adam and Eve stopped after their expulsion from the Garden of Eden.[7] The Bible tells us that Harran is also the place where Terah and his son, Abram (the future patriarch Abraham), Abram's wife, Sarah, and their son, Lot (whose future wife would morph into sodium chloride) settled in for a spell on their oft-delayed journey from Ur to Canaan.[8]

Also nearby is the province's eponymous capital, Şanliurfa, more commonly known as Urfa. According to Turkish Islamic tradition, Urfa was originally the city of Ur, the birthplace of the aforementioned Abraham.[9] (Most archaeologists place Ur in today's Iraq.) According to Jewish legend, it was in Ur that Abraham was thrown into a great bonfire by the nasty Babylonian king Nimrod.[10] God intervened on Abraham's behalf, and he walked out of the flames unscathed. Turkish Islamic tradition explains why: God had turned the flames into water and the firebrands into fishes.[11] To commemorate Abraham's deliverance from the flames, Urfa has for many centuries maintained a sacred pool of fishes next to the mosque dedicated to the patriarch.

Moving on to more verifiable history, the plain just outside Harran was the scene of some of the most significant battles in antiquity. It was on the Harran Plain where the Babylonians defeated the Assyrians in 610 BCE; where Xenophon and his Ten Thousand marched by in 401 BCE, harried by the Kurds along the way;[12] where Marcus Licinius Crassus, the Roman general who had defeated the rebel slaves led by Spartacus, was himself defeated by the Parthians in 53 BCE;[13] where the Roman emperor Caracalla was assassinated by one of his lieutenants during another campaign against the Parthians in 217;[14] where another Roman emperor, Valerian, was decisively beaten by the Persians in 260 and taken captive;[15] and where the crusaders fought the Turks in the Battle of Harran in 1104 (the Crusaders lost).[16]

In short, today's tiny Şanliurfa province has seen the rise and fall of many kings, sultans, shahs, emperors, emirs, and pashas, as well as their respective kingdoms, empires, and emirates. Hardly a day goes by without some farmer discovering an ancient artifact while digging a well or tilling a field.

By the time the archaeologists came to Göbekli Tepe, they had already recorded a number of exciting sites for excavation and were getting a little jaded by this surfeit of riches. Some requisite exploratory digging at Göbekli Tepe uncovered a large number of ancient flint and obsidian tools. However, a few carefully carved limestone blocks poking out of the ground seemed to belong to a much later period. The director of the survey, Peter Benedict, guessed that a Byzantine church and cemetery overlay a more ancient settlement. In a region abundant in important sites, Göbekli Tepe did not appear all that interesting. Benedict noted the geographical coordinates of the hill, wrote a brief description of the stone tools and carved stone blocks found there, and gave the site a name that, while hardly prosaic, was eminently practical for cataloging purposes: "V 52/1."[17] Bennett and the other members of the survey team then moved on to look for other potential sites. None of them would have guessed that the hill would one day be the site of one of most important archaeological discoveries of all time and provide the final piece of an archaeological puzzle that had taken over a half century to put together. For, in addition to its religious and historical claims to fame, Lilliputian Şanliurfa province would be revealed as the birthplace of civilization.[18]

One site that *had* appeared very interesting to the surveyors was located 96 km (*ca.* 60 miles) to the northeast of Göbekli Tepe, at a spot called Çayönü, near the historic town of Diyarbakır. Less than a year later, in 1964, Chicago University's leading Middle East specialist, Robert Braidwood, accompanied by his wife and colleague, Linda, arrived at Çayönü to begin excavating the site.

The Braidwoods were members of that generation of archaeologists who followed the exuberant pioneers of the nineteenth and early twentieth centuries; people like Heinrich Schliemann, discoverer of Troy; and Robert Evans, who uncovered the Minoan capital at Knossos in Crete. Schliemann and Evans were both accused of rushing excavations in feverish attempts to unearth the most spectacular artifact or treasure, while tossing aside seemingly insignificant items that could reveal much about the cultures they were supposedly bringing to light.[19] A gold death mask may be impressive, but it tells us less about the man behind it than do the bones or shells of the animals he ate for supper, or the seeds or pits from the fruits he enjoyed for dessert, or the last surviving threads of the clothes he wore. Evans and Schliemann also brought with them rather romantic notions based on the Greek myths, and this seriously colored their interpretation of the data. To be fair, both men had no manual or established procedures on how to dig up the remains of an ancient civilization, and their spectacular finds did herald the beginnings of a new scientific field that would eventually be called "archaeology." Nevertheless, their work left much to be desired.

The Braidwoods and their colleagues were of a different breed. They designed the rigorous methods now universally used to perform an excavation: the laying out of a string gridline over the site, the shaking of each spade of dirt through a fine screen to capture the smallest artifact fragment or trace of food detritus, and the meticulous recording and cataloging of everything found. Braidwood was also among the first archaeologists to recognize the importance of using radiocarbon analysis to more accurately date artifacts.[20] Developed in 1949 by Willard Libby and others at the University of Chicago (Braidwood's home base), the technology measures the decay rate of the naturally occurring radioisotope, carbon-14, which is found in the remains of all organic material less than sixty thousand years old (little of the isotope remains after that period of time has passed). Like many other important innovations, radiocarbon dating was not immediately or enthusiastically embraced by all those who would most benefit from the technology. Braidwood's contemporary, British archaeologist Stuart Piggott, denounced radiocarbon analysis as "unacceptable," largely because it contradicted the dates and chronologies he had put forward in his work on Neolithic sites.[21] Widespread acceptance of radiocarbon dating did not come about until the 1960s, by which time Braidwood had already been using it with great success for almost twenty years. The innovations advanced by Braidwood and others in the post–World War II period would forever change the science of archaeology.

By the time he came to Çayönü, Braidwood had been surveying, excavating, and studying the artifacts of Middle Eastern archaeological sites for over thirty years, mostly in Iran and Iraq. Like most archaeologists, he believed that the earliest civilization arose somewhere in the Fertile Crescent, that imaginary arch that begins in the Nile delta and runs through the Levant (where today's Israel and Lebanon are located) and then turns east to end in the lower reaches of the Tigris and Euphrates Rivers. During their work in northern Iraq and Iran, the Braidwoods were beginning to come to a slightly different conclusion about the rise of civilization in the Near East than that proposed by other prominent authorities like James Breasted and V. Gordon Childe. While Breasted and Childe believed that the first settlements had sprung up in the lower reaches of the Tigris and Euphrates Rivers, where rich alluvial soil afforded abundant opportunities for agriculture, the evidence the Braidwoods were uncovering instead pointed to the foothills of the Taurus and Zagros Mountains, stretching from southern Turkey to northern Iran.[22] They discovered that the older the agricultural sites were, the farther north they were found. Robert Braidwood called this area the "hilly flanks" of the Fertile Crescent. This made perfect sense. During the earliest stages of agriculture, people were still learning how the planting process worked. And since the first cultivated cereal crops yielded barely more than their wild cousins, these proto-farmers probably needed to hunt game to supplement their diet. The hills were abundant with game, and their forested slopes allowed hunters greater opportunities to sneak up to their prey than did the open plains to the south. Moreover, since farming requires settling down and living at a fixed location, the domestication of some of these game animals also made sense. The archaeological record pointed to sheep and goats—natural denizens of craggy mountains and hills—as the first domesticated farm animals. The critical transition had to have taken place in a hilly or mountainous region. When the Braidwoods came to Çayönü in 1964, they knew that exploratory digs conducted by the previous year's survey had turned up enough interesting material to suggest that this particular Neolithic settlement might be a key transitional site.

The first season's dig offered promise. It was clear that Çayönü was both quite ancient and, for an early Neolithic site, substantial in size. With each passing year, as the accumulated dust of the ages was brushed away and each fresh artifact brought to light, it became evident that Çayönü was indeed one of those critical missing links for which the Braidwoods had spent decades searching. Along with another settlement, Çatalhöyük, which the British archaeologist James Mellaart was excavating at the same time several hundred kilometers to the west[23] (also on the "hilly flanks" of the Fertile Crescent), a solid picture was emerging of the hunter-to-farmer transitional period and of the growth of the first large settlements. Çayönü flourished around 7000 BCE and was the size of a small village. It is the earliest known permanent human community. It is in Çayönü that we see some of the earliest examples of farming, animal husbandry (pigs), woven cloth, copper metallurgy, and fired-clay ceramics.[24] Nevertheless, all these crafts are at their earliest developmental stages. Aside from pork, the meat Çayönü's inhabitants consumed still came from wild game. There is no evidence that people at Çayönü milled grain, much less baked it, and all plants they ate were also gathered from the wild. Stone and bone were still used for virtually all tools. The copper and cloth artifacts discovered are primitive and, not surprisingly, seem like tentative first steps. The dwellings were constructed of sun-dried clay brick (adobe), and the structures were huddled close together, generally sharing a common wall. The ceilings were supported by a row of wood poles—the trunks of smaller trees. The entrance to the dwelling was through a hole in the roof that also served to evacuate smoke from the hearth. People climbed up ladders to reach the roof, then pulled them up to use for the ceiling entrance, a defensive arrangement found to this day in some villages in the Middle East and North Africa.

Çatalhöyük flourished approximately seven hundred years after Çayönü and was about ten times the size of the earlier settlement. It is the oldest known "town" we know of.[25] Besides its far larger size, several important technologies distinguished Çatalhöyük from its predecessor. Grain was stone-ground to produce flour and ovens were used to bake the first known bread, apparently unleavened. The ceramics crafts had also advanced in the seven-hundred-year interval. Instead of the simple fired-clay figurines found at Çayönü, shards of the earliest known kilned pottery were unearthed at Çatalhöyük.[26]

At both Çayönü and Çatalhöyük, the floors of the dwellings were extraordinarily hard. Small pieces of limestone were set into what was initially assumed to be a hard adobe to form pleasant patterns. However, chemical tests established that the foundation material consisted of a mixture of lime with clay—the earliest known examples of artificial stone floors then uncovered. Archaeologists would call these "terrazzo floors," because they resembled the flooring inlaid with marble chips that had originated in Terrazzo, Italy. This lime-clay mixture was also applied as a plaster for the adobe blocks used in constructing the dwellings at both sites. While the Braidwoods and Mellaart were impressed by this early use of lime for building, its true importance was obscured by the other Neolithic accomplishments, such as the evidence pointing to some of the earliest forms of agriculture, ceramics, and metallurgy.

Evidence confirming Robert Braidwood's theory about geographic origins of civilization was growing, and by the late 1970s the pieces were falling neatly into place. Braidwood was now convinced that the foothills of the Taurus Mountains in lower Anatolia, not the Zagros range to the east, was where civilization began to emerge, specifically in or near Şanlıurfa province. He called this region the "nuclear zone," meaning that the area formed the nucleus of where the first civilized crafts arose and where the transition from hunting to agriculture first took place: the "ground zero," so to speak, of civilization.[27] He was convinced that earlier, less developed settlements would be found not far from Çayönü and would thus complete the picture. Sadly, external events would interrupt the archaeologists' work. Şanlıurfa province was about to become the ground zero of a very unpleasant war.

Southeastern Turkey had long been home to the Kurds, a people whose customs and language are closely related to those of the Iranians. When Mustafa Kemal Atatürk came to power in 1919, he brought the long moribund Ottoman Empire to an end and proclaimed his nation a republic and a "Turkey for the Turks." In short, there was no place for the Kurdish language or culture in the new, westernized, secular Turkey.[28] The Kurds had to become Turks—or else. It was not much different from the attempt by Americans of European ancestry in the late nineteenth and early twentieth centuries to make Native Americans "Americans." (In both instances, the objects of conversion had been established on the land thousands of years before the arrival of the now dominant power.) The Turkish government labeled the Kurds "Mountain Turks" and outlawed their language.[29] The Kurds defied the new government and sporadically revolted. Things gradually quieted down after a few years, with the Kurds passively resisting the conversion measures and the Turks generally ignoring the situation as long as their subjects behaved and paid their taxes. In 1984, after nearly six decades of simmering discontent, a Kurdish revolt broke out in Anatolia. It was led by a shadowy figure, Abdullah ("Apo") Öcalan, who called for an independent Kurdistan. Southeastern Turkey exploded in violence. Turkish police were gunned down, and when troops were sent to Anatolia to restore order, they were ambushed as well. The government responded ruthlessly. People believed to be rebel sympathizers were either arrested or assassinated.[30]

Unfortunately for archaeologists, the Kurdish region of Turkey was also where most of the newly discovered or soon-to-be-discovered Neolithic sites were located. Especially affected by the violence was the region around Çayönü (Apo had grown up in a nearby village). The Braidwoods had no choice but to abandon the site and hired a local man to guard it. In an arrangement between the University of Istanbul and the University of Chicago, the former agreed to pay the guard's salary, while the latter offered to cover the cost of the man's food and bullets.[31]

The government's generous-carrot-and-ruthless-stick approach toward the Kurds in Anatolia resulted in a gradual winding down of the conflict in the 1990s, and in 1999 Öcalan was captured in Kenya, where he was remotely directing operations.[32] Sadly, it was too late for the Braidwoods to return to Turkey: both were approaching their nineties, and their health had become too fragile for fieldwork.

Still, the Braidwoods had the pleasure of learning about recent discoveries that confirmed their theories about where civilization had arisen. DNA analysis verified that all the species of wheat grown today throughout the world could trace their lineage back to a wild variety still growing in the foothills of the Taurus Mountains, near Şanlıurfa.[33] Further bolstering their theory was the discovery in the same area of two more Neolithic sites, both far older than Çatalhöyük and Çayönü. Taken together, the four sites would offer archaeologists a near-perfect chronology of how humans made the transition from hunter-gatherer groups to agricultural communities. Strangely, the buildings at both these newly discovered sites also had lime concrete floors.

The principal carrot offered by the Turkish government to the Kurdish population of Anatolia was a vast infrastructure upgrade for the region. Among these projects, which included road improvements and better schools, was the construction of a complex of large dams on the upper Tigris and Euphrates Rivers. These dams would prevent seasonal flooding, provide the local inhabitants cheap hydroelectric power, and offer a steady and controlled source of water for agriculture. It was like America's Depression-era TVA Project on steroids.[34] The downside to all this was that dozens of important archaeological sites—many still unexcavated—would be flooded.[35] One site destined to be lost to the floodwaters was Nevali Çori, a Neolithic settlement that showed promise after an earlier exploratory dig had produced interesting artifacts. The University of Istanbul and the University of Heidelberg quickly organized a rescue project in 1993, and excavation began soon after under the direction of German archaeologist Harald Hauptmann. As with Çayönü, Nevali Çori proved to be another key "missing link" in the story of the rise of civilization.

Dr. Hauptmann and his colleagues established that Nevali Çori flourished five centuries earlier than Çayönü, and although it shared certain common features with the latter site, such as fired-clay figurines and terrazzo lime concrete floors, it was distinctly different. Not surprisingly, some technologies were more primitive, while others were yet to be discovered. Woven cloth and copper metallurgy had yet to make an appearance at that time. In one technology, however, the people at Nevali Çori showed more mastery than their later, near-contemporary Neolithic brethren.[36] Dr. Hauptmann was amazed to discover large, intricately carved limestone blocks and tall monumental pillars shaped in the form of a "T." To cut limestone using flint is a tedious business, yet these blocks and pillars were carved with great care and surprising mastery. The pillars were approximately 3 m (*ca.* 10 ft) tall and once supported a wooden roof. Incised on them were sculptures, including the earliest known relief of human hands.[37] Remarkably, the hands seem to be clasped in prayer. The stonework would not have been out of place in a site dating several thousand years after Nevali Çori. A limestone bust of a man's head—the earliest known life-size anthropomorphic figure—was also discovered. The man's head is bald, except for what appears to be a crawling snake on top—or his hair cut to resemble a crawling snake—and may represent a shaman or priest.[38]

Another thing that set Nevali Çori apart was that it seemed to be more a religious complex than a settlement. Of the twenty-two buildings unearthed, only a few seem to have been continuously occupied dwellings, perhaps the living quarters of priests, priestesses, or shamans.

The rescue operation to excavate and survey Nevali Çori was largely a success, and many tools, utensils, and limestone carvings were recovered before the river waters confined by the new Kemal Atatürk Dam flooded the site.

As one Neolithic site was being submerged, another site was emerging a few kilometers away, at a place that had been cataloged thirty-five years earlier and was filed under the name "V 52/1." Its discovery would rewrite all the books about the late Stone Age.

GÖBEKLI TEPE

Among those who worked under Harald Hauptmann at Nevali Çori was Dr. Klaus Schmidt, another professor of archaeology at the University of

Heidelberg. The carved limestone blocks at Nevali Çori reminded him of those that had been exposed at a site mentioned in the 1963 survey by Peter Benedict. If the stones at Göbekli Tepe were indeed Neolithic carvings—and not the remains of a medieval church—this would explain the flint tools discovered by Benedict, who had assumed they belonged to a far earlier period and stratum.[39]

Plans were made to excavate Göbekli Tepe, and serious work on the site began the following year in a joint project conducted by Turkey's Şanlıurfa Museum and the German Archaeological Institute.

After the archaeologists had set up their grid lines and carefully started their excavation work, they were befuddled with what they began uncovering at Göbekli Tepe. The stonework was similar to that of Nevali Çori but was in some ways even better. Dressed stone masonry was used for walls, and, like Nevali Çori, limestone was carved into "T" crosses that served as pillars for a now-vanished wooden roof. Dr. Schmidt believes that large teams of people were organized to pull massive limestone blocks—some weighing up to fifty tons—from a quarry two kilometers away. For the largest blocks, Schmidt estimates that work crews numbering up to five hundred people were assembled for the task—an astonishing and unprecedented number of individuals cooperating together during this early period.[40] Once the limestone blocks were put into place, the builders then carved them with a dexterity that would not be seen again for many centuries. Bulls, cranes, foxes, boars—the totem figures of a foraging people—were rendered with a deft likeness. And nowhere is there any evidence of agriculture, only the food detritus of hunter-gatherers: wild animal bones and remains of seeds and nuts.[41]

It is conjectured that some sort of common belief system and shared customs and language brought a thousand or more hunter-gatherers each year to this spot, which apparently held some spiritual significance for them. And together they planned and built a massive religious center that has no equal for its time, for the results of the carbon dating tests staggered the archaeologists: the temple complex had been created 11,600 years ago,[42] almost 7,000 years before the creation of the Stonehenge monoliths and predating Nevali Çori by a thousand years. At the time building began at Göbekli Tepe, humans in North America were still hunting wooly mammoths and avoiding sabertooth tigers. The last Neanderthals had only recently gone extinct, and the last members of a far more primitive hominid, *Homo floresiensis*, might have still been dodging modern humans in Indonesia.[43] As far as we know, there was no place on the face of the earth that could even approach the construction mastery of Göbekli Tepe, let alone equal it.

Göbekli Tepe was without doubt a sacred precinct, for there is no sign of permanent human habitation at the site. Apparently, these ancient people reserved their finest structures for the divine and would continue to do so for many centuries, as evidenced by Nevali Çori. As Dr. Schmidt puts it: "First the temple, then the town." (*Zuerst kam der Tempel, dann die Stadt.*)[44] And, as at Çayönü, Çatalhöyük, and Nevali Çori, these people burned limestone to make concrete flooring.

In early 2003, not long after the Braidwoods learned of the exciting data coming from the excavation at Göbekli Tepe, both came down with severe bronchial pneumonia and were taken to University of Chicago Hospital. A few days later, in a poetic denouement that the gods sometimes grant to inseparable lifelong partners, the Braidwoods died within hours of each other. Robert was ninety-five, Linda, ninety-three.[45]

CONCRETE: MOTHER OF INVENTION

Toward the end of the twentieth century, archaeologists who specialized in materials engineering, now called archaeomineralogists, began pondering ancient concrete's importance. They noticed that as one goes further back in time, the technologies begin disappearing one by one: oven-baked food vanishes, then pottery, then woven cloth, then metallurgy, then simple ceramics, and finally, primitive agriculture. At almost twelve thousand years in the past, all the remaining technologies—at least those not related to hunting-gathering—are pretty much the products of limestone: carved reliefs, block masonry, lime plaster, and lime concrete floors. Indeed, this mineral dominated the last years of the Paleolithic Age, which alerted

archaeomineralogists to the possible importance of lime in the technological revolution seen in the succeeding Neolithic period.[46] They also began
to realize that the discovery of lime was far more difficult and complex than
previously believed.

That archaeologists had not realized the importance of lime was largely
due to a simple misunderstanding: almost everyone was under the impression that the substance is easily made; that all you had to do was make a fire
in a limestone declivity to create lime. Some had even suggested that the
discovery of lime was as old as the history of fire. After all, humans had been
manipulating the properties of rocks with heat for many thousands of years.
By simply putting flint in a fire for a couple of hours, the stone becomes
easier to chip—or "knap"—for the purpose of making spear blades, arrow-
heads, and axes.[47] Many authorities have long assumed that the discovery of
lime had similar origins, and one form or another of the scenario noted earlier in this chapter is the most widely believed version.[48]

Unfortunately, few people apparently took the trouble to actually test
this theory with some limestone.[49] If they had, they would have discovered
that it is extremely difficult for a campfire to reach temperatures high
enough (848°C / 1558°F) to extract lime from the surrounding rock. To
put this into perspective, this level of heat is three times higher than what a
modern kitchen oven can produce. Even at this temperature, extracting
lime is a difficult proposition, since that part of the limestone beyond the
point of direct exposure to the heat acts as an insulator. Because of this insulating effect, some limekilns have actually been built of *limestone*. (Modern
rotary kilns heated by blast furnaces do not have this problem, as all sides of
the stone are thoroughly baked.) For this reason, our ancestors had to create
high-temperature kilns to make lime.

Our ancient ancestors made two key discoveries that eventually led to
modern civilization. The first—and most important—of these two discoveries was agriculture. However, agriculture came after the earliest kiln-based
crafts, what we now call "pyrotechnologies." The cultivation of crops came
many centuries after the first major building projects, as well as the
extended social bonds that sustained the cooperative efforts behind such
achievements. It was not farming that brought large groups of people

together but some spiritual yearning, a religious impulse strong enough to prompt the building of a huge temple complex. Göbekli Tepe may have been the focal point of the first "organized" religion—in other words, a Stone Age Jerusalem. It is my suggestion that, in addition to the spiritual yearning and extended societal bonds that led to the creation of the temple complex, another factor was involved. It was a physical phenomenon that seemed magical to these Paleolithic peoples: the chemistry of lime.

Besides underestimating the difficulty in generating the amount of heat required to produce lime, archaeologists also undervalued the impact that the properties of this amazing substance must have had on pre-Neolithic humans. It is difficult to erase from our minds the technologies we enjoy today and supplant them with the very limited knowledge and tools possessed by our Paleolithic ancestors. The curious properties of lime must have upended everything rational to them. It produced heat when it came into contact with water and created—to all appearances—a true rock, not something flimsy like dried clay. Mix it with a little sand or clay, and you could make a larger rock. In a way, it must have seemed like some divine power transferred to humankind, for only the gods could make rocks. The "magic" of lime may have given its discoverer a power that soon transcended the immediate hunter-gatherer group and led to the first intertribal communities based on a particular belief system and set of rituals.

The discovery of lime also seems to have coincided with some of the earliest instances of carving limestone for construction purposes or to create art. It is as if the discovery of lime focused people's attention for the first time on the other attributes of limestone, especially the fact that it is the most malleable of all the hard rocks. As lime was almost certainly considered a sacred substance, so must have limestone been regarded as a sacred rock, for its use— both in construction and art—was, like lime concrete, restricted for many centuries to religious complexes like Göbekli Tepe and Nevali Çori.

It is also likely that the process of making lime remained a closely guarded secret of the priests or shamans, for no limekiln has been discovered within the precincts of these late Paleolithic and early Neolithic sites. (The limestone was probably kilned at a remote location, away from the prying eyes of the profane.) And this high-temperature kilning process was easily the most technologically difficult, most physically demanding, and likely the most resource-intensive craft during this period of prehistory.

Since the art of kilning limestone changed little from remote antiquity until the early industrial age, it would be worthwhile to look at the process as it was practiced for such a long period. The first step was to create an oven in which an enclosed fire could concentrate the heat. This was done by digging a shallow pit and then building a stone structure—eventually brick would be used—around it that could contain and intensify the heat and allow just enough air to enter to keep the fuel burning at a high temperature. Over the centuries, this would assume the shape of a beehive or a Burgundy wine bottle. The "neck" of the bottle-shaped structure functioned as a constricted chimney that helped to keep the interior of the kiln extremely hot during the firing process. A small aperture at the kiln's base, just large enough for someone to crawl inside when it was fully open, controlled the amount of air reaching the flames and enabled the adding of more fuel. Once the structure had been completed, men would crawl through the opening to tightly stacked firewood in the pit and then place pieces of limestone—usually the size of a small fist or smaller—over the wood. More wood was then stacked on top of the layer of limestone. The reason the limestone was no more than fist-size was to allow the heat to completely permeate the stone and fully calcify it. The wood stacking and limestone placement alone usually required a day to complete. The kiln workers would then carefully crawl out of the oven, set the wood alight, and then close the opening with a flat rock, leaving it ajar enough to permit a steady flow of air to feed the fire. Once the fire was started, it needed to be worked regularly; men would poke the embers, fan the flames, and continually add more wood every hour or so.

The firing often lasted three days and two nights. Because the firing had to be carefully managed during this period, the work was almost always performed by at least two people. So much heat would be generated in the kilning process that another two days were required *after* the firing to allow the oven's interior to cool down enough to permit the workers to go inside to remove the calcinated lime. Removing the lime was a hazardous undertaking. As we have seen, lime is very caustic, and handling it can cause skin burns, so gloves or some other protective intermediary were used. (Neolithic people probably utilized animal skins.) The most dangerous aspect of the work was the risk of getting lime dust in the eyes, throat, or lungs. Since water is the activating agent, the effect of calcium oxide settling on moist parts of the human anatomy would be deleterious, to say the least. It is probable that many early limekiln workers suffered from diminished vision, a variety of pulmonary ailments, and, eventually, abbreviated life spans as well. Kilning lime was grueling work and was usually performed by slaves during the Greco-Roman period and in the antebellum American South.[50]

In the early Neolithic, the limekiln most likely was not an aboveground structure but simply a deep pit. Controlling the airflow and adding fuel was probably managed by placing flat rocks over the pit, with two men on each side moving them around with heavy sticks (the rocks would have been too hot to touch with hands).

Since even a modest limekiln required large amounts of fuel, collecting the necessary wood would have involved much effort, especially in late Paleolithic and early Neolithic times because of the then-primitive nature of tools. Repeated blows by flint or obsidian axes against wood frequently

FIGURE 1. Arial view of the excavation site of Çayönü, the earliest known permanent human settlement.

fractured their blades; yet, the blades would be replaced and retied, and the work would be continued. At least a dozen cords of wood (a cord is a 4′ × 4′ × 8′ stack of wood) were probably used in a single firing of a small kiln. As human populations grew, large swaths of forests would be leveled to provide the wood for limekilns.

Kilning limestone represented humankind's first use of complex chemistry. It was also the earliest known industrial process. Contemplating these ancient kiln workers—laboring away, covered with soot, being hit with a stifling blast of heat each time they fed fuel to the flames or adjusted the airflow—one cannot help but recall similar images from old photographs of the grimy laborers who slaved away in the soul-deadening factories of the Victorian Age, a dozen millennia in the future.

FIGURE 2. Archaeologists Linda Braidwood (*left*) and Robert Braidwood (*second from right*) with colleagues at the Çayönü site. Robert Braidwood correctly predicted that the sites marking the human transition from hunter/gatherer to agricultural societies would be found on the "hilly flanks" of the Fertile Crescent.

FIGURE 3. The excavation site of the great temple complex at Göbekli Tepe, built by hunter/gatherers almost twelve thousand years ago.

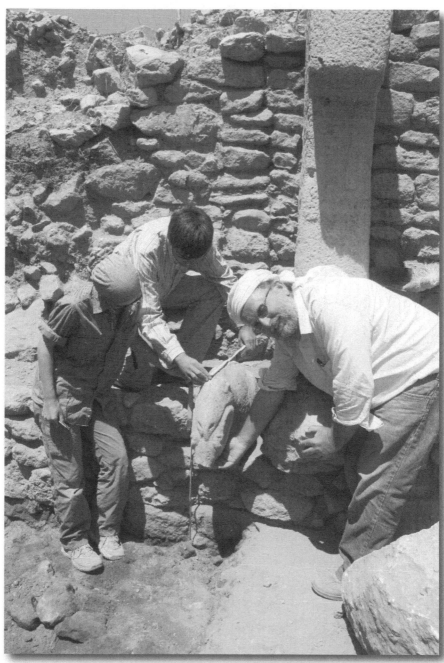

FIGURE 4. Archaeologist Klaus Schmidt (*far right*) working at the site. As at Çayönü, lime concrete was discovered, a technology that required high-temperature kilns.

As odd as it may seem, the archaeological evidence shows that the ear-liest ovens created by humans were not low-temperature affairs for baking or roasting food but rather *high-temperature* limekilns. And this odd frog leap over simpler kilning methods paved the way for the key technologies that followed, as shown on page 49.

It is no coincidence that fired ceramics make an appearance soon after the invention of the limekiln. Sun-dried clay bowls, while good for certain uses, often imparted a bad taste to the food or water stored in them. After the wonders of fired pottery were discovered, the only unkilned clay prod-ucts that would remain in common use would be adobe bricks for building. By Sumerian times (*ca.* 4000–2500 BCE), it was a religious taboo to drink out of a sun-dried clay cup.[51]

Ceramic technology was quickly followed by metallurgy. Copper is typically found as an ore, unrecognizable from its final, processed form. However, on rare occasions, it does pop up in a somewhat pure state called "native copper." One can shape native copper by repeated blows with a very hard rock to form a flat sheet. Still, it is a cumbersome endeavor that often leaves rocky impurities in the metal. On the other hand, by exposing the copper to high temperatures—even a couple hundred degrees short of its actual melting point—it becomes far more malleable. Since copper metal-lurgy arose after the invention of lime and ceramic kilns, it seems probable that the copper was put into a kiln to make it easier to work with. The temptation for a "let's see what happens when we stick this stuff in there" experiment would have been irresistible. By taking a branch and then split-ting it at one end, the Neolithic smith had a convenient set of tongs with which to hold the native copper. Covering the branch with wet clay mud would have been sufficient to insulate it from the fire for the brief time needed to make the copper malleable enough for pounding.

As the efficiency of limekilns improved, it is possible that they could reach temperatures high enough to completely smelt the copper. Unlike limestone, where the insulating properties of the stone delay calcination, copper quickly liquefies once it reaches its melting point. After the advent of ceramics, a fired-clay bowl could now serve as a fireproof crucible in which the copper could be fully smelted. Fireproof clay molds would also

TECHNOLOGICAL PROGRESSION AFTER THE DISCOVERY OF LIME

Note: Italicized items are those that require high-temperature (>190°C) ovens. Italicized and underlined items require *very* high-temperature (>500°C) ovens. The dates below refer to when the respective technologies (**boldfaced**) first came into use according to the present archaeological evidence. *Lime products* include plaster, mortar, and concrete, all of which utilize kilned lime.

Göbekli Tepe	Nevali Çori	Çayönü	Çatalhöyük
ca. 9600 BCE (preagricultural)	*ca.* 8600 BCE (preagricultural)	*ca.* 8000 BCE (partly agricultural)	*ca.* 7400 BCE (agricultural)
Limestone carving	**Limestone carving**	**Limestone carving**	**Limestone carving**
Lime products	*Lime products*	*Lime products*	*Lime products*
	Fired-clay figures (no pottery)	*Fired-clay figures* (no pottery)	*Fired-clay pottery*
		Copper metallurgy	*Copper metallurgy*
		Woven cloth	**Stone-ground wheat**
			Baked bread
			Woven cloth

Temperatures for Crafts Requiring Kilning (in chronological order of discovery)

To extract lime from limestone:	848°C	(1,558°F)
To fire ceramics:	500°–600°C	(932°–1,112°F)
To bake bread:	190°–204°C	(375°–400°F)
To smelt copper:	1,085°C	(1,985°F)

allow the smith to cast a specific tool, like a sword or axe. However, these last innovations would come centuries later, and they probably coincided with another invention: the use of charcoal and the blacksmith's bellows.

However, the discovery of lime presents us with a very engaging mystery. Why would our Neolithic ancestors go to such tremendous effort to build and service a high-temperature kiln for the purpose of creating lime unless they first knew what the end product would be—and how would they know that?

Clearly, some extreme natural agency must have provided the inspiration, a geologic or atmospheric phenomenon demonstrating that limestone heated to a very high temperature will produce a "magical powder." Only three natural forces on the face of the earth can reach temperatures high enough to calcify limestone: volcanism, lightning, and, on rare occasions, a forest fire. Let's look at them. Volcanism comes in two basic forms: explosive and effusive. An example of the explosive type would be the eruption of Mount St. Helens in 1980. A vast subterranean volcanic chamber of hot gases and molten rock builds up so much pressure that it eventually explodes with the force of a nuclear bomb. An example of the effusive type is the ongoing, languid flow of lava from Kīlauea on the island of Hawaii. Volcanism is unlikely to have served Paleolithic peoples as a demonstrative agent for calcination. No one can closely observe the effect of an explosive eruption without being blown to smithereens or instantly carbonized by a pyroclastic avalanche. Slow lava flows, while far less dangerous (they typically advance at the speed of a crawling infant), almost always travel on older igneous (volcanic) rock, not limestone. Lava reaches temperatures high enough (700–1,300°C / 1,300–2,400°F) for calcination, but even if a fresh effusive flow were to crawl across limestone, the heat of the molten lava would obscure whatever effect it was producing on the rock directly beneath it.

Another candidate for calcinating limestone is a forest fire, which reportedly can produce temperatures approaching 1,200°C (2,192°F), but this is very rare even today and was much less common in ancient times. For example, active suppression of wildfires in North America for the last century has caused an unnatural buildup of vegetation that has resulted in very hot "high-intensity" fires. Only recently have we awakened to the fact that fires are a natural and essential part of all but the dampest environments. They enrich the soil and control noxious plant species, and the burnt vegetation serves as a chemical "wake-up call" to seeds buried beneath the surface. Natural fires tend to be "low-intensity" affairs that quickly sweep through an area, burning mostly grasses and brush and only lightly charring tree bark. The burn effects of low-intensity natural fires usually disappear within a few years.

However, high-intensity fires, while less frequent than now, certainly took place in ancient times. Vegetation could escape a natural low-intensity fire for a number of years and gradually accumulate. If a severe drought then occurred, followed by a natural (lightning) or human-made (campfire) ignition event, things could have gotten rather hot. Still, the fire itself, which burns at 380°–590°C / 720°–1,100°F, is not hot enough to create lime from limestone (it is not contained). Rather, it is the air *above* the fire that is capable of reaching the necessary temperatures, but only briefly. Another problem persists with this scenario. The dense growths necessary for the formation of a high-intensity fire are generally found where there is more soil than rock. However, it is just conceivable that ancient high-intensity fires took place below a limestone outcrop and burned hot enough and long enough to calcinate the limestone.

Was a forest fire the source of inspiration? It is possible, but unlikely. Hunter-gatherers would avoid burned-out areas because there was no longer anything to hunt or gather. The discovery of a freshly precooked carcass of, say, a deer might be the exception, but that would argue for a low-intensity fire, since a high-intensity blaze would have carbonized the carcass. Also, the evidence of calcination would have been eliminated with the first rainfall after the fire, since water transforms lime back to limestone; calcium oxide becoming calcium carbonate once again. Finally, limestone, like most rocks, acts as an insulator as well, one reason why it must be broken up into chunks no larger than a small fist. A forest fire burning on top of, or next to, a limestone outcrop will have little effect on it.

This leaves us with lightning, which, at first glance, seems as improbable as the other choices but is actually the best candidate. Satellite data show that lightning occurs roughly one hundred times *per second* around the globe.[52] In some places, the number of lightning strikes per square kilometer (roughly 0.4 sq miles) of land is well over one hundred *each year*, and one hundred and fifty is not unheard of.[53] And lightning can kill—and does so in a spectacular sound and light show. In 2010, one lightning bolt killed thirty-two people in two separate villages during a single storm in Pakistan.[54] In 2005, another strike killed sixty-eight dairy cows in Australia.[55] In the earliest literature, as well as in the later writings of the classical period,[56] being struck by a lightning bolt was near the top of the list of deadly natural forces to be feared. For people who were outside much of the time, like our hunter-gatherer ancestors, the prospect of getting hit by lightning was particularly dreaded.

Yet, lightning was also considered a divine force. It was some sky god *in action* at the most direct and immediate level. Getting hit by lightning demonstrated that the victim had no doubt offended some deity—one reason why it was probably more feared than, say, a fatal illness. And since it was a divine force, the spot where lightning had struck the ground was considered holy by many early peoples.[57] Adding to this imagined sanctity were the unusual artifacts discovered at the point of impact. For instance, if lightning strikes sand, a fulgurite is usually found. A fulgurite is created when a lightning bolt vitrifies the silica of the sand and forms a hollow glass tube. Fulgurites range from a few inches to over twelve feet long. They often resemble the shape of a lightning bolt, which accounts for another name given them: petrified lightning.

When lightning hits limestone, lime is created. The temperature of a lightning bolt is over 27,000°C (about 50,000°F), hotter than the face of the sun, so the "bake period" is instantaneous. While lightning is usually accompanied by rain—which would return the lime to limestone—sometimes it is not. At least one news story is filed each year about some golfer who, though seeing an oncoming storm, is little worried and decides to go for another hole and, as a result, is struck dead by a lightning bolt in broad daylight.

So, how did the people of the late Paleolithic discover that lightning produced lime? Let's revisit that ancient shaman we met earlier. This time he is gathering curative herbs in the mountains. Menacing clouds gather against the slopes, obscuring the sun, and the sky grows darker by the minute. The shaman knows that storms come fast and furious in the mountains, and so he decides that *now* would be a good time to return to his camp.

As luck would have it, as soon as he starts back home, rain begins to fall and the sky flashes with lightning that is quickly followed by a very loud thunderclap. Even primitive people understand that the period between a lightning's flash and its accompanying boom denotes how close the strike is, and so the shaman knows that this deadly force is very near. Suddenly he sees a lightning bolt strike the ground about fifty feet away. He squats down to make himself a smaller target for the obviously angry gods above and decides to ride out the storm. On this occasion, the storm quickly passes, and the shaman breathes a sigh of relief and perhaps offers up a prayer for his deliverance.

As we have seen, shamans are naturally curious about things; they probably possess the same strands of DNA held by scientists and good police detectives. The shaman walks over to look at the spot where the lightning struck the ground and finds that it landed on a limestone outcrop, scorching out a hole in the rock. The charred hole is surrounded by radiating cracks and a strange white powder (the sheltering trees prevented the brief rain from reaching this spot). The shaman picks up a pinch of the powder and makes the same painful discovery as our previous hypothetical primitive. The shaman experiments with it and, as in the earlier scenario, discovers that the powder is an amazing substance.

The shaman knows that lightning is hot and fiery. Accounts have no doubt been passed down of people being killed by it (and the resulting charred flesh and clothing), as well as its propensity for starting fires. The shaman understands that the powder was created by the especially intense heat and fire of the god-wielded lightning. He asks himself: "What if I could bring such heat to the rock? Would I also be able to produce this divine powder?" At this point, superstitious apprehension prompts a rationalization, and so he rephrases the question: "What if this god wanted me to witness the effects of the sky fire, so that I could also make this magical substance?" (It is easier to believe that you are following a god's will than usurping its power.) Inspired by this seemingly mystical experience, the shaman then spends perhaps ten or twenty years in trial-and-error experimentation in an attempt to get limestone to yield its strange white dust. Eventually, his efforts are rewarded. He wows his fellow tribesmen by showing them how he can make a hot, flameless heat using... *water!* And then, *presto*, a rock! Soon thereafter, he teaches his son and future shaman the secret of how to make the powder. The rest of the story is easy to guess: limestone comes to be regarded as a divine stone possessed with both hidden powers (its magical lime) and more commonplace qualities (its strength and ease of carving). Eventually, lime and limestone are used to create the first stone temple.

It is certainly within the realm of possibility that the shaman or his descendents chose the potbellied mount to build a temple complex because it may have been here that the revelatory lightning strike took place. Certainly, the hill would have been the highest point in the immediate vicinity and, so, especially vulnerable to lightning bolts. As with all ancient preliterate people, we will never know the full story and can only speculate about the nature of their gods or the inspiration for building such a grand religious center.

It would be centuries later, at the time of Çayönü and Çatalhöyük, that the lime would finally be used for secular dwellings as well. The once magical powder has now become nothing more than an everyday construction tool. Nevertheless, people would remain captivated by the counterintuitive properties of lime and its products.

Was the technological evolution in ancient Anatolia unique? Probably not. Excavations of the Dadiwan culture in China have revealed a large settlement that is roughly the same age and size of Çatalhöyük and that possesses the same crafts: agriculture, concrete, ceramics, and metallurgy.[58] Neolithic sites in Serbia, such as Lepenski Vir,[59] suggest a similar technological progression, including the use of concrete and carved limestone in a preagricultural, hunter-gatherer society.

However, what makes the excavations in Anatolia especially interesting is their extreme age and the fact that we have a fairly comprehensive picture of the progress of those social and technological revolutions. Excavations of more recent sites in the area also demonstrate that knowledge of these respective crafts were not lost to humanity through social upheavals or natural disasters. Agriculture continued on its course and was constantly refined and made more productive. Ceramics craftsmen would soon create products of great beauty and utility, while the blacksmith's forge would eventually produce both plows and lethal weapons. Lime and limestone construction would also remain in use. Eventually, they would be applied to constructing the most enduring and beautiful buildings of all time.

As for concrete, its story had just begun.

Chapter 2

TOWERING ZIGGURATS, CONCRETE PYRAMIDS, AND MINOAN MAZES

While foothills of the Taurus Mountains saw the development of the earliest stone temples, furnace-based crafts, permanent settlements, and agriculture, it would be to the south, in the delta region of the Tigris and Euphrates Rivers, where all these technologies and societal developments would coalesce to create the earliest human civilization: Ki-en-gir, "Land of the Lords of Light." We know it today as Sumer. What really distinguished Sumerian civilization was its written language, which is the true dividing line between prehistory and history. The beginnings of the Sumerian writing system were simple enough. The burgeoning growth of agriculture required enumeration beyond the easy reference of the digits of both hands, so a written transcription of small inventories was first conceptualized as numerical scratches or hash marks under a picture of the item being tallied. This became more complex as more products and services needed to be depicted. Soon, symbols representing specific actions and ideas were added. Eventually, the Sumerians used letters made of wedge groupings called cuneiform (Latin for "wedge shaped") to represent the different sounds of human speech. A simple reed stylus was used to impress the wedges onto a wet clay tablet that was then baked or left to dry. This phonetic cuneiform script became one of the world's most successful writing systems and was adopted by many of the peoples whose cultures and kingdoms would follow the Sumerians in the Tigris-Euphrates delta. It endured for almost four thousand years, surviving until the second century of the Common Era. Tens of thousands of cuneiform tablets have been uncovered and translated, providing archaeologists and historians with a rich source of information about the everyday lives of the people, both commoners and royals, who trod the earth of ancient Iraq and Syria.

Clay not only formed the basis of the Sumerians' revolutionary writing system; it was also their principal construction material. The river deltas that gave birth to the earliest civilizations were also rich in mud clays. These clays were pressed into a simple rectangular wooden mold and then dumped out to dry in the sun. The clay mud also provided a cheap mortar and plaster. Primitive adobe masonry can create relatively sturdy one-story structures whose thick walls keep the interior cool in summer and retain heat in the winter. If regularly maintained, an adobe structure can last centuries. Adobe remains one of the most popular building materials in many of the warmer climes of the Third World.

The kilning of limestone to produce lime continued: having long lost its seemingly supernatural charms, it was probably looked upon as a fine, but terribly expensive, building material. There was less limestone in the lower Tigris-Euphrates region than in the mountains to the north, and the wood needed to fuel the kilns was also a scarcer resource in the delta. Straw and wood from riparian trees would serve as cooking fuel and for kilns, but the latter were restricted mostly to pottery making, since metal-forging ovens used more efficient charcoal as fuel. Although metal forging required higher temperatures, the smiths needed to concentrate their heat in only one small spot, charged by their bellows, rather than across a large interior space that was the size of a small room. The end of the second millennium BCE saw increased use of kilns to make a new building material: baked bricks. Like pottery, bricks required a shorter bake time and lower temperatures than did lime production. However, because the large number of bricks required for building purposes was substantial, their use was mostly restricted to public buildings or palaces financed by kings. As the grand, conical ziggurats of Babylon rose to the sky, the preferred building materials were cut stone or brick masonry, with adobe used within their thick walls as cheap filler.

It was not only the huge fuel requirements that limited lime's use as a mortar during the Sumerian and later Mesopotamian empires but also the discovery of two other materials that seemed more practical. The first was gypsum, the material now used to make plaster of paris. Just as calcium carbonate is the chief constituent of limestone, so calcium sulfate is the principal element of gypsum. To the ancients, gypsum must have appeared to be a good alternative to limestone. It calcinates at a far lower temperature (120°C / 248°F), and the resulting powder is far less caustic (you can scoop it up with your hands with no ill effects). Add water to the powder, and though the resulting mixture is very warm, it does not generate the scalding temperatures of its chemical cousin. Once thoroughly hydrated, the substance sets after a few hours, returning to its original state as a hard mineral. Gypsum did not create a mortar as hard or adhesive as the lime-based one, but it was probably "good enough" for the ancient contractors, who no doubt appreciated the fact that they did not have to worry much about their sons, apprentices, or themselves being seriously injured while handling the material.

Another useful substance locally abundant was tar, which bubbles up in some spots from the vast subterranean collections of petroleum that lie beneath the crust of this part of the Middle East. Unlike limestone or gypsum, tar requires no processing; it can be used *as is*. As a result, the petroleum that would one day bring great wealth to the region was initially used as a convenient and satisfactory masonry adhesive.

Lime concrete and mortar did not disappear during the Sumerian, Akkadian, Babylonian, or Hittite Empires of Mesopotamia. In places where trees and limestone remained plentiful, the old white magic continued to be practiced. It was still the best stuff around.

LAND OF THE PHARAOHS

Rich clay deposits surrounded the Egyptians of the Nile Delta as they did the people on the Tigris and Euphrates. Naturally, sun-dried blocks served most of their building needs and would continue to do so for thousands of years.

The exceptions were grand tombs and monuments. From the time of the earliest dynasties, Egyptian rulers wanted to be buried in lavish tombs

FIGURE 5. The pyramid of Pharaoh Djoser, reportedly designed by the renowned architect and artist Imhotep.

that would serve their imagined needs in the afterlife. The first of these tombs were mastabas. *Mastaba* is Arabic for "bench"—an appropriate name for these rectangular structures, the interiors of which replicated the ruler's own domiciles, with the same rooms set aside for various purposes, complete with furniture and artwork. The tomb chamber was hidden underneath the main structure. The earliest mastabas were built of adobe bricks; centuries would pass before the pharaohs switched to the more permanent limestone. Interestingly, the walls of the first limestone mastabas

FIGURE 6. Arial view of pyramids at Giza; the Great Pyramid of Khufu is in the right foreground. There is a contentious controversy about the materials used for the Giza pyramids.
Some authorities claim that, instead of limestone, a special concrete was poured into molds
that formed the blocks used to construct them. If true, it would be the largest use of concrete before the construction of the Panama Canal in the early twentieth century.

were carved and painted to resemble clay bricks. In other words, the early
Egyptian rulers still lived in adobe structures, but, knowing the material's
limited life span, they sought a substitute that would endure for the eternity
of an afterlife: limestone. Still, it had to resemble the walls of their own
palaces and so was made to appear like clay. The mastabas gradually became
larger and more elaborate, paralleling the pharaohs' own increasingly opu-
lent lifestyles. Stories (called "steps"), each with sloping walls, were added to
the later mastabas. The Egyptian word for mastaba seems to have meant
something like "place of ascension." Evidently, the additional steps meant a
higher mastaba, and so the less one would need to ascend.[1]

When the Third Dynasty pharaoh, Djoser (2667–2648 BCE), came to
power, he wanted to build a mastaba grander than those of his predecessors.
Work on the structure began early in his reign at a place now called Saqqara,
then known to the ancient Egyptians as *kbhw-ntrw*. (Like all ancient

Semites, the Egyptians were somewhat averse to expressing their vowels in writing.) The largest previous mastaba had three steps, so Djoser, of course, wanted four. His architect was the renowned Imhotep, who would be deified centuries later as Egypt's "Father of Architecture, Sculpture and Medicine."[2] (Poor Imhotep would later be vilified in American and British "mummy movies" in which he appears as the resurrected heavy. The plot equivalent would be that of an undead Leonardo da Vinci rising from his tomb to terrorize modern Florence.)

Either Djoser kept changing his mind, or Imhotep kept urging his master to build on a grander scale, for there are three distinct modifications to the tomb. The first version is a typical mastaba, though grander than earlier ones and featuring a fourth step. The structure was then enlarged, and a fifth step was added. Shortly after this phase, all the stops were pulled, and a *sixth* step was added, as well as a dazzling temple precinct surrounding the pyramid. The temple precinct incorporated not only every architectural form then known to the Egyptians but several others never seen before, which would serve as archetypal designs for Egyptian buildings over the next three thousand years: the first colonnade (using the first fluted columns); the first columned hall, called a "hypostyle"; the first cavetto cornices; the first porticos; the first life-size statues; and, of course, at the center of it all, the first pyramid: the famous step pyramid of Djoser. Instead of just a large burial chamber underneath the pyramid, Imhotep built an underground warren of hallways, tunnels, and shafts that stretch almost four miles in length, connecting some four hundred rooms and galleries. To fool any would-be tomb raiders, much of the underground labyrinth was designed as a series of mazes that led to multiple dead ends and false entrances. (Djoser's mummified remains and treasure would remain safe for a few centuries, but like all the pyramid tombs, it would eventually be found and robbed of all valuables.)

Djoser's six-step pyramid is just above 60 m (*ca.* 197 ft) high and echoes the older mastabas in that it is rectangular: 173 m by 107 m (*ca.* 568 ft by 531 ft). Limestone was used to a greater extent here than in all previous Egyptian buildings combined. (One could make a case that the amount of limestone used in Djoser's pyramid and tomb district exceeded all previous *human* construction efforts with this stone *combined*.) Surrounding the tomb complex was a ten-meter-high (*ca.* 33 ft) wall that was 1.6 km (1 mile) long made from pure white limestone from Tura, Egypt. The temples inside the district were also built of limestone, and their walls were masterfully carved with depictions of Djoser enjoying such bucolic pursuits as hunting, fishing, and dispatching enemy soldiers with club blows to the head.[3] No longer were the walls made to appear like adobe bricks, for by this time pharaohs lived in limestone palaces. Clay bricks were now for the common folk.

Shortly after Djoser's reign, a new family came to power, which we call the Fourth Dynasty. As before, each pharaoh of this new dynasty would compete with his antecedent and build tombs larger and more perfectly proportioned than the previous ones. These remarkable structures would reach their apogee during the reign of the pharaoh Khufu (2589–2566 BCE), who would build the most magnificent of all these monuments, the Great Pyramid of Giza. The finished result rose 147 m (*ca.* 482 ft), and its four-sided base was a perfect equilateral square, 230 m (*ca.* 755 ft) per side. The pyramid was built with solid blocks of limestone, with the finest white limestone reserved for its exterior cladding, or overlay. Gypsum[4] was used to cover the masonry joints, and the stone was then polished to a smoothness that reflected the sun with an almost mirror-like intensity.[5] The best stone—the exterior Tura limestone cladding—was removed centuries later to build palaces and mosques in medieval Cairo. Even with its magnificent exterior stripped away, the Giza pyramid is staggering to behold. Barring multiple nuclear attacks, it will probably outlast any other human-made structure—and probably the human race as well.

But was limestone really used to create the pyramids? The blocks *appear* to be limestone, but according to two materials engineers, a substantial part of the pyramids—most notably Khufu's Great Pyramid—actually consists of cast concrete. If this is true, the pyramids represent the greatest volume of concrete manufactured and applied to a single engineering project until the construction of the Panama Canal some twenty-four centuries later.

THE GREAT CONCRETE PYRAMID CONTROVERSY

The Egyptian pyramids have always been magnets for people proposing *fringe* theories—to put it politely—about their creation. Despite the fact that a dozen pharaohs spent over a century perfecting the building techniques required for the construction of the pyramids, some people still cannot believe that the ancient Egyptians built these remarkable monuments without some sort of mysterious assistance. Some suggest that extraterrestrial aliens (their most popular locus being the Pleiades star group) were involved, while others opt for tech-savvy survivors of the "lost continent" of Atlantis. Many respected Egyptian authorities would prefer to term such theories as "unhinged" rather than "fringe"; the latter suggesting that these imaginative scenarios have one foot grounded in truth and the other in speculation. "Unhinged" seems to be the better adjective for describing the alien- or Atlantis-based theories.

Somewhere in the middle ground between the unhinged and the more widely accepted theories about pyramid construction is the cast-concrete-block hypothesis proposed by two men with respectable scientific credentials. Neither Michel W. Barsoum, a professor of materials engineering at Drexel University in Philadelphia, nor chemical engineer Joseph Davidovits, head of the Geopolymer Institute in Saint-Quentin, France, come across as fringe—or unhinged—theorists. Yet these two men have lit the fire of perhaps the most hotly debated controversy in Egyptology today: that major portions of the pyramids were built of concrete, not limestone, as many archaeologists and historians believe.

The Frenchman started it. Davidovits suggested back in the early 1980s that some of the rock he examined at the Great Pyramid looked more like concrete than limestone.[6] And not just any concrete: an eco-friendly version using a "geopolymer." To refresh your memory of high school chemistry, a polymer is a long molecular chain that lends strength and/or stability to a chemical compound. One example of a geopolymer—a natural polymer arising from the earth—is petroleum. (Davidovits has trademarked the preexisting term, converting it to his proprietary Geopolymer™) According to Davidovits, his Geopolymer concrete—

which he claims the Egyptians were the first to discover—uses an alkali solvent, natron (sodium carbonate), in water to dissolve clay rich in aluminosilicates. (Aluminosilicates are the mineral form of aluminum and silicon that are needed to serve as a concrete binder.) Crushed limestone and a little lime is added to this soup, which is then thoroughly mixed and allowed to dry to a thick mud form, during which time the dissolved aluminosilicates recondense to form a stronger crystalline structure. The resulting "mud" is then rammed into wooden forms in a process similar to that used to create adobe bricks—although with far larger molds, of course. The result is a block with a look and texture very similar to stone. A demonstration of the process can be viewed at http://vimeo.com/1657432. (To more clearly make his point, Davidovits dressed his workers in ancient Egyptian garb.)

Davidovits's hypothesis is both ingenious and intriguing. He believes that soft limestone with a high clay content was quarried near the Giza Plateau, then dissolved in a water, lime, and natron solution (natron was used by the Egyptians for mummification) and held in large pools or holding tanks fed by the Nile. The pools were then left to evaporate, leaving behind a moist mud, the ancient equivalent of wet concrete. This concrete was carried to the pyramid site in baskets where it was tamped into molds. After a few days, the concrete would cure, forming a building block for the pyramid. Davidovits claims that a work crew of just ten people could make a dozen large blocks within the span of several days. Since the blocks were cast in place, no elaborate hoisting equipment or levers were needed to build the pyramids. By using their smarts, and not their sweat, the Egyptians saved millions of man-hours in constructing the pyramids. Davidovits explains his theory in detail in a book coauthored with Margie Morris titled *The Pyramids: An Enigma Solved.*[7] His revisionist account of Egyptian engineering is compelling and disturbing. If Davidovits's explanation is accurate, Egyptian history would need to be reevaluated, countless books about Egypt's architectural splendors rewritten, and, of course, numerous History Channel documentaries scrapped, as well. And could anyone ever again watch the Hebrew slaves struggling to build Pharaoh Rameses's monuments in Cecil B. DeMille's *The Ten Commandments* without shaking their head and muttering disapproving smart-aleck remarks? (Or, at least, *more* smart-aleck remarks than the film already provokes?)

Putting aside Egyptian building techniques for a moment, the low-lime concrete discovered—or rediscovered—by Davidovits seems to offer great promise. The resulting concrete has great compressive strength and uses only a very small amount of lime and no kilned clay. If we assume that its bugs can be worked out—it is still non-hydraulic and begins to fall apart when immersed in water for more than two weeks—this Geopolymer concrete could substantially cut the energy required for, as well as the pollution generated from, cement manufacturing.

Egyptologists hotly contest Davidovits's theory. They point out that there is no historical data supporting this form of construction and that engineers have already examined the limestone quarries near Giza and calculated that the amount of removed stone was roughly equivalent to that used for all monuments in the area. As for the resemblance—at least to the human eye—of Geopolymer concrete to limestone, even standard concrete resembles a variety of limestone and even some granite.[8]

There the matter stood until 2004, when Egyptian-born Michel Barsoum took a trip to Khufu's pyramid while visiting his native country. Hiking around the monument, he noticed that a few of the stone blocks looked more like concrete than limestone. Selecting several different blocks, he knocked off a few pieces with a rock pick, put the fragments in a plastic bag, and brought them back to Drexel University to study.[9] Barsoum should not have taken samples in this manner, but he knew that applying for official permission would be a long, drawn-out affair. There was also a good chance that his request might be denied. The former minister of state for Egypt's Antiquities Affairs, Zahi Hawass, ruled his fiefdom with an iron fist and was very cautious about granting site permits. (Hawass, who has never been accused of being shy, pops up in seemingly every recent television documentary about ancient Egypt.)

Examining the fragments under a microscope, Barsoum discovered a kind of crystallization commonly found in concrete, but not in limestone. A Columbia State University geologist, David Walker, agreed that the microscopic examination "certainly revealed things you wouldn't expect to find in normal limestone."[10] To Barsoum, this suggested that some twenty-five hundred years before the Romans began using concrete, the Egyptians had been using it on an even more massive scale to construct their pyramids. Unlike Davidovits, Barsoum believes that concrete construction was not used on most of the pyramid's blocks but only the outer portions beneath the limestone sheets covering the structure.

Because of the controversial nature of his findings, Barsoum had trouble getting his paper published in a scientific publication. Eventually it was printed in the December 2006 issue of the *Journal of the American Ceramic Society*.[11] As the journal started reaching subscribers in late November, Barsoum began talking to the media. He also put together a slideshow to explain his work,[12] in which he shows microscopic images that contrast the different crystallization of concrete fragments and limestone. Barsoum also points to photographs that show that the pyramid blocks fit together so precisely, a thin sheet of paper cannot be placed between them. This, he contends, typifies cast concrete but would be almost impossible to achieve with stone carved with the relatively soft copper tools that the Egyptians used. A materials science professor at MIT, Linn Hobbs, was intrigued by Barsoum's work and assigned his students the task of building a small-scale pyramid using Geopolymer concrete. The project was completed without any difficultly, demonstrating that the technology was certainly feasible.[13]

Naturally, Joseph Davidovits was happy to see his original hypothesis apparently vindicated. A masterful and indefatigable publicist, Davidovits churned out a new flurry of press releases, papers, and online videos about his theories, and pointed to Barsoum's data and Hobbs's model pyramid as collaborative evidence.

One cannot help but be impressed by the amount of evidence Davidovits has marshaled in support of his theory. Studied in isolation, the case he presents in his books and papers seems almost unassailable. Unfortunately, almost all the evidence he and Barsoum point to is either misleading, wrong, or very wrong. It is a theory that, while extraordinarily clever, ignores a mountain of conflicting data. There are no limestone deposits rich in clay (called "marl") near Giza—although it's possible some existed thousands of years ago—and no pools have been uncovered that could have been used to process the concrete. There are no nearby sources of natron[14] or any other archaeological evidence that might support Davidovits's hypothesis. Independent scientists have exhaustively tested core samples from the monument stones and compared them to

Davidovits's Geopolymer concrete. They found no similarities.[15] Many of the blocks on the pyramid still show, even after the weathering of centuries, clear quarry markings. They are unquestionably limestone—and unquestionably the same limestone found at the quarry near Giza. As for the precisely fitted masonry blocks seen at the Egyptian pyramids, this kind of masterful carving is hardly unique. Mayan and Inca stonecutters in Mesoamerica achieved similar precision using tools more primitive than those of the Egyptians. Finally, it must be remembered that Khufu's tomb chamber within the pyramid is constructed with granite,[16] and granite is far harder to carve than limestone, especially with primitive tools. Yet, the Egyptians managed to do so.

What about Barsoum's sample from the Great Pyramid that seems to identify it as artificial stone? It may well be artificial stone, and concrete at that. The Egyptian monuments have been subjected to major restoration projects since pharaonic times and during the Roman period, as well. Concrete blocks were used in the modern era to fill in some gouges left by quarrying performed in medieval times. Zahi Hawass believes that this was the source of Barsoum's fragments: "I would ask Dr. Barsoum the question: where did he get the samples he is working with, and how can he show that the samples are not taken from areas that have been restored in modern times?"[17]

Barsoum and Davidovits have apparently decided to defend their position by mounting a good offense. Do an Internet search for "concrete pyramids," and you will get a blizzard of hits that originate from their websites, or from the regular interviews the men grant reporters, or from mainstream media stories that simply recycle their press releases (apparently written by research-averse reporters who simply ask the opposing side for a few brief remarks). It is the cyber version of a debate in which one advocate is soft-spoken while the other uses a megaphone. Unfortunately, Davidovits's refusal to yield on this long-settled controversy is obscuring his exemplary work in developing an eco-friendly concrete that also possesses high compressive strength. The potential energy savings and pollution-curbing attributes of Davidovits's Geopolymer concrete is more vital to the public interest than his attempts to uphold a fascinating but deeply flawed theory on how the pyramids were constructed.

THE MINOANS AND THE GREEKS

Toward the end of third millennium BCE, a seafaring people from the island of Crete began showing up at ports throughout the Eastern Mediterranean, eagerly looking for items to trade. They had mastered the art of constructing small ships capable of sailing long distances and sturdy enough to carry tons of cargo. This gave them a decisive trading advantage over other peoples who lived on the Levantine coast. These seafarers eventually created a mercantile empire with colonies or trading posts throughout the region.[18] We call the people of this lost kingdom the Minoans for the legendary King Minos, who reportedly once ruled the island of Crete.[19] In truth, we do not know what they called themselves.

Trading not only brought the Minoans wealth but, more importantly, exposed them to the arts and sciences of the many societies with which they came into contact. These skills and arts were adapted and then refined by the Minoans. By 1700 BCE, the Minoans were inarguably one of the most culturally advanced peoples in the world. They built huge palaces with a labyrinthine series of rooms and halls, which probably provided the basis of Greek myths about the labyrinth of the Minotaur, a half-man/half-bull monster that pursued the Athenian hero Theseus. The Minoans possessed a remarkably sophisticated drainage and sewer system and enjoyed hot and cold running water in their homes and indoor toilets. And these amenities were not restricted to their royalty but were also enjoyed by a relatively large middle class.[20] As in modern Western societies, women wore clothes that accentuated their beauty instead of hiding it. Women also seemed to have shared equal status with men, and they certainly dominated the island's powerful priestly class. The surviving mural paintings of the Minoans are beautiful, too, and are quite unlike the stiff and usually menacing religious and political art seen in the Mesopotamian and Egyptian civilizations. In these murals we see charming scenes of everyday life. It appears that their artists loved creating beauty for beauty's sake. It is not unrealistic to assume that the Minoans would have become the dominant Western culture had they not been subjected to one of the most colossal volcanic eruptions to occur in the history of the human race.

Around 1640 BCE, a major outpost of the Minoan civilization, the island of Santorini (called Thera by the Greeks), was annihilated when its volcano exploded in a super eruption, ejecting an estimated 603 cubic km (*ca.* 143 cubic miles) of rock and ash into the atmosphere, six times more than that ejected by the 1883 Krakatoa eruption.[21] Santorini was instantly transformed from a large island to a few small ones. No living thing within sight of the island could have survived. The blast generated a massive tsunami perhaps a hundred or more feet high, far larger than the one produced by the 2004 Indonesian or 2011 Japanese earthquake.[22] Since Crete forms the major part of the southern flank of islands that surround the gulf in which Santorini lies, the Minoan ports bore a large brunt of the tsunami and were obliterated. Most people who did not escape to high ground were likely drowned or crushed to death by the debris of smashed ships, docks, and warehouses. Although the surviving Minoans were able to rebuild their culture, their population had been severely reduced and materially weakened. Obviously, their military strength was also greatly diminished. Not long afterward, Greeks from the north began settling on the island, whether by force or peacefully, we do not know. The Minoan culture gradually disappeared, and only the names of a few of their gods and settlements survived into Greco-Roman times. Their language, preserved on a few inscriptions called "Linear A" by archaeologists, remains undeciphered.[23]

However, although the extinction of the Minoan civilization undeniably altered the course of history, it does not relate to our main inquiry here. What does concern us is the fact that the Minoans created a reasonably strong concrete. The volcano that wiped out this civilization had already erupted several times in the past, long before modern humans lived on Crete, and each eruption blanketed the islands of the eastern Mediterranean with a thick layer of pumice and ash. This volcanic "earth," rich in aluminosilicates, would later become known as "pozzolanic soil," a name derived from the town of Pozzuoli, Italy, where nearby Mount Vesuvius had been periodically depositing volcanic pumice for thousands of years (its most famous eruption wiped out the Roman towns of Pompeii and Herculaneum in 79 CE). Engineers now refer to such ingredients in concrete and mortars as "pozzolans" or "pozzolanas." Fly ash (ash from coal-fired power plants) and most kilned clays are also rich in aluminosilicates and are also classified as pozzolanas. High aluminosilicate volcanic powder does not need to be baked first, for Nature has already kilned it, and so it can be used straight from the ground. When pozzolans are mixed with lime, a remarkable transformation occurs when water is added: the two active ingredients combine to create a far harder and more durable material. This material is not only highly impervious to water and weathering but can actually set underwater, unlike strictly lime-based mortars or concretes for which exposure to air is necessary for the setting process. For this reason, concretes and mortars with these properties are called "hydraulic," and the pozzolanic portion is referred to as the "hydraulic element" or "hydraulic ingredient." Pozzolana is what separates Roman, natural, and modern concretes from the lime concrete used for thousands of years since the Neolithic period.

That the Minoan concrete was not as hard as later concretes is due to the fact that their building craftsmen were still experimenting with the material, sometimes choosing clay instead of sand as the mixing medium. No doubt they liked the more plastic qualities of clay, a malleability made more convenient by its slower setting period. The Minoans used their various concretes for floors, for foundations, and as a water-resistant mortar. They also laid some of their terra-cotta drainage and sewer pipes in this concrete,[24] perhaps to prevent their breakage during earth movements (Crete is regularly subjected to strong earthquakes). Sadly, just as the Minoans' use of concrete began to expand in the seventeenth century BCE, their empire suffered the shock of the Santorini eruption.

A SIMPLE CISTERN

Sometime around 700 BCE, a large, rectangular stone cistern was constructed on the island of Rhodes in the city of Kamiros (*Kameiros* in Greek). Kamiros was the principal city of Rhodes at this time, and was famed for being the birthplace of Peisander, a poet whose epic *Heracleia*, about the labors of Hercules, was ranked just behind the works of Homer and Hesiod by the ancient Greeks. The cistern held 605,600 L (*ca.* 160,000 gallons) of water, enough to support four hundred families. What makes this cistern unique is that the bonding agent was a hydraulic mortar utilizing lime, pozzolanic earth, and sand. It is the earliest known true hydraulic mortar, essentially *pre-Roman* Roman concrete without the heavy aggregate. It is not known whether or not the local masons understood the unique properties of this mixture, but its hydraulic nature was certainly ideal for a cistern. Down through the centuries, brickmasons and stonemasons have been known to be notoriously reticent about disclosing their mortar formulas.[25] This secrecy about their tradecraft, related in numerous stories, has endured up to the present. If the masons in ancient Kamiros truly understood that they had something special, it's likely they kept it to themselves, for we do not see this kind of mortar used in any of the pre-Roman sites in the Greek world. It seems to have disappeared after the cistern was constructed. Kamiros was twice destroyed by earthquakes before the Common Era and was eventually abandoned. Perhaps the last stonemason or two who were privy to the formula were killed when the earth shook and their shops collapsed on top of them. We will never know. What we do know is that a similar product was discovered centuries later on the Italian peninsula, near the Bay of Naples. The ancient Romans were the first people to recognize the full potential of this novel material. And they used it to create some of the most spectacular and enduring edifices in the world.

Chapter 3

THE GOLD STANDARD

...go to Rome and try to break old Roman concrete with an axe; you will only dent the steel.

—French architect Auguste Perret, 1950 interview

Japan's Society of Civil Engineers, concerned over the limited lifespan of modern concrete, is forming a committee to investigate why Roman concrete has endured for so long.

—News item from 2004

There has been an explosion of interest in Roman concrete, and it is not hard to see why. The Romans used concrete to build edifices capable of lasting thousands of years, while most modern concrete structures are incapable of lasting two centuries—and many are unlikely to endure beyond just one. Did the Romans, ignorant of modern chemistry, just happen to latch onto the right formula, while we, armed with several orders of magnitude more knowledge, accidentally chose the wrong mix? Or did the Romans, who used concrete for over a century before applying it to their most ambitious building projects, better understand the long-term environmental impact on certain formulations? The answers to these questions are complicated, but it seems that the Romans did indeed understand key aspects of concrete that we would not wake up to until the twentieth century. Because of our ignorance, we will have to spend over a trillion dollars in the coming years to either repair or replace our crumbling infrastructure. And, even though we have improved the formulation and application of modern concrete to improve its longevity, it still has the shortest life span of any major building material. For this reason, the way the Romans used concrete is of critical importance to us, and it provides our thematic focal point.

THE GREATEST ENGINEERS OF THE ANCIENT WORLD

When I was going to school some decades ago, we were told that ancient Rome's principal contribution to history was as conveyor and disseminator of the Greek culture. This would be like claiming that the United States' principal contribution to history was as a conveyor and disseminator of European culture. It's partly true, but mostly not. The Romans have also not fared well in today's media-driven, minimum-content society, in which most people learn their history from stirring but inaccurate movies and television miniseries. Pretty much all they can remember about the Romans are the duels in the arena and the dissipated lifestyles of some members of its upper class. Relying on these sources for our knowledge of the Romans is akin to making a moral appraisal of contemporary Americans based solely on tapes of Jerry Springer's show and the accounts of torture at Abu Ghraib prison.

In fact, the Romans and the Greeks were cocontributors to Western civilization. Together they represented a complementary confluence of cultures—a sort of European yin and yang. The Greeks brought art and literature to an unprecedented level of mastery. More importantly, they intensively explored both the scientific *and* ethical questions of the world: the first by improving and expanding the existing mathematical systems (which also led to deductive reasoning), and the second through drama and moral philosophy. However, the Greeks were also a remarkably quarrelsome people. The Greek mind was quicker but less stable than that of the Roman. The Greeks gloried in the minutiae of counterarguments, verbal obfuscation, and confusing paradoxes, while the Romans sought to discover the moral heart of an issue and find grounds of commonalty.[1] Greek identity was often tied to their city-states, which constantly warred against, or made alliances with, other city-states. Today's enemy was tomorrow's friend, and then an enemy again. Few Greek governments enjoyed a stable political framework, and classical Greek history is essentially the dreary account of one long internecine conflict after another.[2] The Romans, on the other hand, absorbed their conquered peoples into a system that offered real benefits. By the second century BCE, the Italian peoples who belonged to tribes that once fiercely opposed Rome, like the Samnites and the Etruscans, had become Roman citizens, many of whom were members of the equestrian or senatorial class. Later, the Punic North Africans—descendents of Rome's once mortal enemies the Carthaginians—along with the Syrians, the Greeks, the Britons, and the Germans, would eventually hold high power in the empire as well. This is not to say that the "absorption process" was painless (no one likes being conquered), but it worked for many centuries, and, with a couple of exceptions—like the ancient Judeans—the people of the empire grew to recognize the benefits that the *Pax Romana* offered them, and they prospered.

The Romans were a practical people who abhorred instability, which was one reason why they worked hard to come up with just and enduring laws. Importantly, the Romans held an almost transcendental respect for their legal system, which was not only accessible to all Roman citizens but to noncitizens as well. For example, the people of a province, though they might not be Roman citizens, could bring suit against their ex-governor (invariably a well-connected Roman aristocrat) for corruption or the arbitrary use of power. Based on the surviving accounts of such cases, the provincials stood a good chance of winning such suits.[3] Consequently, a rich patrician could suddenly find himself impoverished by the steep penalties imposed for his malfeasance, or banished from Italy, or both. The emperors were technically above the laws, but the better ones were loath to act outside them. By the early third century, all freeborn or manumitted men were officially Roman citizens, blurring or eliminating the old distinctions between "provincial" and "Roman."

Another corollary to the Romans' sensible nature and love of stability were their engineering skills, which they exercised toward practical ends.

Before the Romans, major engineering efforts were usually directed toward creating awe-inspiring tombs (the Egyptian pyramids or Mausoleus's famed sepulcher) or making gaudy power statements (the Colossus of Rhodes). Of the Seven Wonders of the Ancient World, the only one that offered real practical benefit to its peoples was the Lighthouse of Alexandria. The Romans rightly held that most of their engineering achievements, while generally less thrilling to behold, were more significant in that they provided a higher standard of living to the inhabitants of their empire.[4] Aqueducts brought an unprecedented amount of fresh water to town and city dwellers, sometimes exceeding the volume now enjoyed by their modern descendents. Expertly designed and well-bedded Roman roads connected a realm that stretched from Northern Europe to the Middle East. Discerning travelers could peruse the equivalent of today's *Michelin Guide*[5] to check the quality of the inns along the way and note the attractions of each town they passed through. Civil servants generally made sure the needs of the local people were addressed, kept the infrastructure repaired, and saw that private disputes were settled equitably in court and not by vendetta. It was a world that respected and rewarded education, encouraged daily bathing, and whose medical knowledge and technology would not be surpassed until the early industrial age. Although they were ignorant of germ theory, Roman surgeons sterilized surgical instruments before an operation[6] and sometimes closed wounds with biocidic silver staples to avoid later infection.[7] Most people would be surprised to hear that the Romans possessed steam turbines, odometers, analog computers (using finely milled brass gears), coin-operated machines, and a host of other technologies that made life easier or more entertaining than ever before.[8] The cartoons and graffiti uncovered by archaeologists in ruins of taverns and bordellos have shown that literacy in the Roman Empire, while not universal, was fairly widespread, since most of the graffiti appears to have been written by slaves and members of the lower classes.[9] The English historian Peter Salway observed that the literacy rate in Roman Britain was higher than any subsequent British government was able to attain for the fourteen centuries following the empire's fall.[10] The Romans also fostered a meritocracy that was unequaled until modern times. Hard study and diligent work allowed many

Roman citizens—including "barbarians" and former slaves—to reach high administrative posts. By the second century, most of Rome's rulers were no longer Italian natives, and most provinces would eventually see at least one of its native sons become emperor. It is with good reason that we refer to the period that followed the fall of the Roman Empire as the "Dark Ages."

Of course, the engineering achievement that chiefly interests us is the Romans' rediscovery of hydraulic concrete. However, unlike before, when the formula was lost or forgotten, the Romans recognized the potential of this material and would use it with gusto throughout their empire until its fall in the fifth century. They systematized its production and application and were the first people to utilize concrete as we do today: putting it into large molds to create a strong monolithic architectural unit. Even after the Industrial Age gave us the tools and machines to surpass most Roman engineering efforts, we still scratch our heads and wonder about what methods the Romans employed to build some of these concrete structures, and how these buildings have managed to endure for so long.

Until the middle of the twentieth century, it was generally assumed that the Greeks had discovered concrete and used it in their major building projects before passing it on to the Romans. (The unspoken subtext of this belief was that the Romans had not been clever enough to invent something so remarkable.) The concrete remains of the platform at the summit of the Pnyx (pronounced *pnüks*) in Athens, the little hill from which speakers addressed the assembled citizens below, seemed to support this theory. Concrete was also discovered in the harbor works of Piraeus, Athens' port city. However, later investigative work revealed that the platform on the Pnyx did not date from the classical period but was instead a second-century Roman restoration. The same is true of the harbor emplacements at Piraeus, which were also renovated in Roman times. Aside from the cistern in Rhodes mentioned previously, no other structure incorporating hydraulic concrete or mortar from the archaic or classical period has yet been discovered from the Greek mainland or islands. We find no mention of concrete from the surviving Greek literature, and just one reference to lime mortar. The latter comes from a book dating from the fourth century BCE, called *On Stones* by Theophrastus of Lesbos.[11] Theophrastus speaks of the properties of gypsum, but it is evident that he has confused it with limestone, for he reports that, when water is added to the "gypsum," the mixture is too hot to touch. It is clear that he is actually referring to calcium oxide (lime), not to calcium sulfate. No mention is made of sand or any other filler. Most authorities agree that Theophrastus based this information on secondhand, or possibly thirdhand, accounts.

If the Greeks did pass on the knowledge of pozzolanic concrete to the Romans, it was definitely through an indirect route. The Greeks began establishing coastal colonies in Italy around 1000 BCE, but the ones who settled in the area near Vesuvius—where Roman concrete was first discovered—were Euboeans, and not the Doric settlers of the Greek islands with pozzolanic soil. Finally, no remains of Roman concrete have been discovered in Italy that can be confidently dated earlier than the third century BCE. Roman concrete piers have been uncovered at the site of the ancient Italian port of Cosa (Portus Cosanus), which, based on the pottery fragments used for some of the aggregate, might be dated to the middle of the third century. However, most of the concrete work at Cosa seems to date from the late second century or early first century BCE. (The broken pottery could have been taken from an old rubbish heap.) Nevertheless, even if the material dates from this later period, the ancient concrete found at Cosa is among the earliest examples of Roman concrete so far discovered, and the hydraulic ingredients came from the area around Mount Vesuvius, more than 300 km (*ca.* 187 miles) to the south.[12] Perhaps a sailor arriving on a ship from Naples told the Cosans of the remarkable cement being used back home. Barring some dramatic discovery—an ever-present possibility in archaeology—the current data suggest that the people in the Naples region independently discovered what would later become known as "Roman concrete" in the early third century BCE. However, by this time, its discoverers were either Latin- or Samnite-speaking Italians of Greek ancestry or Romans. Consequently, it is probable that the Romans, unassisted by Greek artisans or architects, independently discovered concrete through a process of trial-and-error experimentation that spanned several generations.

THE EVOLUTION OF ROMAN CONCRETE

It is frustrating tackling the story of Roman concrete from the few literary sources that have survived on the subject. The earliest, though indirect, reference comes from a book published around 200 BCE by Cato the Elder,[13] when the Romans were mostly using a nonhydraulic lime mortar and concrete. Our next—and most important—source comes from a book published almost two centuries after the first by the famed architect Vitruvius.[14] In the latter, we find the first detailed reference to true Roman concrete. Unfortunately, this source dates from a period just before the Romans had perfected the material and began using it in volume. Another, though quite brief, allusion to hydraulic concrete is found in two of the volumes of a Roman encyclopedia published around 78 CE by Pliny the Elder.[15] Although these literary references are mostly sparse and chronologically scattered, they have proven invaluable in our understanding of Roman concrete.

CATO

Although lime had been around since Neolithic times, the first detailed description of its manufacture and use in Western literature is Marcus Porcius Cato's *On Farming* (*De Agricultura*, also known as *De Re Rustica*—*On Rural Affairs*), written around 200 BCE. Cato, known to us as "Cato the Elder" to differentiate him from several of his descendents with similar names, was notorious for his extreme and often heartless penny-pinching (he advocated selling off old or infirm slaves instead of caring for them) and ruthless nationalism (in the last twenty years of his life, he ended every speech with the closing remark "Furthermore, Carthage must be destroyed," until that end was finally achieved by the Roman army). Even to his austere countrymen, Cato came across as a bit "over the top" in his severity. A little ditty was written about him:

Porcius snarls at everyone and at every place
With bright gray eyes and flushed red face.
Even after death, one can imagine well
That he'd scarce be admitted to Hell.[16]

A surviving portrait bust of Cato exemplifies Roman mastery of subtle character delineation, for the sculpture shows a sour-faced, unrepentant reactionary. Today he would probably be a popular talk show host.

In his book, Cato talks about using lime mortar for masonry walls and—as was done for the previous ten thousand years—spreading it on the ground of a dwelling to serve as an artificial stone floor. Cato's only aesthetic concession for his lime flooring is the use of tiles. However, being a tightwad, he thought that broken shards of pottery would serve just fine. Waste not, want not. Reading *On Farming*, one cannot help but wonder what it must have been like being Cato's wife.

His description of a limekiln—the earliest such reference to have survived—is especially fascinating. In keeping with his skinflint ways, Cato suggests having all necessary implements and materials ready for the contractor doing the lime kilning to avoid incurring any additional expense. Those things to have prepared in advance include limestone, a kiln for baking the limestone, the wood for the kiln, and the sand for the mortar.

Build the limekiln ten feet across, twenty feet from top to bottom, sloping the sides in to a width of three feet at the top. If you burn with only one door, make a pit inside large enough to hold the ashes, so that it will not be necessary to clear them out. Be careful in the construction of the kiln; see that the grate covers the entire bottom of the kiln. If you burn with two doors there will be no need of a pit; when it becomes necessary to take out the ashes, clear them through one door while tending the fire with the other. Be careful to keep the fire burning constantly, and do not let it die down at night or at any other time. Charge the kiln only with good stone [limestone], as white and uniform as possible. In building the kiln, let the throat [chimney or smokestack] run straight down. When you have dug deep enough, make a bed for the kiln so as to give to it the greatest possible depth and the least exposure to the wind.... When it is fired, if the flame comes out at any point other than the circular top, stop the orifice with mortar.[17]

Centuries of practice and experimentation before Cato's time had led to a number of advances. The grate described above was unquestionably iron, and this allowed more limestone to be kilned, as pieces could now be stacked right over the flames. Another refinement was the shape of the limekiln, the so-called Burgundy bottle form, which would be used until the beginning of the twentieth century. It allowed the maximum amount of stone to be calcinated with the least amount of fuel. Nevertheless, a phenomenal amount of fuel was still required: perhaps a couple dozen or more cords of wood for a single firing of the limekiln described by Cato. While the lime kilning methods used by Cato might seem primitive, he was using very up-to-date technology. However, he mixed and applied the *caementis* in a rather unsophisticated manner, at least in comparison to later Romans. Cato was a successful farmer and a formidable politician, but he was a crude architect.

Aside from using lime to create artificial stone flooring, Cato mostly talks about using lime for wall building. He instructs the reader to mix two parts of sand with one part of lime. This proportion is odd, and, while certainly functional, it goes against the miserly mien of the old Roman. Why not recommend the equally usable and thriftier mix of three or four parts of sand to one of lime? Perhaps because the quality of the sand Cato used to mix with lime was subpar. For him, "sand" was any kind of naturally pulverized rock—and it was most likely adulterated with dirt to varying degrees, as well. Using more lime was probably the only way of overcoming the sand's poor quality and/or contaminants. Another reason for the two-to-one proportions could be the fact that less sand makes a slightly harder mortar, although this marginally harder mix comes at an economic cost that Cato would have likely shunned. Consequently, the high amount of lime Cato used was likely due to low-quality sand.

Cato does not provide any instructions about constructing a wall. He gives only the dimensions, material requirements, and costs. To save money, he naturally recommends that the farm owner provide the rock rubble (*cae-menti*), sand, and lime for the contractor's use to avoid any markups. Cato calculates that each linear foot of a five-foot wall should require one *modius* (*ca.* two dry gallons) of lime and four *modii* (*ca.* eight dry gallons) of sand. If the contractor provides the lime, Cato tells the reader exactly how much it should cost, so he won't get cheated.

Why did Cato provide details about constructing a limekiln but write nothing about how a wall should be built? The likely answer is that, while many farmers were unfamiliar with lime kilning, almost everyone knew how a Roman wall was constructed. And the way the Romans built their concrete walls was unique for their times.

ROMAN CONCRETE WALLS

Even though the temples built on the summit of the Acropolis in Athens— especially the Parthenon—were marvels of engineering and the pride of all Athenians, their own homes were of far humbler construction and were usually built of adobe brick. While adobe was certainly serviceable, it had a major security flaw: one could dig a hole through an adobe wall in less than an hour. Indeed, the ancient Greek word for *burglar, toicorucos* (τοιχώρυχος), means "wall burrower." And while adobe holds up well in sunny regions, like Greece and the Middle East, it does less well in damp areas, such as northern Italy, France, and Germany. In short, adobe is not a universal building material.

Most Romans had no worries about wall burrowers. By Cato's time, all but the lowliest citizens built their walls with lime concrete, and these were among the best in the ancient world. The Romans recognized that lime concrete—barely different from lime mortar—possessed limited strength. And even though lime concrete held up much better to the elements than did adobe, it was still subject to gradual weathering. To address this problem, the early Roman mason would lay down two courses of mortared stone running parallel to each other, a half meter (*ca.* 20 in) or more apart.

Once the courses were about two or three feet high, a layer of rock aggregate was laid down between them and then lime mortar was dumped on top of it and strongly tamped down to fill in all the cavities and crevices. The Romans used only enough water to make the mortar pliable, and so it was very thick. Their concrete construction was quite different from the modern method of thoroughly mixing rock aggregate with a less glutinous concrete before pouring the whole into a form. This wall-building process of laying aggregate and ramming in the mortar was repeated until the top of the courses was reached, whereupon additional masonry courses were laid and more aggregate and mortar was tamped between them until the desired height of the wall was attained.[18] The ramming process ensured a maximum rock-to-mortar ratio, thus giving the wall substantial compressive strength—more due to the voluminous rock aggregate than anything else—while the masonry of the outside courses protected the lime concrete core from the elements. Once the wall had set, a burglar would have a very difficult time trying to burrow through such a barrier.

While the Greeks and Etruscans occasionally built rubble-cored walls, the Romans perfected the method and would use it in most of their construction work from the late Republic to the end of their empire.

Around Cato's time, the Romans began using molds for their concrete work. In the beginning, these were often just wooden planks spaced far enough apart to form the outline of the wall to be built. As an aesthetic concession, the sides of the outward-facing stones were often filed down to a square or diamond shape, and together they formed a pleasing netlike pattern on the wall's surface. (Cato would have avoided such frivolity, since it increased costs.) Eventually, brick was used for the outer courses of the wall, and this style would become the preferred construction method.

At this point, it would be helpful to first explain a few Latin technical terms. A misunderstanding of these terms, as well as misinterpretation of the physical archaeological evidence, has caused much confusion about Roman concrete.

The Romans called their lime mortar *arenatum* ("sandy stuff") because it mostly consisted of sand (*arena* or *harena*). They knew that the active ingredient was lime, but in their naming conventions, the Romans often referred to a material's principal constituent. Likewise, the Romans referred to their lime concrete as *caementis* ("rocky stuff"), because, even though it was essentially lime mortar mixed with stone aggregate, it was mostly composed of the latter, which was called *caementa* (plural *caementi*) in Latin. *Caementi* were small, sharp stones that ranged from broken pebbles to fist-sized rocks. Oddly, it is from the Latin *caementis* that we derive the modern English word *cement*, which we frequently—and mistakenly—call concrete. (Cement is the "glue" that, together with water, rocks, and sand, creates the finished product, concrete.) The term *concrete*, though derived from the Latin *concretus* (meaning "brought together" or "congealed"), was never used by the Romans to describe the material. What historians and engineers today call "Roman concrete" is the hydraulic version, for which, as far as we know, the Romans did not even have a name. In the surviving literature, the hydraulic component is described as an additive to standard *caementis*. This additive was called *pulvis puteolis*, pozzolanic soil or powder (*Puteoli* being the ancient Latin name for the modern Italian city of Pozzuoli, near Vesuvius). This volcanic powder was mixed with the *caementis* to make it either impervious to, or to allow it to set under, water.[19] Besides these hydraulic properties, the Romans discovered that *pulvis puteolis* also made *caementis* harder and more durable.

To keep things simple, we will refer to the non-hydraulic version, already ancient in Cato's day, as lime concrete or *caementis*. For the hydraulic version, we will follow convention and call it Roman concrete. For many years, archaeologists examining Roman ruins could tell no difference between walls using *caementis* and those using Roman concrete, as both looked the same. For this reason, they assumed that Roman concrete had been used centuries earlier than it actually was. Only with the recent advent of sophisticated techniques for mineral analysis has this controversial issue finally been settled.

Back to Roman walls. Even though Cato does not directly provide construction details in his book, the remains of numerous Roman walls, plus the specific instructions he gives for the materials to have ready for the contractor, tell us that he was referring to a classic lime concrete wall. The first clue is rather obvious: the chapter in his book is titled "Walls Made of Lime

Concrete and Stone" (*macerias ex calce caementis silice*). The less obvious clue is the previously mentioned amounts of lime and sand he recommends for each linear foot of a thick five-foot wall. These amounts are correct for the high-aggregate lime concrete used by the Romans at this time. (Later Romans would use better-quality sand, and so they could increase the measure to three or four parts sand to one part of lime.)

The Romans liked *caementis* because it allowed them to build thicker, sturdier walls for less money than a pure masonry wall of the same dimensions. If the wall was a crude affair, like the ones Cato built, it could be plastered over for aesthetic purposes and as a further safeguard against weathering. In the case of important temples, government buildings, or lavish villas, cut and polished sheets of limestone or marble were laid across the wall's surface to make it appear as if superior stone had been used for its construction. The remains of many of these walls, though not made with hydraulic concrete, can be found throughout Italy. In Cato's time, the superior *caementis* using pozzolana was still not widely used and was probably unknown to most Romans. A century after Cato's time, hydraulic *caementis* had come into more general use, and the first surviving mention of it is found in a book written by a remarkable man.

VITRUVIUS

The earliest and only known detailed reference to Roman concrete in the ancient literature is by the renowned Roman architect Marcus Vitruvius Pollio in his *De Architectura* (*On Architecture*), also known as the Ten Books on Architecture, written a couple of centuries after Cato's work.[20]

Vitruvius began his career as an artillery specialist in Julius Caesar's army. His occupation required expertise in the construction and maintenance of ballista: devices that used torsion-springs instead of gunpowder to hurl sharp iron bolts or heavy stone balls with great force at enemy troops or over the walls of a besieged city. Vitruvius's book shows that he also had keen knowledge of other aspects of military engineering, such as siege emplacements and the rapid construction of sturdy wooden bridges, as well as larger-scale works that we would today call civil engineering projects, like town planning, municipal drainage, building aqueducts and harbor emplacements, and so on. For the average Joe, Vitruvius also addresses the comparatively simple issues involved in home building and maintenance, like plumbing basics and what kind of stucco to use.

It seems somehow appropriate that the earliest reference to concrete comes from a Roman military engineer. Some authorities have suggested that Rome's lime-based technologies—and her adversaries' lack of such— contributed to her success in conquering a large part of Western Europe. The ancient tribes of Gaul, Britain, and Germany were ignorant of lime mortar or concrete, and so built their forts with thick earthen walls reinforced by logs. Once completed, these defenses were formidable and immune to ramming. However, after a couple of decades these once stout barriers became gently sloping mounds that offered little defense against a determined army.[21] Vitruvius was probably well acquainted with these earth-wood walls from his time in Gaul.

We know little of Vitruvius's life, but it is generally assumed that he became a civil engineer after Caesar had completed his military exploits and had assumed the title Dictator for Life (it would be a short term of office). Vitruvius mentions having built a basilica in Fano, Italy, during this period, but all traces of it have vanished. Like most Roman buildings, it may have served as medieval quarry where local inhabitants could freely obtain precut stone. Vitruvius eventually received a generous pension from Caesar's grandnephew and adopted son and heir, Augustus, Rome's first emperor. It was during this comfortable retirement that Vitruvius wrote *On Architecture*.

On Architecture is counted among the most influential books on the topic of architecture ever written. Part of its influence comes from the fact that it is the only detailed manuscript on architecture to have survived from the Greco-Roman world. Vitruvius provides useful bibliographic references on each of the subjects he explores, but the works of most of these authors have survived only in fragments or vanished completely during the Dark Ages. Consequently, when *On Architecture* was rediscovered in the fifteenth century, it had a profound impact on all the architects of the

Renaissance, especially Andrea Palladio, whose own *I Quattro Libri dell'Ar-chitettura* (*Four Books on Architecture*) could be viewed as an updated and appended version of Vitruvius's work.[22]

On Architecture is encyclopedic in its breadth. Besides the topics mentioned above, it also covers sundials, water mills, pneumatics, crane and hoisting technologies, geometry, and even a little astronomy. This was the time of the great ancient encyclopedists, such as Vitruvius's near contemporaries, Pliny the Elder and the famed North African scholar King Juba II, two authors who tried to address as many topics as possible in their books and, in the process, perpetuated a number of myths.

Fortunately, Vitruvius does not often stray far from those subjects with which he was personally familiar. It must be remembered that the word *architecture* in the ancient sense referred to the construction of any structure or device, whereas today it refers only to designing buildings. This fact, along with Vitruvius's stated belief that a good architect should also possess a strong grounding in the sciences and liberal arts, explains why the book covers such a wide variety of subjects, including the author's occasional philosophizing. As for philosophy's own relationship to architecture, Vitruvius explains that "it makes an architect high-minded and not self-assuming, but rather renders him courteous, just, honest, and ungoverned by greed. This is very important, for no work can be rightly done without honesty and incorruptibility. Let him not be grasping, nor have his mind preoccupied with the idea of receiving excessive fees, but let him maintain his position with dignity and by cherishing a good reputation."[23] This sounds almost like a building contractor's version of the Hippocratic oath.

Many of Vitruvius's pronouncements are simple yet profound. For example, his remark that a structure must be durable, useful, and beautiful (*firmitas, utilitas, venustas*) holds true for any age. Look around sometime and ask yourself how many of the buildings you see fit all three of these criteria. And such a mandate is applicable not only to buildings but to any manufactured product. It is what sets a fine mechanical watch apart from a cheap electric model, or a beautifully crafted pen from one made of plastic and designed to be disposable; or, more appropriate to our theme, it is what distinguishes a beautiful Roman arched bridge capable of lasting millennia from one made of modern concrete that is subject to disintegration after a century or less.

Vitruvius was also capable of making some very shrewd observations. In his chapter on "Aqueducts, Wells, and Cisterns," he reflects on the toxic properties of lead plumbing:

> Clay pipes for conducting water have the following advantages: In the first place, regarding construction issues: if something happens to them, anybody can repair the damage. Secondly, water from clay pipes is much more wholesome than that which is conducted through lead pipes, because lead is found to be harmful for the reason that white lead is derived from it, and this is said to have deleterious effects on the human system. Hence, if what is produced from it is harmful, no doubt the thing itself is not wholesome.
>
> Of this, we can draw an example from plumbers, since the natural color of their bodies has been replaced by a deep pallor. For when lead is smelted in casting, the fumes from it settle upon their extremities, and daily burn away all the beneficial properties of the blood from their limbs. Hence, water ought by no means to be conducted in lead pipes if we wish for it to remain wholesome.[24]

This was written some two thousand years before lead pipes for plumbing were finally banned in the United States in 1989.

Besides being better organized than the Greeks, the Romans possessed another attribute that contributed to their empire building: a fascination for any technology that had a practical purpose. Greek intellectuals held a strong prejudice against craftsmen. A potter or sculptor was unlikely to be invited to one of Plato's symposiums, no matter how well read he might be. (Socrates was a *retired* stonecutter.) Indeed, the Greeks called the manufacturing trades the Bausotic Arts, from the Greek word *bausos*, meaning "vulgar." Romans, especially those of the patrician class, adopted this prejudice because they generally held Greek taste in high regard. Nevertheless, the Romans' innate practical streak prevented them from completely embracing this elitist snobbery. Romans remained fascinated by technologies that could serve useful purposes, and since they were among the most conscientious administrators in history—building roads, bridges, aqueducts, and public baths wherever they went—many of them felt it was necessary to know how things worked. Reading *On Architecture*, it is clear that one reason Vitruvius wanted to share his knowledge was that he knew many of his readers would be government administrators confronting the same infrastructure challenges addressed in his book.

Book I of *On Architecture* covers fundamental subjects: the proper background and knowledge an architect should possess; how to find a proper site for a city and its walls, public buildings, and so on. Book II covers building materials, and describes mortar in far more detail than Cato's work. Book II also demonstrates just how far construction knowledge had advanced in the intervening two centuries. Vitruvius explains the variety of different sands that can be used in lime mortar, and discusses their relative strengths and weaknesses. Vitruvius advises against using sea or river sand (possibly because the granules of both are likely to have been worn smooth by water and are thus less capable to form strong bonds). Sea sand is particularly unsuitable because the dissolved salts within it will cause unsightly splotching. The highest-grade sand he calls "pit sand," and of this variety there are four kinds: black, red, gray, and carbuncular. From Vitruvius's description, the last seems to be of volcanic origin, but it is not clear if it is of the pozzolanic variety; the other three apparently derive from rock erosion. Of pit sand, the best is dry, sharp-grained, and unadulterated with dirt. Vitruvius helpfully provides a simple, do-it-yourself test to determine the sand's quality: "Of these the best will be found to be that which crackles when rubbed in the hand, while that which has much dirt in it will not be sharp enough. Try this: throw some sand upon white cloth and then shake it out; if the cloth is not soiled and no dirt adheres to it, the sand is suitable."[25] If one has no choice but to use sea or river sand, Vitruvius knows of an additive that will ameliorate their defects: "[P]otsherds ground and sifted through a sieve, and added in the proportion of one-third part, will make the mortar better."[26] This seemingly innocuous comment contains within it profound possibilities, as we shall later see.

Vitruvius then goes on to discuss limestone and lime, and here we find the earliest surviving reference to hydrated lime, although its use probably predates Vitruvius's book by some centuries. To remove lime's caustic properties, a small amount of water is added to it. Craftsmen performed the process of hydration in a dozen different ways. Often, the powder was laid out over a smooth, dry surface, sprinkled with water, and then thoroughly mixed with rakes or trowels. In very humid environments, the lime is simply exposed to the air for a period of time, although this method was probably not employed in the mostly temperate climes of the Roman Empire. Vitruvius suggests that the lumps caused by the moisture be thoroughly mixed with the rest of the powder. Once it was hydrated, the lime was sealed in a waterproof container. In Roman times, this was usually a ceramic amphora, though wooden barrels may have been used on occasion. The Romans held that the older the hydrated lime was, the better its quality, and so they specified in their building codes that it be aged several years before use.

After finishing with hydrated lime, Vitruvius moves on to discuss something that evidently strikes his fancy, and, from its description, we perceive that it is an interesting novelty.

> There is also a kind of powder (*pulvis*) that naturally produces admirable results. It is found in the area of Baiae and among the farming communities around Mt. Vesuvius. This substance, when mixed with lime and rubble, not only lends strength to various buildings, but even when piers of it are constructed in the sea, they set hard under water.[27]

Vitruvius is describing concrete that is created by using pozzolanic earth, the latter being a granulated version of volcanic pumice (*pumex*) that the Romans also called sponge-stone (*spongia*) for its many holes. It is essentially the same material used in the mortar for the cistern in Rhodes. It is the earliest surviving reference to the material that would one day dominate the visual landscape of our modern world: hydraulic concrete.

At the time that Vitruvius wrote his book, toward the end of the first century BCE, true Roman concrete had only recently emerged from being an intriguing waterproof mortar to becoming a building material that, by itself, could be used in new and creative ways. Instead of the simple wooden planks used for wall molds, more elaborate forms—called "shuttering" by today's engineers—were fashioned. The same method of ramming together layers of lime concrete and stone aggregate was followed, though the concrete was now the hydraulic version known today as Roman concrete. There are the partial remains of a ceiling vault, and a largely intact Roman concrete dome over a bathhouse dating from the century in which Vitruvius lived. Unsurprisingly, both the vault and dome were constructed in the immediate vicinity of Mount Vesuvius, which had a virtually inexhaustible supply of pozzolanic soil. Both architectural forms are related to the Roman arch: stretch an arch along its side, and you have a barrel vault; two intersecting barrel vaults form a cross vault; rotate an arch around its center axis, and you have a dome. The Romans would go on to perfect all these architectural forms, utilizing the plastic nature of concrete to create them. The bathhouse dome, the earliest known monolithic concrete dome, is a curious affair, however. Apparently, the pioneering architect who built it was still a bit nervous about the material and uncertain about its strength. Consequently, he made its walls very thick. Even though the bathhouse was constructed roughly around the same period that Vitruvius wrote his book, he does not mention the use of concrete for building domes or vaulting. The master architect probably felt that these recent and relatively rare experiments had yet to meet the test of time, which, to him, was the final arbiter of any building's worth.

Vitruvius makes one more mention of Roman concrete in book 5, chapter 12, where he discusses building breakwaters and harbors:

> Take the powder that comes from the country extending from Cumae to the promontory of Minerva [pozzolanic earth from the vicinity of Vesuvius], and mix it in the mortar trough in the proportion of two to one. Then, in the place previously determined, a cofferdam, with its sides formed of oaken stakes with ties between them, is to be driven down into the water and firmly propped there; then, the lower surface inside, under the water, must be leveled off and dredged, working from beams laid across; and finally, the concrete [*caementis*] from the

FIGURE 7. Bust of Cato the Elder (aka *Cold, Cold Heart*), Roman author and reactionary politician. Cato's book *On Farming* contains the earliest surviving literary reference to *caementis*

FIGURE 8. Figure 8. Image from a seventeenth-century book depicting Roman engineer Marcus Vitruvius Pollio presenting his magisterial work *On Architecture* to Emperor Augustus. Vitruvius was the first to describe the uses of true Roman concrete.

mortar trough [*mortario*]—the stuff having been mixed as prescribed above—must be heaped up until the empty space which was within the cofferdam is filled up by the wall.[28]

These are the earliest instructions on how to use Roman concrete to create a dock or the piers of a bridge. Note that the sand has been entirely replaced by pozzolana. The Romans would go on to use a variety of concrete mixes, including some incorporating both sand and pozzolana, but for underwater work, little or no sand was recommended.

By Vitruvius's time, the use of concrete for piers and jetties was undergoing explosive growth. Vitruvius was almost certainly aware that Roman concrete was being used on an unprecedented scale in a major construction effort where its hydraulic properties were being put to the ultimate test. Since he had dedicated his book to Caesar Augustus, it was perhaps impolitic for Vitruvius to mention a bold civil engineering project that dwarfed anything Rome's first emperor had yet instigated. And, as if to rub salt into the wound, the man who had authorized the project and was eagerly following its progress was neither a Roman nor even a Greek but the king of a widely despised people: the Jews.

A HARBOR WHERE NO HARBOR SHOULD EXIST

The first large-scale use of Roman concrete did not take place in Rome, or even in Italy, but 2,300 km (1,400 miles) to the east, in Judea. And it was the largest application of hydraulic concrete in a single construction project until the early twentieth century. Roman concrete, formerly a specialized mix used in perhaps a couple dozen projects, became a mass-produced commodity because it was the essential ingredient used to fulfill one man's obsession: to build a magnificent harbor in a place where a harbor should not exist. The man was Herod the Great, king of Judea, and his pet project was the Harbor of Caesarea.

Like his father, Antipater, Herod had come to power with the help of Rome and at the expense of the previous royal house of Israel, the

Hasmonean Dynasty. Herod was a realist. Rome called the shots, and the Romans wanted a friendly client king in Israel, which, along with other buffer states, would keep in check the powerful and hostile Parthian empire to the east. Antipater and, later, his son Herod, cultivated friendships with powerful Romans, and thus began the Herodian Dynasty. The Roman Senate recognized Herod as king of Judea. He did not owe his position to popular acclaim or royal connection, though he did marry a Hasmonean princess to give the appearance of continuity of the royal bloodline. (He later executed this wife and their two sons.) Herod was king because Rome said he was king. End of story.[29]

Herod is mostly remembered today for the rebuilding of the Temple of Jerusalem, but it was his construction of the city and harbor of Caesarea that really defined his reign in the eyes of the world. Caesarea was a far more ambitious project, and, to some of his contemporaries, it must have seemed like one of his craziest ideas.

The site Herod chose for his city and harbor was simply a long stretch of beach that connected the desert with the Mediterranean Sea. Here were the ruins of an old stronghold, called Strato's Tower, a fortress originally built by the Phoenicians, but which had been successively captured (and lost) by the Greeks, Jews, Romans, and Ptolemaic Egyptians. Thanks to Roman support, the land now belonged to Herod. Strato's Tower originally had a small wharf and breakwater that were formed by dumping boulders of the local sandstone called *kurkar* into the water. Much of this modest wharf/breakwater had largely vanished by Herod's day, due to the strong currents and silting. Its few remaining inhabitants probably pulled their small fishing boats onto the hot sands of the beach to protect them from notorious storms that plagued the Levantine coast and to keep the barnacles off their hulls.

Strato's Tower was not a place to build a major international harbor of the kind Herod envisioned: one to rival Alexandria in Egypt or Athens's Piraeus in Greece. The main difficulty was the local geography. There were no nearby offshore islands or promontories that could offer shelter against the winds and currents, or which could serve as starting points for the dumping of stone into the sea to create a major breakwater. Nor was there a navigable river that could provide protection to ships during the harbor's construction, or from which fresh water would constantly flow out to keep marine pests like barnacles and shipworms in check, or even to provide enough drinking water (meager wells had to suffice). Nor was there any suitable rock in the immediate vicinity that could be used for building a permanent mole, or jetty; there was only the soft *kurkar* that had proved so useless previously and that was vulnerable to breakup when submerged. Even if local rock had been suitable, the bed of the proposed harbor was composed of deep sand, which had already demonstrated an annoying tendency to swallow up the rocks used to create the earlier mole. Additionally, the strong current coming from the southwest, intensified by the great volume of water pouring from the Nile into the Mediterranean—and bringing with it countless tons of silt—would have made the project difficult under the best of circumstances. Combined with the other natural obstacles, the engineering problems seemed insurmountable. Even to Herod it was clear that, despite all his material and manpower resources, he was going to need some help.[30]

Of all Herod's Roman friends, his most powerful and influential was Marcus Agrippa, Augustus's right-hand man. They met in 40 BCE, when Herod made his first trip to Rome. Herod's father, Antipater, had recently passed away (poisoned by court enemies), and Herod no doubt thought it wise to make friends with those in power to help secure his position as king of Judea. Undoubtedly, palms were greased and lavish gifts bestowed. Nevertheless, bribes were not enough. He would have to make a strong, logical case for Rome to support his dynasty's claim to Judea. Herod was probably required to provide the Romans military and material assistance in the region. The Jewish historian Josephus writes that Herod had struck up a warm friendship with Agrippa during this lobbying junket. Though they came from very different worlds, Herod and Agrippa apparently found some sort of rapport. Both were intelligent men who, like most educated people of their day, spoke fluent Greek, and it was probably in this language that they conversed. Also, both men were builders. Augustus liked to say that he found Rome a city of brick and left it a city of marble, but it was Agrippa who was responsible for much of the city's transformation.[31]

In 23 BCE, Herod and Agrippa arranged a meeting in the city of Mytilene, on the Greek island of Lesbos. Though few details of their talks have survived, most historians believe that it was at this meeting that Herod brought up his plans to build a city and harbor at Strato's Tower. It is likely that Herod put forward all the strategic reasons why a large port should be built there. The huge Roman grain ships—the supertankers of their day— would have a safe haven if a storm should arise and they found themselves too far from Alexandria to turn back. The only other major port in the region, Antioch, was too far north to be of any assistance, and Antioch had its own problems: its harbor was fast silting up, and it was strategically vulnerable. The Parthians could send troopships by boat up the Euphrates then disembark east of Antioch and march west to take the city. On the other hand, Strato's Tower enjoyed the protection of the vast and merciless Syrian Desert to the east. Even if Antioch were to fall, the Romans would still be able to use Strato's Tower to deliver the troops and supplies to counter such an incursion. Finally, Strato's Tower would be a Hellenistic city, like Alexandria. It would have a forum, theater, temples to the gods, and public baths fed by an aqueduct that would also bring freshwater to its inhabitants—in short, all the things to make a Greek or Roman feel at home. It would be mostly populated by the local people—Hellenized Syrians who spoke Greek as a second language—as well as Roman and Greek merchants. Naturally, there would also be a Jewish community— after all, Herod was *king of Judea*—but he'd keep them in line (non-Hellenized Jews already had established a reputation for being unwelcoming to pagans and their religious practices.) Of course, Herod's real reason was that such a harbor was a necessary prerequisite for a major expansion of his kingdom's economy. The overland caravans bringing the silks, spices, and other luxury goods from the East would no longer need to divert to the north for the port of Antioch or south to Alexandria to ship their goods on to Rome and the other wealthy cities of the empire.[32]

Once Herod had convinced Agrippa of the project, it was time to examine the engineering details for the site. The Roman had likely brought engineers with him to discuss the tricky problems involved in creating a great harbor in a place where none should exist. Agrippa had some experi-

FIGURE 9. Marcus Agrippa, power broker, patron of the arts, builder of the original Roman Pantheon, and friend of Judean king, Herod the Great.

ence in harbor building, having built Port Julius in the Bay of Naples for the Roman fleet. Of course, Port Julius was far smaller than the harbor Herod wanted, and the Naples area had everything that Strato's Tower lacked, especially a sheltered bay with a good supply of building materials nearby. It is probable that Agrippa had used Roman concrete in the con-

FIGURE 10. Herod the Great, who, with assistance from Marcus Agrippa, built Sebastos Harbor at Caesarea, easily the largest single application of concrete in all antiquity

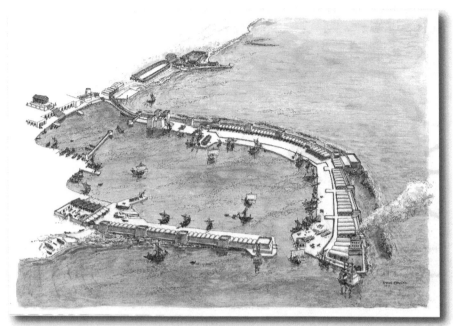

FIGURE 11. Reconstruction of Sebastos Harbor at Caesarea in Judea. The logistical challenges faced by its builders were extraordinary: hundreds of thousands of trees were felled just to provide the fuel needed for the lime kilns to make the Roman concrete used in its construction. Equally challenging was the transportation of the thousands of tons of pozzolana from Italy, which likely required the use of the ancient equivalent of today's "super tankers."

struction of Port Julius, since all the pozzolanic soil he needed was close by. One can easily imagine Agrippa or one of his engineers telling Herod, "You know, we have this special *caementis* that we used at Port Julius. It sets underwater. You could probably use it to create your harbor. However, you would need to use an *awful* lot of this stuff. I mean, no one has used pozzolanic concrete on this scale before. It's *theoretically* possible, but frankly, I don't see any other way you could pull this thing off."

In fact, without Roman concrete, there was no other way for Herod to build his magnificent harbor. Concrete solved all the logistical problems that would have normally doomed such an enterprise. No suitable rock to build the harbor? Use concrete. No sheltering promontory to use as a starting point? Build your own with concrete. Once that issue was settled, the next hurdles to overcome were the other formidable logistical challenges. Herod had no doubt brought along a scale model of the harbor—

the usual preliminary step in a major building project in the Greco-Roman period—which was probably accompanied by calculations of the volume of material needed, based on the breakwaters' proposed length, breadth, and the depth of water where the material was to be laid. Also likely included were the local tide tables and the number of available working days (seasonal storms probably restricted the construction effort to less than two hundred days a year).

The harbor's design also had a very clever feature. The bane of all artificial harbors is the danger of silting. Agrippa's own Port Julius was already beginning to silt up at the time of his conference with Herod and would eventually have to be abandoned. Herod's solution—or that of Agrippa's engineers—was to have channels at the top of the moles that would be open only at high tide, ensuring a flow of silt-free water through the harbor. It was a well-thought-out, state-of-the-art design. The trouble was, it would be located in the worst spot imaginable, and even Herod probably conceded that his chosen location was less than ideal. Unfortunately, the whole coastline of his kingdom was pretty much the same. To help grease the wheels, it was probably Herod who suggested naming the city Caesarea, and its harbor Sebastos[33] (the Greek word for *Augustus*). Such things do delight monarchs.

At this point, Agrippa probably gave the plans to his engineers who were present at the meeting and asked them to figure out the logistical requirements of building such a harbor with hydraulic concrete. ("Can you have it ready by tomorrow?") After the engineers left to mull over the figures, Agrippa and Herod probably moved on to discuss the political situation in the region, to delve into the latest intelligence from Parthia, to exchange court gossip, and perhaps to make the stock inquiry "How are your kids doing?" (Although in Herod's case, that might not have been such a prudent question.)

One can imagine the expression on Agrippa's face when his engineers returned with logistical requirements for Herod's harbor. Only now, after much underwater archaeological surveying has been performed, are we beginning to understand how colossal those logistics were.

Roman engineers were thorough and fastidious in their planning, and they certainly knew how much lime, pozzolana, and aggregate would be needed for a certain measured volume, as well as the amount of wood needed to kiln a specific quantity of lime. Just as a CEO today is given a thick binder with all the details concerning a proposed project but usually decides about whether to go forward based on the information in a flashy slide how, so Agrippa's chief engineer must have delivered a thick scroll but then provided a verbal sum-up. My guess is that it went something like this (using modern measurements for the convenience of the reader):

"Your Excellency, building a harbor in the place King Herod desires will be a formidable undertaking. If we forget for a moment the amount of aggregate needed, the harbor emplacements will require 24,000 cubic m of pozzolana (*ca.* 847,552 cubic ft) and 12,000 cubic m (*ca.* 423,776 cubic ft) of lime. Let's deal with the pozzolana first. Getting that much to Judea will be tricky. That's many times the amount we used at Port Julius, and we had the advantage of enjoying an inexhaustible supply directly at hand. The number of normal ship cargo loads this represents boggles the mind. I mean, we're talking about almost *23,000,000 kg* (*ca.* 63,566,399 lbs). How do you move that much pozzolana to Judea?"

Having been in the corporate world for a quarter century, I have had the pleasure of listening to many expositions by engineers. The good ones first present the problem in such a manner that it seems insoluble. Then, after giving you a few seconds to ponder the imponderable, they smile and then explain their clever solution. Agrippa's chief engineer must have enjoyed seeing his boss draw a long face before announcing their clever scheme. "We do have an idea, however. If you could borrow the giant grain ships after they have off-loaded their cargo at Ostia, and then divert them south to Naples, you could load them up with volcanic soil there; then, on the way back to Alexandria, they make a stop at the construction site in Judea to drop off the powder. The old mole at Strato's Tower, just south of the planned harbor, could be extended with concrete and sandstone blocks to provide refuge for a few ships during construction of the main harbor. Once the latter is completed, Strato's mole can serve as a subsidiary break-water to lessen the impact of the currents on the southern jetty.

"Still, the toughest nut to crack is the 12,000 cubic m (*ca.* 423,776 cubic ft) of lime that's also needed for the concrete—and that's just for the jetties alone. That much lime will weigh around 29,000,000 kg (*ca.* 63,800,000 lbs). Unlike pozzolana, which can be simply scooped up, lime needs to be manufactured. To produce that much lime, you're going to need hundreds of limekilns, which will have to be manned twenty-four hours a day, every day for the five or six years this harbor will be under construction. Once you have the lime, it needs to be slaked, then put into amphorae— many thousands of them—which will then need to be carried by cargo ships with the ropes and drilled storage decks to handle them without breaking—you definitely do *not* want amphorae of lime shattering on the wet deck of a ship. For efficiency's sake, it would be best to have the pottery ovens and limekilns near the limestone outcrops and the fuel sources, but that's the biggest problem of all. Figuring that one oak tree is needed for the fuel to kiln the limestone to produce 190 kg of lime, we will need approximately one hundred thousand to two hundred thousand trees. Where are those trees going to come from? The coastline of the Mediterranean Basin has been pretty much denuded of trees. Remember a few years ago when we needed to get permission from Augustus to cut down the sacred grove that surrounded the Sibylline Shrine at Cumae? As you recall, it was perhaps the last forest of virgin oak trees in Italy near the sea, but we had no choice because that lumber was necessary for building the ships we used for the war against Mark Antony and Queen Cleopatra. And that amount of wood was *nothing* compared to the massive volume needed for this project. And here's the kicker: in addition to the trees needed for the limekilns, you will need almost as much wood for the pottery ovens to make that many amphorae, and to construct the large concrete forms. The proposed size of the forms will require long planks, so we'll need conifer wood for those. However, where are these thousands of pine trees going to come from?"

Here Agrippa probably drew another long face before his chief engineer smiled and came once more to the rescue with a clever idea.

"As Your Excellency knows, Moesia on the south bank of the Danube River recently became a Roman province. Next to it, on the same river, is Thrace, ruled by a king who, like Herod, is loyal to Rome. On the north bank is Dacia, an independent kingdom with which—at least for now—we also enjoy cordial relations. All are rich in trees and limestone. We can set up a group of limekilns every few miles on the banks of the Danube, which can be used in turn as the logging work progresses. I'm sure we'll be able to find clay deposits somewhere close to the river for the amphorae, but since there will be so much timber at hand, we could use a new technology recently imported from Gaul: wooden barrels. These can carry more lime than amphorae, are less difficult to handle, and far less susceptible to breakage. Both the timber and lime can be sent downriver on boats to the Black Sea port of Troesmis, where they can then be loaded onto larger cargo ships destined for Judea. Of course, since this lumber and lime-making enterprise will no doubt be a state monopoly, the revenues to the treasury will be substantial." (Always bring up the cost benefits—you want your boss's head swimming with *denarii* signs.)

"In conclusion, the construction of the Port of Sebastos and the City of Caesarea is not only feasible but doable. Both will be magnificent monuments to Augustus, just as Alexandria will forever memorialize Alexander the Great."

All we know for sure is that Herod must have presented a strong case for the harbor to Agrippa, and Agrippa in turn must have persuaded Augustus that it was a project worth supporting, because Roman assistance on a large scale began shortly after the meeting in Lesbos. Of course, there was an additional motivator for the Romans. If Judea should ever become a Roman province, Caesarea would make a wonderful capital. The proconsul would enjoy the comforts of a cosmopolitan Western city, something Jerusalem definitely was not.

Putting aside the phenomenal resource requirements involved, the construction of the harbor was a marvel of ancient engineering. It faced unique challenges that had never been grappled with before, and so served as a massive test bed for new building technologies. Before the city of Caesarea could be built, the harbor of Sebastos had to be in place; and before the harbor could be constructed, its southern breakwater needed to be built. Without this seawall to blunt the powerful northward flowing currents, the water would have been too turbulent to permit construction of the rest of the harbor.[34]

Archaeologists have uncovered three different containment methods employed by the ancient engineers in constructing these concrete moles, making it clear that they were learning as they went along. The first method involved using a pile driver to ram wooden beams into the seabed, their positions defining a rectangle. Divers—probably sponge divers who could hold their breath for several minutes at a time—would then nail long planks of spruce or pine to the upright beams. The sandy seabed had been prepared in advance by laying down a thick layer of *kurkar* rocks to prevent the currents from undercutting the sand beneath the finished jetty, a practice still followed today by modern engineers constructing breakwaters. A thick layer of concrete was dumped into the form and then tamped down into the rubble bed. Once this was accomplished, *kurkar* aggregate was dumped in and raked to create a flat surface before another layer of concrete was added and tamped down. The lime, pozzolana, and sand were probably mixed on a floating platform next to the form and then put into a large basket that was maneuvered into place by ropes attached to a small crane. Once correctly positioned, one of the ropes would be pulled, upending the basket and dropping its load of concrete into the water. Within the still water of the form, the lump of thick Roman concrete would drop straight down. The divers would then go down to check whether the concrete load had dropped into the right place; if not, it would be correctly positioned before being tamped down with a wooden ramming device, perhaps weighted with a lead core to overcome its buoyancy. After the concrete had been tamped down, more *kurkar* aggregate was laid down and again raked to form a level surface before another layer of concrete was rammed on top of it. This process was followed until the top of the form was reached. If one side of the form was to be part of the seaward flank of the planned jetty, more *kurkar* rock was dumped against this side to further secure it against the forces of currents and waves.

This method of construction was arduous, to say the least. A sponge diver, despite the ability to hold his breath for up to five minutes, probably needed two or three dives to hammer just one nail into the planks because of the increased water resistance (dealing with bent nails underwater must have also thrilled him), and many more dives were required for laying down and compacting each layer of concrete and aggregate.

Clearly, another approach must have been considered early on, for we see a transition to a less cumbersome process. The form was soon being constructed on land, with the planks making up the floors and walls incorporating the same mortise and tenon joints used by ancient ships to ensure a watertight fit. To lend additional strength, the interior of the form was heavily braced by a series of wooden ties that crossed at right angles.[35] This floating caisson was then ballasted with enough concrete to keep it steady in the water while it was moved into place with ropes. The same process of loading and tamping the concrete and aggregate was followed, but its efficiency and speed were greatly enhanced now that the work was performed in a relatively dry environment. Hoist operators managed the ropes, ensuring that the form would slowly sink into place, snug against the previously laid caisson.

Archaeologists discovered a third method of concrete-form construction at the northern jetty, which was probably built after the completion of the sheltering southern jetty. This form was built on a base of four heavy wooden beams, their ends notched with axes. These were then slotted into each other, forming a rectangle not dissimilar to the base of a log cabin. Instead of a single wall hull, double walls were constructed, also incorporating the mortis and tendon joints. The 0.23 m (9 in) space between the double walls was then filled with Roman concrete with a *very* high lime content—almost 35 percent—and small aggregate of various stones.[36] Archaeologists have theorized that the cavities between the double walls were carefully filled with the concrete, the gradually increasing weight causing the platform to slowly sink into the water. Strangely, no wooden floor was uncovered. If that had been the case, the divers would have had to once again perform the tedious task of laying and tamping the concrete and aggregate underwater. Another possibility is that the base of the form was constructed of concrete, its remains obscured by the concrete dumped on top of it. It seems difficult to imagine that the filling of the double hull with concrete would alone counter the buoyancy of the wood, especially the large beams at the form's base. Perhaps it had a concrete floor, resting on the inside lip of the base beams and reinforced by the intersecting wooden ties at the bottom. This arrangement—my own theory and one to which I am not wedded—would have allowed a dry working environment for the laborers.

The one attribute common to all the forms is that they were quite large, some ranging up to 11.5 m wide by 15 m long (*ca.* 38 ft by 50 ft). Some were rectangular, some square, depending on their placement in the jetty and whether or not they were stacked. Their height ranged from 1.5 m to 4.5 m (*ca.* 5 ft to 15 ft).[37]

After enough concrete forms had been put into place, their flat tops would have risen several feet above the water. Nicely dressed blocks of the ever-abundant *kurkar* were then laid over the concrete surface, which perhaps caused the ancient Jewish historian Josephus to claim that the harbor was constructed of cut stone, not concrete, a belief that was held until underwater investigations conducted by archaeologists in the second half of the twentieth century proved otherwise.[38]

After eight years of construction, including the arduous preparatory work of securing and manufacturing the building materials, Sebastos Harbor was inaugurated in 15 BCE. It was an unparalleled engineering achievement and would still be considered a remarkable accomplishment by today's standards. Sebastos was larger than Athens' facility at Piraeus and rivaled the port of Alexandria in Egypt, the largest harbor then existing on the planet. Roughly two millennia would pass before another concrete harbor would match its size, let alone surpass it. The crown jewel of Sebastos Harbor was its southern breakwater. Instead of directly blunting the powerful northwest flowing current, the southern breakwater extended in a gentle west-northwesterly direction to guide the stream farther out in the Mediterranean. Its left bank continued in this direction, while its right assumed a more northerly route so that the breakwater grew in width as it reached its terminus. The southern breakwater's terminus was opposite that of the smaller—though still massive—northern jetty, which followed a straight westward trajectory from the land. The gap between the termini of the jetties was approximately 20 to 30 m (*ca.* 66 ft to 98 ft) wide and formed the entrance to the harbor.[39]

Although these structures are described as breakwaters, they were much more than that. The completed edifices were flat, rigid artificial stone peninsulas that trumped the features of any naturally formed promontories. The southern jetty was 40 m wide (*ca.* 131 ft) at its shore-end, and 60 m wide (*ca.* 171 ft) at its finishing point almost a half kilometer (over a quarter of a mile) away in the Mediterranean Sea.[40] It had a road and walkway, and it supported a series of large, vaulted stone warehouses. At its seaward end was a massive lighthouse, the highest and brightest beacon outside of Alexandria. Two stone and concrete towers, each supporting three colossal statues, were positioned to each side of the harbor's entrance. Josephus does not tell us who or what the statues represented, certainly not the Jewish god, for such images were forbidden by Hebraic law. Perhaps they represented Olympian deities whose favor was no doubt sought by the harbor's builders.

One major engineering concession made for both breakwaters was the use of *kurkar* for the aggregate. Just as a mason never uses a mortar that, when set, is harder than the masonry blocks it binds together, so must an engineer never use an aggregate weaker than the set concrete. The Roman concrete used for Sebastos's jetties must have mauled its porous *kurkar* aggregate, but the mix held together. And that was all that mattered to Herod.

Caesarea itself would take another five years to finish, and it became the largest and most beautiful city in Judea, with a population of 120,000, roughly the same size as Athens during this period. It is hardly surprising that the sponsor of this amazing project, King Herod, would soon thereafter become known as "the Great."

The Herodian Dynasty did not last very long. After the construction of Caesarea and the new temple in Jerusalem, Herod felt politically secure enough to have his Hasmonean wife, Mariamne, executed on trumped-up charges of adultery. A few years later, their two sons, Aristobulus IV and Alexander, would suffer the same fate, allegedly for treason. Despite all this, Herod retained considerable affection for his sons' children. One, the son of Aristobulus, Herod Agrippa (named for his grandfather's friend), was sent to Rome to be raised in Augustus's own household, undoubtedly a more congenial family environment. There he made friends with young men who would go on to play important roles in Roman history, including the future emperor Claudius. Herod Agrippa, who was as much Roman as Judean, was widely respected in Rome for his levelheaded views, so his political advice was often sought after. A couple of years after Herod Agrippa assumed the title of king of Judea in 39 CE, his old friend Claudius became Roman emperor. Claudius ceded more lands to the Judean king so that his territory was now larger than that of his grandfather; indeed, it probably encompassed more land than any other Jewish king in history. Agrippa continued the building work of his grandfather, as did his own son and heir, Agrippa II. In 66 CE, a revolt forced Agrippa II and his wife to flee for their lives. The rebels, who belonged to various dissatisfied factions, slaughtered the entire Roman garrison at Jerusalem, and a few months later they defeated a Roman army. Unfortunately for the rebels, they were divided into several mutually antagonistic political and religious groups and thus were unable to develop a coherent military strategy. Their bid for independence faced long odds, but the murderous infighting that arose after the first successes effectively doomed their cause. The Romans soon returned with a force of sixty thousand men, ably commanded by the general (and future emperor) Flavius Vespasian. Caesarea had remained in Roman hands, and from here Vespasian's legions marched out to take one town and city after another. When Vespasian left Judea to assume power in Rome after the death of Nero had plunged the empire into chaos and civil war, he turned over his military command to his son Titus, who supervised the siege of Jerusalem, which fell after starvation and a months-long heavy artillery barrage had sufficiently reduced its population and fighting strength. Many of those captured alive were enslaved and sent to Rome.[41] Another Jewish revolt took place in 130 CE—coinciding with an earthquake that damaged much of Caesarea and its harbor—but the emperor Hadrian squashed it with the same grim efficiency as the earlier insurrection, and Judea was incorporated into the Roman province of Syria. Another eighteen hundred years would pass before Jewish self-determination was again restored with the creation of the modern state of Israel in 1947.

Caesarea and its harbor Sebastos enjoyed mixed fortunes over the centuries. No human edifice, especially those built on or near the ocean, is exempt from Nature's power, for sea levels and seabeds rise and fall, and coastlines can change dramatically over the centuries, especially if, as we have seen, another natural force comes into play: earthquakes. A major seismic fault runs along the coast of the Eastern Mediterranean, and Sebastos Harbor sat on it (the fault is now 150 m from the present shoreline). By the time of the earthquake in 130 CE, the harbor's seabed had likely sunk a foot or two, and the tremor probably caused even more subsidence. The Romans, who did a fine job of keeping their infrastructure in good order, probably repaired the damage to the town and harbor, but they could not stop the slow and relentless subsidence of the seabed and coastline, which was accelerated by major earthquakes every few centuries (another large tremor struck Caesarea in 363 CE). By the beginning of the sixth century, much of Sebastos, now called Portus Augusti, was probably waist-deep in water, for a contemporary historian reports that it was no longer usable. The Byzantine emperor, Anastasius, restored the harbor around 505 CE, no doubt by adding more *kurkar* blocks on top of the submerged jetties. A little over a century later, the Arabs swept through the Levant, and Caesarea became part of the Rashidun Caliphate. Crusaders took the city in 1099, but by then most of harbor had once again sunk beneath the waves. The Christian knights used *kurkar* blocks to create a small harbor and surrounded its land end with stout defensive walls. The knights managed to hold onto Caesarea for almost two centuries before losing it to forces under the command of the Mamluk sultan, Baibars al-Bunduqdari, who razed the fortifications (the harbor had already silted up by this time[42]).

Caesarea faded from history until the twentieth century, when archaeologists began conducting underwater surveys and started excavating the ruins that still remained on dry land. The scuba-diving scientists were staggered by the size of the concrete blocks, which still remain in remarkable shape after two millennia. Caesarea soon became a popular tourist destination in Israel, and it now has a small modern marina, most of which is situated over what had been the western edge of the ancient city. The remains of the greatest harbor on the eastern coast of the Mediterranean, host to countless war galleys and merchant vessels, now lies under 12 m of water, a refuge for small fish and octopi.

The lessons learned from the building of Caesarea's harbor were applied to the dozens of concrete wharves and jetties the Romans would build throughout the Mediterranean over the following three centuries.

The remains of these later structures generally show better workmanship and materials. The concrete appears to have been mixed more thoroughly, and the rock aggregate is almost always of a better grade than the *kurkar* sandstone of Judea. (Because of geologic changes, most of these edifices are now either underwater or stranded on dry land.) Pozzolana from the Vesuvius region would go on to become a major Italian export and has been found as a secondary cargo in sunken Roman vessels, where it probably also served as ballast.[43] Interestingly, the knowledge that the volcanic soil of Santorini and the other nearby islands was just as good for making a hydraulic mortar or concrete, as demonstrated by the cistern in Rhodes constructed five centuries earlier, had been lost by this time.

Herod Agrippa's friend, the emperor Claudius, would use concrete to expand the harbor of Ostia. Claudius's nephew and imperial predecessor, the barmy Caligula, brought a massive 25-meter-high Egyptian obelisk to Rome.[44] To transport it, he had a special-purpose cargo ship constructed that carried the obelisk to the naval base at Misenum, near Naples (Port Julius had probably silted up by then). There it was off-loaded and transported—no doubt by a huge special-purpose built wagon—up the Via Popilia to the Via Appia and then north to Rome. Once the gigantic ship had delivered its cargo, it just sat in the harbor, its specialized design making it unsuitable for any other purpose. Apparently, it was a local tourist attraction, for Pliny the Elder writes in his encyclopedia *Natural History* that "it was the most incredible floating vessel ever seen."[45] Eventually, someone figured out a useful purpose for it. It was loaded with pozzolana from nearby Puteoli and then sailed to Rome's port of Ostia, which was undergoing the expansion program initiated by Claudius. Apparently, lime was mixed with the pozzolana in either Puteoli or Ostia (the text isn't clear), for, as Pliny tells us "the Emperor Claudius had it sunk there and used as a base for three breakwaters that rose as high as the ship's towers that were built on it. These breakwaters were constructed using Puteolian powder [pozzolana], especially dug and taken there for this purpose."[46] It is possible that separate ships brought the powder after the special-purpose transport vessel had been moved, but that would not have been practical, so the latter must have conveyed the material. Thus, the craft used for transporting Caligula's obelisk served the same purpose as the floating caissons used for constructing Sebastos, but on a considerably larger scale. With the sinking of the obelisk ship, a major part of the construction effort was taken care of in a single stroke. And the rest of the project appears to have gone smoothly, for another half century would pass before the harbor was again renovated and expanded (silting from the River Tiber was always a problem).

However, it was not the Romans' use of concrete for port construction that has so captured the attention and imagination of people around the world but rather their application of this material toward the creation of some of the most beautiful and enduring buildings in history.

Unfortunately for us, Vitruvius wrote about concrete before Rome's use of the material had reached its greatest level of sophistication and its composition and manufacturing techniques had been further refined. For this reason, much of what has been written about Roman concrete has been inordinately influenced by Vitruvius's book. An analogy would be our distant descendants uncovering the Wright brothers' design plans for their first airplane and using this document to draw a host of assumptions about the operating characteristics of World War II aircraft.

Fortunately, we can again turn to the archaeological data, which shows us that the Romans gradually used an increasing variety of concretes and did so with a greater assurance and sophistication. Soon their architects would achieve a mastery of the material that we would not see again until the twentieth century and that, in some ways, have still not been equaled.

THE ARCHITECTURAL MASTERPIECES OF ROMAN CONCRETE

After Vitruvius wrote *On Architecture*, the next time concrete appears in the surviving literature is approximately ninety years later, in Pliny the Elder's previously mentioned *Natural History*. The elder Pliny—to differentiate him from his equally famous nephew and adopted son, Pliny the Younger—compiled his encyclopedia shortly before his death in 79 CE (the scientist had ventured too close to study the eruption of Vesuvius that buried Pompeii and Herculaneum and was overcome by sulfurous gas).

Pliny mentions concrete just once, in his reference to the cargo ship that was to form the jetty at Ostia. His references to lime mortar and stucco seem to be mostly lifted from Vitruvius's book. Like so many people over the years, Pliny cannot suppress his amazement about the properties of lime: "It is something truly marvelous, that quick-lime, after the stone has been subjected to fire, should ignite on the application of water!"[47]

After Pliny's encyclopedia, no surviving reference to Roman concrete is found, aside from two inconsequential books written toward the end of the empire, but both simply plagiarized Vitruvius's text.[48]

One important development was the use of crushed and sifted pottery shards. Mentioned by Vitruvius as being a component of waterproof stucco, it did not take the Romans long to recognize that the red powder had properties similar to pozzolana. Indeed, a major component of modern concrete is kilned clay, and pottery is just that. Soon, pottery and brick dust became a major ingredient of Roman *caementis* walls, giving them the enduring qualities so long admired by engineers down through the centuries.

THE GOLDEN HOUSE

In 54 CE, Empress Agrippina, Emperor Claudius's third wife, tired of waiting for her husband to die a natural death and decided to put some poisonous mushrooms in his food to help the process along. The toadstools had the desired effect, and her son by a previous marriage, Nero, assumed supreme power. Since Agrippina was a dominating mother, and since Nero hated to be told what to do, he ordered her execution a few years later, thus completing a tidy what-comes-around-goes-around karmic circle.

About a decade into his infamous reign, Nero decided that he did not like the imperial mansion he was living in. Although the existing palace was impressive, Nero felt it was not sumptuous enough. He concluded that new and more lavish living quarters needed to be built. Unfortunately, he wanted to build the new palace in the center of Rome, which had long since been developed and was now crowded by such pesky things as apartment complexes, temples, and government buildings. What was Nero to do?

The Great Fire of 64 CE destroyed much of central Rome, killed or injured thousands of its residents, and left perhaps as many as a hundred thousand more homeless. Roman historians count Nero as the chief suspect in this unprecedented arson, as men with torches were seen deliberately setting fire to buildings, unhindered by the local authorities. Nero blamed the Christians, members of a new religious sect, and executed hundreds of them in a number of grisly ways (the morbidly curious can Google® the information). According to the Roman historian Tacitus, the persecution only served to highlight Nero's cruelty and gain sympathy for the Christians. This did not much trouble Nero, who was now delighted that the fire had freed up the 80 ha (ca. 198 acres) of land on which he wanted to build his new residence and surrounding parkland. Construction on Nero's pleasure palace began almost before the last embers of the fire had cooled. Five years later, the residential portion was finished. (It would consist of several separate buildings, some completed after Nero's death.) The palace complex was called the Domus Aurea, the "Golden House," for its extensive use of gold leaf on the building's decorative flourishes. The palace utilized brick-clad concrete walls that were mostly veneered in marble (some walls were covered with ivory panels, which must have cost a few hundred elephants their lives). In addition to the gold leaf, the Golden House featured beautiful frescoes and elaborate stuccowork embedded with jewels and semiprecious stones. The land surrounding the palace was extensively landscaped to create a bucolic setting: large trees were transplanted to create a small forest, and there were gently rolling hills of pastureland (dotted with grazing sheep), a small lake stocked with fish, and even a tiny vineyard—all this so the emperor could reside in Rome and yet feel as if he were living in the Campanian countryside. When Nero finally took up residence in the main building, he exclaimed that at last he had a house that allowed him to "live like a human being!"[49]

What makes the palace so interesting are its Roman brick-faced concrete walls, some of which have survived. While brick-faced walls predate Nero's time, their use grew after the conflagration. The reason for this is simple: brick is fireproof, while stone is not. Exposed to high temperatures, stone flakes off in a process called exfoliation. Once the outer stone of a

Roman wall is damaged in this manner, the structural integrity of its concrete core is compromised as well. Concrete is even more susceptible to
exfoliation: it literally crumbles away when subjected to extreme heat for a
sustained period. A concrete wall faced with fireproof brick can far better
protect its core, which constitutes up to 80 percent of its volume. Consequently, the majority of concrete walls in Rome constructed after Nero's fire
used brick facing. Sometimes we see a combination of rock and brick
facing, but these were probably walls that originally had veneers made of
marble or limestone sheets, their attachment points being at the brick
courses, so that heat would be transmitted to the brick. The knowledge of
concrete's vulnerability to fire is just one of many aspects of the material
that we would not relearn until the twentieth century.

Another major feature of the Golden House is a concrete dome over a
large octagonal dining room. The dome, parts of which have survived, is
not a true geometric dome—a perfect half sphere—but is rather an eight-
sided vault, each side rising from each sectional wall of the octagonal room.
The dome is the earliest surviving example of such sophisticated vaulting. It
is a kind of prototype for the larger and more elaborate Roman concrete
vaulting that would be used in later basilicas and public bathhouses.
Although such vaulting can be constructed in stone masonry—as seen in
medieval cathedrals—the Romans realized that concrete was more ideally
suited for the purpose. The dome itself was 13.48 m (ca. 44 ft) wide, and at
its apex was a six-meter-wide (ca. 20 ft) circular opening called an oculus
(Latin for "eye") that also allowed light to enter the room, supplementing
the light streaming in from windows beneath the dome's base. However, it
is almost certain that only indirect light came from the oculus, as the
external top of the dome had a flat concrete base that once supported what
is now called a *tempietto* (Italian for "little temple"), a circular, lantern-
shaped structure. According to contemporary accounts, the vault of the
dome was painted to resemble the sky and dotted with numerous crystal
gems that served as "stars."[50] The *tempietto* above the oculus was also
domed and decorated in a similar manner and—as reported by the Roman
historian Suetonius—continually revolved "night and day,"[51] probably by
waterpower, since the palace possessed a sophisticated hydraulic system that

also fed elaborate fountains. Although the Golden House was an example of wretched excess, it also represented the most complicated and elegant application of Roman concrete up to that time.

The fire, and the construction of the Golden House and its park afterward, caused long-simmering discontent to finally explode into open revolt. Armies rebelled and the people rioted. Nero suddenly found himself abandoned. Rather than face public execution—or being literally torn apart by an angry mob—he decided to commit suicide. His reputed last words were: "Jupiter! What an artist the world loses with my passing!"[52] Nero—a decent poet, a middling singer, and an awful ruler—then had a slave assist him in cutting his own throat.

THE ROMAN COLOSSEUM

A brief but bloody civil war followed the death of Nero, and Flavius Vespasianus, known to us as Vespasian, defeated his rivals and emerged triumphant. Vespasian's reign was a much-needed tonic for the Romans. Unlike Nero, Vespasian was a sensible, even-tempered man who worked hard to restore moral integrity to Roman government and bring its finances back under control after Nero's reckless spending had plunged the empire into insolvency. (Among other measures, Vespasian instituted the first public pay toilets to help balance the budget deficit he inherited, a fiscal measure that survives in the modern Italian word for a public urinal: *vespasiano*.)

Unlike the thin-skinned Nero, the new emperor possessed a natural and imperturbable equanimity. The historian Suetonius wrote that Vespasian endured "the frank language of his friends, the barbs of attorneys, and the impudence of philosophers with the greatest patience."[53] Like most Romans, Vespasian found Nero's palace an embarrassment and opened up most of the park surrounding the palace for public and commercial development. Though Vespasian refused to live in the Golden House, archaeological evidence shows that work continued on many of the buildings. It is likely that most of them went on to serve as government offices, with, of course, the gaudier bits of decoration removed. Vespasian set aside the land around Nero's lake for a major building project close to his own heart: a massive amphitheater like no other on earth.

We know the Flavian Amphitheater today as the Roman Colosseum (sometimes called the Coliseum). This is a misnomer, for that term originally referred to the colossal statue (*colossus*) that once stood nearby. The colossus was a super-sized bronze representation of Nero, a remnant of the Domus Aurea. The statue's facial features were modified to remove Nero's visage, and a "halo" embossed in gold leaf and sporting radiating flames was attached to the head. The colossus was then rededicated to the sun god Helios.

The Colosseum was not the first Roman stadium. There had been earlier ones, but these were usually nothing more than open-air wooden bleachers. The exception was the largely stone amphitheater built by Statilius Taurus on the Field of Mars in 29 BCE. Although it was called an "amphitheater," the Taurian structure was probably like a classical Greek theater, with most of the audience seating to one side. This amphitheater was destroyed in Nero's fire (perhaps indicating that the seating and stage portions remained of wood construction). The Flavian Amphitheater was far more ambitious, and it is evident that much thought went into its design, because it has served as a blueprint for almost every major Roman— and modern—stadium built since. It makes no difference whether the modern "colosseum" is used to host football games or rock concerts, the Roman design provides a maximum seating capacity with full view of the arena or playing field below, while also allowing the fairly rapid ingress and egress of thousands of people.

Much of the cost of building the Colosseum came from the booty taken in the Judean war, which also provided most of the cheap manpower needed for the project (an estimated hundred thousand Jews were taken back to Rome as slaves). Work began on the Colosseum in 72 CE, and Vespasian's son and successor, Titus, opened it to the public in 80 CE. (Titus's brother and successor, Domitian, would spend another two years revamping the stadium.)

Its original dimensions were impressive. The Colosseum was 189 m (*ca.* 615 ft) in length at its longest point—the structure was elliptical, not circular—and its outer wall was 48 m (*ca.* 157 ft) high, and its perimeter was 545 m (*ca.* 1,788 ft) around. Ringed along the base of the outer wall were eighty numbered entrances (inside, they were numbered as exit points), four of which were set aside for VIPs. Three of four VIP entrances were reserved for members of the senatorial class, and one—the entrance facing true north—was set aside for the emperor and his guests (all four VIP entrances were positioned axially and faced the four cardinal directions). The attendee was given a ticket with the entrance, row, and seat number. For additional convenience—and as a further preventive against crowd congestion—the corridors and stone staircases were also marked to help a person quickly find his or her place. Once comfortably seated, the spectators could then watch death-row prisoners killed by wild animals or gladiators duel in the arena below. The men or animals would pop up from trap doors hidden beneath the sand (*arena*) connected to winch-driven elevators that arose from a *hypogeum,* an underground complex of limestone chambers and corridors. (Despite all its historical inaccuracies, the 2000 movie *Gladiator* faithfully portrays the Colosseum and its workings.)

Most gladiators did not fight to the death—the sport would have ended after several games if they had. The spectators knew this and settled for a good show of swordsmanship and stunts instead.[54] Gladiators were to sword fighting as the Harlem Globetrotters are to basketball, or as television wrestlers are to Olympic wrestlers. The gladiators did get cut often, and the crowd enjoyed this, and, yes, sometimes there were "grudge matches" that led to deaths. Occasionally, a gladiator would show up with a hangover and perform badly, and would get the "thumbs-down" (the real signal is not clear), if he was decked and had a sword held to his throat. Although technically slaves, gladiators were also celebrities who enjoyed considerable freedom and were worshipped by many of the crowd. Like rock stars, they had a following of "groupies." (One of their nicknames—with all due apologies to the New Testament—was "fishers of women.") Not a few members of the equestrian class voluntarily surrendered their social reputation to become slaves so they could attend gladiatorial schools and go on to stardom, so to speak.[55]

According to some modern commentators, it was Roman concrete that made the construction of this magnificent stadium possible, while others

FIGURE 12. The Roman Colosseum. The huge stadium employed a tremendous amount of concrete in its construction, although the vast majority of it was principally used for its foundation.

assert that it could have been built without concrete. And almost no one can confidently say—although many do—how much concrete was actually used in the Colosseum's construction. Estimates vary from 6,000 metric tons to 653,000 metric tons. There is obviously a great disparity between the two figures, and it is a contentious issue.

In truth, the Colosseum could have been built without concrete, but it would have taken more time and manpower. The Colosseum is basically a masonry structure made of travertine limestone and brick.

Of the latter, the brick masonry was cored with concrete. In places, there is more brick than concrete; in other places, more concrete than brick. It is as if the builders slowly began putting more trust in the material as they went on. The vast majority of the concrete used for the Colosseum, perhaps 80 percent, was used for the stadium's foundations, something that earlier historical architects mostly ignored. The high figure for the amount of concrete used, 653,000 metric tons, is probably closer to the mark.[56]

The Colosseum, while hardly representing a ringing endorsement of concrete by its builders, does signify an important transitional phase in the story of the Romans' use of the material. In the years following its completion, Roman builders seem to have had more faith in concrete's structural strength.

Still, Roman concrete did play an important role in one event held in the Colosseum, a spectacle that could not have been staged without it.

DON'T GIVE UP THE SHIP!

According to Roman historians, Emperor Titus decided to celebrate the inauguration of the Flavian Amphitheater by staging a massive wild animal hunt in which "hunters" chased some nine thousand creatures around the arena. The hapless animals were then dispatched by sword, lance, trident, or arrow. Titus is also reputed to have staged a sea fight called a *naumachia* (from the Greek word for "naval battle," *naumakhía*—ναυμαχία) within the new stadium. *Naumachia* had been staged in the past—Julius Caesar, Augustus, and Claudius had each sponsored one—but they were rarely held because of the great expense involved. Usually, a massive basin had to be excavated near the Tiber and then flooded with water (Claudius had his *naumachia* performed on a natural lake outside Rome). Real warships—at least a dozen and usually more—were used, many of which were permanently damaged in the melee. Since a *naumachia* was a real contest to the death, condemned convicts and prisoners of war—probably including many from the recent Judean revolt—were used instead of gladiators. Since the spectators could expect to see real carnage, *naumachii* were extremely popular. The combatants were especially motivated to win, since the survivors could expect a pardon afterward, although this may not have been the case with all the *naumachii* staged.

Many historians were skeptical of accounts that a *naumachia* staged by Titus was held in the Flavian Amphitheater. They reasoned that such an event would have flooded the hypogeum and destabilized the Colosseum's foundation. However, a subterranean aqueduct that leads to the stadium could have been the source of the water. The aqueduct was built using stone heavily mortared by hydraulic concrete. Archaeologists now believe that Titus's brother and imperial successor, Domitian, built the hypogeum several years after coming to power, so flooding would not have been an issue at the time the reported *naumachia* was held.[57] To pull off such an event, Titus must have covered the Colosseum's arena with flat stone paving generously mortared with Roman concrete. The walls of the Colosseum's lower tier also appear to have been strong enough to contain the lateral water pressure. Once the fight was over, the same conduit that flushed out the Colosseum's numerous public urinals likely directed the water and blood out to the already heavily polluted Tiber River. Domitian reportedly held a *naumachia* in his reign, but it likely was held at an artificial pool (*stagnum*) like the earlier imperial shows.

The Colosseum has not fared well over the centuries. Damage caused by lightning strikes (it was the tallest structure in Rome until the modern age) and powerful earthquakes in the eighth and fourteenth centuries, and the even more grievous harm caused by people who used it as a stone quarry for over a millennium, took their toll on the venerable structure. Indeed, it is remarkable that so much still remains. By the nineteenth century, the base of the Colosseum was totally buried under earth brought in by regular flooding of the Tiber River. By then, most of the edifice had become overgrown with weeds and a host of other flora (botanists have counted over six hundred plant species living among its ruins). Serious excavation—and weed eradication—did not begin until 1871. It is now one of Rome's chief tourist attractions, bringing millions of visitors to the Eternal City each year.

Roman concrete continued to improve, and about fifty years after the completion of the Colosseum, it would reach its apogee with a building that would be much imitated but never equaled.

THE PANTHEON

Tho' splendid ruin round you lies,
The proud Pantheon time defies,
Nor yields to Nature's law;
Rome's mighty Genius rear'd the dome,
To give man's conquer'd Gods a home,
And strike the world with awe.

John Courtenay, "Congratulatory Ode" (1792)

". . . angelic, and not of human design."

Michelangelo Buonarroti on the Pantheon's dome

American engineer David Moore was visiting Rome with his wife in the late 1980s when the desk clerk of the hotel where they were staying suggested they take a stroll over to the nearby Pantheon. Mr. Moore describes the visit:

After a brief walk, we found ourselves facing an unusually large, round-shaped building covered by uniform brick and neatly capped with a massive dome. Something out of the modern architectural world that could have been created by the famous Frank Lloyd Wright. But surely this could not be the right building for the ancient Pantheon. It was too big, too new, and complete in every detail for an ancient Roman building some 1,800 years old.

The high colonnade porch with large marble columns beckoned us to enter, and we did, through two impressive tall metal doors. The interior view came as another shock. We stood on a highly polished marble floor with an interesting pattern. Amazingly, the Romans had given this floor a slight camber from the middle to facilitate drainage. Gazing about revealed several large niches in the walls with carved purple marble columns lining each side. At one time these niches possessed the important statues of Rome. The ceiling held a large, open skylight in the center, but what was really unusual to an engineer was the waffle-like indentations which made up the lower portion of the massive dome.

This building looked very complicated for anyone to build with only Roman hand tools. I asked the guard at the door for some assurance that this building was really built by the Romans near the time of Christ. He promptly responded to counter any doubts about his countrymen. The building was indeed built by the Romans some 1800 years ago, and it had not been rebuilt. Yes, it was built with Roman concrete. The official tourist pamphlet said its dome spanned 143 feet. I was amazed. How could anyone build such a large, elegant structure with hands using some mysterious concrete?[58]

Moore was so staggered by the building that he would spend the next ten years studying the Pantheon and Roman concrete, resulting in his book *The Pantheon: Triumph of Roman Concrete*. And he is not alone. A number of unprepared people visiting the Pantheon for the first time, especially architects and civil engineers, come to similar conclusions, mistakenly assuming that the building is really a magnificent early twentieth-century structure or that the building is indeed Roman, but the dome is a modern addition or enhancement. If such visitors are also acquainted with the limited life span of contemporary concrete, they are even more befuddled. Not only is the design and workmanship too modern, but how could it last two thousand years without disintegrating?

One can easily spot the first-time visitors to the Pantheon. Like David Moore, there is brief astonishment, then awe (many English-speaking visitors cannot suppress a "Whoa!" or "Wow!"), followed by the question "How did they do that?" For close to two millennia, the Pantheon possessed the largest true dome (a halved geometric sphere) in the world. Popes, princes, kings, caliphs, Byzantine emperors, and German electors have pushed their architects to create buildings with domes larger than that of the Pantheon, and all failed to match its size, let alone surpass it. Only modern technologies and material sciences have allowed us to build domes larger than the Pantheon's. Even to this day, despite occasional earthquakes and constant exposure to the elements, it remains the largest unreinforced concrete dome in the world.

FIGURE 13. Architect and engineer Apollodorus of Damascus. Apollodorus made the mistake of making sarcastic remarks about Emperor Hadrian's fascination for domes.

FIGURE 14. The brilliant, but thin-skinned Emperor Hadrian,
the probable designer of the new Pantheon.

FIGURE 15. The Pantheon. An artistic and technical tour de force built of Roman concrete. Afterward, Hadrian ordered the dome-skeptic architect Apollodorus to commit suicide.

How did the Romans do that?

It might be best to pause here and recount the fascinating history of the Pantheon. Like all great human-made edifices, the story behind the Pantheon's construction is almost as interesting as the building itself.

The first couple of centuries of the Roman Empire represented a generally happy period for its inhabitants. Tiberius and Domitian may have terrorized the Senate, but, by and large, they ruled well. Even under the reign of such nutcases as Caligula and Nero, the empire's exemplary civil and legal institutions, while disrupted in Rome by the whims and ravings of mad emperors, usually ran smoothly in most of Italy and the provinces. The assassination of Domitian in 96 CE inaugurated what most historians consider the empire's "Golden Age," also known as the "Reign of the Five Good Emperors." Historian Edward Gibbon refers to this time as a "happy period of more than fourscore years" when the empire "comprehended the fairest part of the earth, and the most civilised portion of mankind."[59] The Roman

Senate chose one of their own to succeed Domitian: Nerva, who soon initiated the finest method of hereditary succession ever known, before or since. Chosen by the Senate because he was an intelligent, humane, honest, and hardworking individual, Nerva in turn adopted a younger man of similar attributes, Trajan, and designated him as his successor. Trajan did the same by adopting Hadrian, and Hadrian did the same by adopting Antoninus Pius, and Antoninus Pius did the same by adopting Marcus Aurelius. Unfortunately, Marcus Aurelius broke this wise and long-standing tradition by designating his natural son, Commodus, as his imperial heir. Sadly, Commodus was a disturbed megalomaniac.

Of the five good emperors, Nerva is remembered for his gentle nature, Trajan for his martial prowess, Hadrian for his many building projects, Antoninus Pius for his quiet and efficient administration of the empire, and Marcus Aurelius for his philosophical musings. Hadrian was the most complex, fascinating, and intelligent of these five highly intelligent men. He was also the most unsavory of the group. Shortly after coming to power, Hadrian earned the Senate's enmity when he ordered the execution of four members of that body, men he did not like or could not trust. Starting off your reign by executing four senators was not something a "good" emperor did. Still, despite a dozen instances of summary "justice," Hadrian was generally a good emperor. One time, while he was traveling through the provinces, a woman approached him with a petition. When he tried to brush her off, saying that he had no time to review her case, the woman replied, "Well, then stop being emperor!" Stung by the remark, Hadrian stopped and patiently heard her petition. Hadrian's policies were farsighted and sound. He abandoned many of Trajan's Eastern conquests and withdrew the Roman forces behind more defensible natural barriers and passes. To prevent raids by the barbarian Picts, who would swoop down from what is today Scotland to ravage Roman Britain, he built a formidable wall across northern England that was interspersed with forts every mile or so. Large portions of "Hadrian's Wall" remain, although—as usual with the majority of Roman structures—most of the stonework was removed centuries later for other buildings, mostly local abbeys, churches, and farmhouses.[60]

There seems to have been no subject that Hadrian did not know some-

FIGURE 16. Interior view of the Roman Pantheon with light streaming in from the dome's oculus (eye) at its apex.

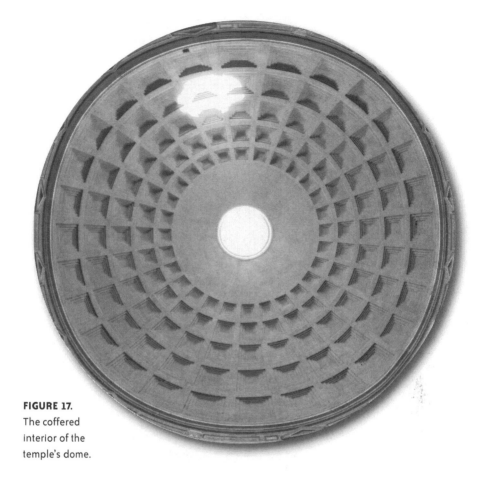

FIGURE 17.
The coffered
interior of the
temple's dome.

thing about, and very few he did not know well. He would invite learned men to dinner, discuss a wide range of topics with them, and then proceed to point out—in detail—each of his guests' errors. Once, when the famed grammarian Favorinus of Arelata (Arles) was Hadrian's guest, they discussed the etymology of words. Hadrian disputed Favorinus's theory of the origin of a particular word, and the latter conceded that the emperor was probably right. When Favorinus later told his friends the story, they upbraided him for giving in to the emperor, since he, Favorinus, was almost certainly correct about the word's origin. Favorinus shook his head and said, "You are urging the wrong course, my friends, to suggest that I not regard the man with thirty legions as the most learned of men."[61] This story

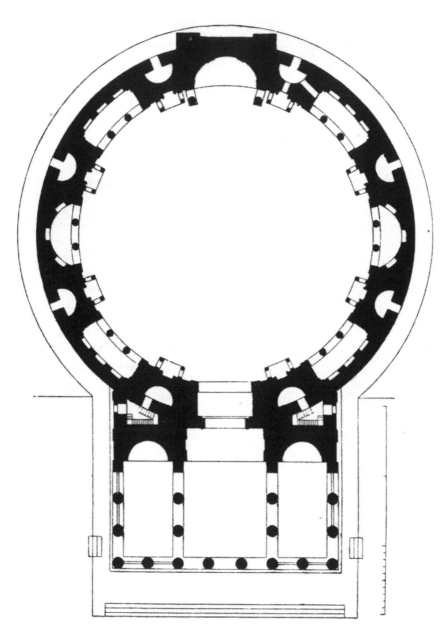

FIGURE 18. The floor plan of the Pantheon, showing the massive
walls built to contain the awesome weight of the concrete dome.
It remains the largest unreinforced concrete dome in the world.

made the rounds, and it was not long before the emperor exiled Favorinus to the Greek island of Chios. Hadrian was that kind of guy.

Besides Hadrian's voluminous knowledge of history, mathematics, and philosophy, he was also an amateur architect who particularly liked domes. There is a story that typifies Hadrian's character and, perhaps, provides a significant clue to the Pantheon's creation. Once, while Trajan was discussing a construction project with the famed architect Apollodorus of Damascus, Hadrian ventured a few suggestions. Apollodorus told him to mind his own business and go back to drawing his "pumpkins," a sneering reference to the young man's fascination with domes. This was a dumb thing for Apollodorus to say, for Hadrian was the kind of person who never forgot or forgave an insult. When Hadrian became emperor, Apollodorus, in an effort to make amends, dedicated a book on field artillery to the new ruler. It didn't work. While Apollodorus likely waited for building commissions that would never come his way, Hadrian decided to build a magnificent temple, one that sported a "pumpkin," the likes of which the world had never seen. The architectural motif scorned by Apollodorus, domes, would at last be vindicated with a building that would take everyone's breath away. Since real estate was limited in downtown Rome—as Nero had discovered—Hadrian decided to realize his dream by rebuilding a temple that had gone up in flames almost a half-century earlier: Agrippa's Pantheon.[62]

Marcus Agrippa, Herod's pal and Augustus's best friend, was also a wealthy patron of the arts. In 27 BCE, he decided to build a magnificent temple dedicated to all the gods, a *pantheon*. Unfortunately, fire completely destroyed the temple in 80 CE, sometime in the middle of Domitian's reign.[63] We have no idea what it looked like, but it was most likely a standard rectilinear temple. Domitian probably never got around to rebuilding it because he was too busy fighting the Germans and Dacians. Nerva never got around to rebuilding it because he died just sixteen months after becoming emperor. Trajan never got around to rebuilding the temple because he was too busy fighting the Dacians and Parthians, and he first wanted to finish constructing his new forum and building complex known today as Trajan's Market (it's true function was more likely governmental than commercial), which Apollodorus was building for him in Rome. By the time Hadrian came to power, the site of the old temple had been vacant for so long that it had probably become a long-established farmer's market.

Instead of simply restoring the previous structure, Hadrian wanted to create something different and much grander: a building that would dazzle everyone who saw it. As we noted, he was partial to domes, but Hadrian's genius was that he wanted to realize the full aesthetic possibilities of concrete and its potential for expressing not just beauty but also *power*. And he wanted to do it in a way that the finest and most carefully laid masonry could not.

Some authorities have suggested that Apollodorus designed the Pantheon. This proposal seems highly unlikely, given Hadrian's dislike of Apollodorus and the latter's apparent preference for vaults over domes. Apollodorus was also apparently wary of concrete's strength, because most of the vaults at Trajan's Market were brick-ribbed. This method, long used and reliable, involved placing mortared bricks on their sides around a wooden half-barrel form to create a vault. Indeed, until the nineteenth century many architects assumed the Pantheon's dome to be of brick-ribbed construction—using a hemispherical wooden form—that was then covered with concrete. Apollodorus did use concrete to create a large cross vault at Trajan's Market, but it was likely constructed in sections and possessed none of the size or complexity of the Pantheon's dome.

Hadrian was probably familiar enough with concrete's attributes to know that it would support a large dome. Of course, concrete domes had been built in the past, but never on such a spectacular scale. Hadrian's dome would be over twice as wide as any previous one, twice as high, and certainly more beautiful. No one in antiquity would ever put so much trust in the strength of concrete as Hadrian did in his design of the Pantheon. Even today, you will find most engineers unwilling to attempt such an enterprise without the use of steel reinforcement.

Naturally, Hadrian's ambitious project demanded that extreme care be exercised. We know of at least one compromise—or change of design—that took place during the construction of the Pantheon. The columned portico in front of the Pantheon was originally intended to be much taller. Above the existing portico one can just make out a slightly flattened triangle that precisely matches the roofline below it. Evidently, the original size proposed for the portico's granite columns was simply too large. Granite is extremely hard and heavy; quarrymen would have had to find and cut out a massive block of granite—one without flaws that might later jeopardize the column's load-bearing capabilities—and then lower it with great care onto a sled so it could be moved to a place where the arduous task of carving this obdurate material into a perfectly round pillar could be performed. (Fluting the columns was probably also avoided because of the difficulty of working with this kind of stone.) The present columns are still quite large: the granite portion of each is 12 m (*ca.* 39 ft) high, 1.5 m (*ca.* 5 ft) thick, and capped with carved Corinthian capitals that are 2.5 m (*ca.* 8 ft) high, bringing the total height of the columns to 14.3 m (*ca.* 47 ft).[64]

There may be another possible reason for the modification. If one looks at an image or model of the Pantheon as originally planned, the larger portico appears to be a more harmonious fit for the rotunda behind it. However, from the perspective of a person on the ground looking up, the smaller portico seems large enough. Hadrian must have realized that the dome would appear slightly less impressive if one entered the temple through a larger portico. Hadrian may have occasionally lashed out at people he felt threatened by, or simply disliked, but no one has ever questioned his aesthetic sensibilities or his willingness to use the material and political power at his disposal for their realization in construction projects. For example, instead of using the granite from the nearby Alps or the Calabria range for the Pantheon, Hadrian chose the granite from Mount Claudius (Djebel-Fateereh) in eastern Egypt. The logistics and expense involved in transporting these pillars from Egypt must have been substantial, but Hadrian was probably struck by the pale gray-blue color of this particular granite and thought it a perfect match for the temple's heavenly theme. From the standpoint of aesthetics, Hadrian would have been right in choosing to make the portico smaller.

If there had been no rotunda, and the portico continued to form a classic rectilinear temple, it would still be considered a wonderful achievement of the Greco-Roman era, but Hadrian instead wanted to push the envelope further than anyone before him.

It has been estimated that a number of geometric and material stress calculations were made before work could begin on the temple. Were the dome
capped at its base with an upside-down twin below, its volume would
describe a perfect sphere, its base precisely touching the rotunda's floor.
Approximately fifty years earlier, the remarkable Greek scientist Heron of
Alexandria wrote a book on the mathematical solutions to building vaults
and domes, and it is possible that Heron described such a configuration. We
will never know, because Heron's treatise, like the vast majority of the books
published in antiquity, has not survived to our times. Recently, another
mathematically derived feature of the dome was discovered, one that perhaps
explains the dimensions chosen for the hole at its apex: the *oculus* (eye). In
2005, Robert Hannah of the University of Otago in Dunedin, New
Zealand, visited the Pantheon while performing research for his book *Time
in Antiquity*. He soon realized that the layout of the Pantheon suggested that
it was more than just a temple. Between the fall and spring equinox, the light
of the noonday sun traces a path across the inside of the domed roof; while
between the spring and fall equinoxes, the higher sun shines along the lower
walls and floor. At each of the equinoxes, the sunlight coming in through the
dome's oculus strikes the junction between the roof and wall, above the Pantheon's grand northern doorway, allowing a single ray of light to pierce the
grilled window above the portal and fall on the courtyard outside. Hannah
points out that this is no coincidence, as a type of sundial common in
Roman times that incorporated a dome with a hole—although obviously
much smaller—was used to indicate the time of year.[65]

Still, as the old saying goes, it's 1 percent inspiration and 99 percent per-
spiration, so after the calculations were performed and the design for the
Pantheon completed, the really hard work began: constructing an imposing,
extraordinarily complicated, and unprecedented edifice. The Romans did
not reinforce their concrete with steel rods as we do today, so the tensile
strength of the concrete was very limited, and the compressive strength of
Roman concrete—its ability to support heavy weight (including its *own*
weight)—was also less than modern concrete. A large concrete dome places
tremendous stresses, both downward and outward, on the walls supporting
it. In short, domes made engineers nervous, and a dome of this size would

have made all the people connected with this project—the architects working under Hadrian's direction, the operating engineer, contractors, and material suppliers—especially anxious. Hadrian wanted his dome bare, so it had to have a smooth surface. A number of challenges had to be overcome, and if the solution to any of these daunting obstacles were to fail the final test, the whole building could come tumbling down.

The downward and outward stresses caused by the dome were handled in several ingenious ways. To help contain the dome's outward stresses, a thick ring of brick masonry was laid around the exterior base of the dome. However, the most difficult part was building walls that could contain both the tremendous downward stresses, and the remaining—but still substantial—outward pressures. As we have seen, the typical Roman wall consisted of masonry on the inside and outside of a concrete core. To handle the stresses caused by the Pantheon's heavy dome, the concrete core of the wall was especially thick, and the bricks used for the exterior and interior portions were especially large. To provide additional compressive strength to the walls, the bricks were laid in an arch formation, with the arches filled in with more masonry. These embedded arches are called relieving arches, and they had been used in earlier Roman buildings to handle the stresses of concrete vaulting. However, the most interesting method used for reinforcing the rotunda's walls was integrating within it eight 6.4 m (*ca.* 21 ft) thick barrel vaults, each additionally supported by two pillars that flank the interior entrances to the huge circular galleries within the Pantheon. It was in these galleries that large statues of the various gods were once placed (they are now devoted to sculptures representing various saints). These eight barrel vaults are directly opposite each other, supporting the dome as if it were a succession of intersecting arches, which, theoretically, is what a dome is. And so an engineering necessity is used toward an aesthetic end.[66]

Another major problem the architects faced was that if a standard dome—such as the ones seen in the bathhouse in Baiae or in Nero's Domus Aurea—had been built to this scale, its weight would have been such that even the thick walls and barrel vaults of the rotunda would have been unable to support it. This was partially remedied by the twenty-eight vertical rows of beautiful inlaid coffers that not only lessened the dome's concrete mass and its subsequent load but also added a stunning artistic adornment. Once again, engineering requirements were imaginatively used to add beauty, emphasizing the great deal of thought and planning that went into the design of the Pantheon.

To reduce weight even more, lighter aggregate—mostly volcanic pumice—was used for the dome, unlike the walls, which featured harder rocks with better load-bearing capacities. Indeed, one sees a variety of different aggregates used in the construction of the Pantheon: one kind for the foundation, another for the walls, and another for the dome, each perfectly suited for a special application and/or load-bearing need. This shrewd use of various aggregates was something that would not be fully appreciated until the twentieth century.

The dome of the Pantheon would not have been so dazzling if the form into which the concrete was poured had not also been a model of perfection. The mold for the Pantheon's dome was the most artistically advanced example of concrete shuttering used until the modern age. Imagine a convex reversal of the interior surface of the Pantheon's dome, with the four-stepped coffers sticking out instead of in. Essentially, a mathematically perfect wooden mold with a near flawless surface had to be created onsite, with dozens of workmen laboring away on the top and bottom of it, and with every step in its manufacture closely supervised to ensure the exacting precision required. The mold had to have been made of strong wood, like walnut or live oak, and it had to have been quite thick to hold up the hundreds of tons of concrete laid upon it. And supporting it all was a thick forest of really stout scaffolding beneath, probably consisting of timbers carved from whole trunks of massive pine trees.

Nevertheless, the tamping of Roman concrete was an essential element of the material's durability. Not only did it fill all the voids of the shuttering, but it also compressed the concrete itself, making it denser and more compact, ensuring a minimum of micro air cavities within the material that might allow the ingress of water or chemicals that might damage it. This compaction method was not rediscovered until US Department of Interior engineers noticed its benefits in concrete dam construction in the 1980s.[67] It is now widely employed today.

The finished temple is stunning in every sense of the word. You first enter the portico between the tallest granite columns that have ever existed in Rome and then walk through the original 6.4 m (*ca.* 21 ft) high double bronze doors into a massive open space that is awe-inspiring in its dimensions. The huge dome high above you was designed to represent the heavens, and the stunning concentric rows of vaults rising toward the oculus at its center seem to convey some great natural force, both divine and intelligent, frozen in the act of becoming. However, there are as many impressions and interpretations of the Pantheon's dome—mathematical, aesthetical, and even spiritual—as there are human beings. One person described the Pantheon as a kind of time portal, in which the visitor is immediately transported back to ancient Rome in all its glory. It's true. Here, more than anywhere else, a visitor can experience the power and grandeur of imperial Rome.

Once the temple was finished, Hadrian did something curious. He had Agrippa's original dedicatory inscription incised on the pediment. It reads "M. AGRIPPA L. F. COS. TERTIUM FECIT." Most monumental dedications at that time were abbreviated in a form instantly recognizable to the Romans, who would read it as M[arcus] AGRIPPA L[ucii: *of Lucius*] F[ilius: *son*] COS[ul: *consulship*] TERTIUM [*third*] FECIT [*built this*]. "Marcus Agrippa, son of Lucius, built this in his third consulship." Unlike most Roman emperors, Hadrian did not like leaving his name on every edifice he built, and this has led to some confusion. Hadrian restored several buildings in Athens and built several more. Recall that speaking platform on the small hill in Athens called the Pnyx? The restoration was performed in Hadrian's reign, but for a long time the structure was assumed to be a product of the classical period. The same was true of the Pantheon. Just seventy-five years after Hadrian had constructed this magnificent domed structure, the Roman historian Cassius Dio would number it among the buildings that Marcus Agrippa had erected.[68] This belief continued to be held for most of the succeeding centuries, with everyone assuming that the Pantheon had been constructed one hundred fifty years earlier. Fortunately, the Romans usually stamped their bricks with the year of the then-current consuls, as well as the manufacturer (often the bricks were government- or army-issued). In the early twentieth century, several bricks from the Pantheon were removed and examined; they confirmed that Hadrian was really the builder. The bricks date from 118–125 CE,[69] about six years into his reign, once more confirming that much time, planning, and preparation took place before the actual work began. It is even possible that Hadrian began designing the Pantheon during Trajan's rule, and maybe it was this design that Apollodorus alluded to in his cutting remark about "pumpkins." Hadrian eventually ordered Apollodorus's execution (he was probably ordered to commit suicide). My guess is that Hadrian first gave the Greek architect a tour of the completed Pantheon before issuing his death warrant. Hadrian's message to Apollodorus would have been clear: "So, what do think of this *pumpkin*?"

Concrete vaulting and domes became popular in the Roman Empire in the second, third, and fourth centuries, though none of these later models ever reached the size of the Pantheon's. Cross vaults were especially utilized in the massive public baths. All the baths have fallen into ruin with a single exception: one section of the fourth-century baths built in Rome by the emperor Diocletian. This section was once the *frigidarium*, the wing for cold-water bathing. It was saved from destruction by being converted to a church: the basilica of *Santa Maria degli Angeli e dei Martiri* (St. Mary of the Angels and Martyrs). It is worth a visit, for it is the best-preserved example of Roman concrete vaulting.

Another example of Roman concrete that has survived the ravages of age and the hand of man is the Roman Senate House, called the Curia Julia (named in honor of Julius Caesar). It had burned down twice since Julius Caesar's day but was rebuilt each time as it was before. The one that stands today was rebuilt at the beginning of the fourth century, also by the emperor Diocletian. Like the Pantheon, the marble sheets that once covered the Curia Julia's brick-clad concrete walls were removed and used elsewhere. Few buildings in Europe can lay claim to so much history as the Curia Julia.

Following the reigns of the five good emperors, the Roman Empire—at least its Western half—endured for another couple of centuries, despite almost unrelenting civil war and barbarian invasions. The Eastern Roman

Empire survived in truncated form until the fifteenth century, when its capital, Constantinople, was captured by the Turks. However, long before that time, innumerable books of literary and historical importance had vanished, as well as the secrets to dozens of different technologies, including the formula for making Roman concrete.

With a couple of curious and isolated exceptions, almost fifteen hundred years would pass until concrete was slowly rediscovered, and another century or two until we realized that we were making the same errors that the Romans had already learned to avoid in the ancient past. Sadly, by the time we discovered our mistakes, the planet had already been covered in a material that was made and applied in the wrong way. Unlike the Pantheon and the Roman Senate House, virtually all the concrete structures one sees today will eventually need to be replaced, costing us trillions of dollars, pounds, euros, yen, and yuan in the process. But that is another story, and one that will be covered later.

Chapter 4

CONCRETE IN MESOAMERICA AND RENAISSANCE EUROPE

Between 1500 and 1600 BCE, at a time when the Egyptian and Minoan civilizations were thriving, the first major civilization in the Americas, the Olmec culture, was consolidating its power in southeastern Mexico, near today's city of Vera Cruz. By the twelfth century BCE, the Olmecs were producing outstanding art, both stylistic and naturalistic, that was on par with the best being produced in the Old World. The Olmecs are mostly remembered today for the numerous carved stone monumental heads, evidently of their rulers, that now grace museums throughout the world.

The influence the Olmec civilization had on succeeding Mesoamerican cultures was profound. These contributions include writing, a sophisticated calendar, number system (with the earliest known representation of zero), and a ball game similar to soccer that would endure until the Spanish conquest. Human sacrifice may have been another cultural legacy of the Olmecs, but the archaeological evidence is still hazy. Some have speculated that the pantheon of Olmec gods, though under different names and guises, also survived right through the later Mayan and Aztec civilizations.

The Mesoamerican civilizations are especially fascinating because they developed in isolation from the rest of the world. Though many of the cultural contacts between the disparate societies that stretched across Eurasia were indirect, many ideas and technologies were carried back and forth. One can find many cultural similarities between the Han Chinese and Roman Empires, though each mostly knew of the other only through secondhand and thirdhand sources. In the Mesoamerican cultures, there was something unique, and—at least to us—considerably more alien. The Mesoamericans saw blood sacrifice as a necessary element to their religion and worldview. As they saw it, since their gods had used their blood and body parts to create the human race, so must humans return the favor by sacrificing their blood and lives to the gods. For them, offering blood to the gods was as important as water was to irrigation, for if the blood offerings stopped, the world would become a wasteland. Besides sacrificing humans to their gods in great public ceremonies, the individual could also beseech divine assistance by offering his or her own blood. This usually required a great deal of pain—simply opening a vein was evidently for sissies—and involved such grim exercises as passing the barb of a stingray through one's tongue or genitals. Some people have trouble reconciling the advanced scientific achievements of the Mesoamericans with the apparent barbarity of their religions, but to the Olmecs, Mayans, Aztecs, and other societies of the region, such practices were a normal aspect of their lives.

The Olmecs evidently burned lime to make plaster, but they did not use it to the same degree as the cultures that succeeded them. As the Olmecs were declining, grander cultures were arising in the valleys of Mexico and Oaxaca (the Teotihuacán and Monte Albán peoples) and to the southeast, in what are today the nations of Guatemala and Belize and the Mexican state of Yucatán. By the time Commodus was mismanaging the Roman Empire, the Mesoamericans were living in grand and beautiful cities and building pyramids, a few of which would equal those in Egypt. By the first centuries CE, the Mesoamerican cultures were using a tremendous amount of lime for plaster (including types that served as a mural base) and stucco, which they lavishly applied to their public buildings. The Mayans occasionally made concrete, usually in the form of a simple beam, in post and lintel structures, but surviving examples are somewhat crude. Since some of the limestone in Mesoamerica was adulterated with clay—ranging from 10 to 20 percent—the Mayans occasionally produced what is called "natural concrete cement." Based on a few surviving examples, some of this concrete was among the best in the ancient world, though the Mayans never came close to mastering the material to such bravura effect as had the Romans.

Estimates about how much lime was used varies, but of one thing we can be certain: it was a *lot*. It appears that the limestone was burned in pits, with the kilning process alone requiring at least thirty-six hours to complete. Since twenty full-grown pine trees were needed as fuel to create just a cubic meter (35.3 cubic feet) of lime, the amount of deforestation caused by the need for farmland *and* plaster and stucco probably tipped the environmental balance deep in the red.[1] Toward the end of the Mayan civilization, during the period called the Late Classic, the Mayans were using lime more sparingly and applying thinner coats of plaster to their walls. Centuries before the Spanish arrived, most of the once-great cities of the Mayans had been lost to the jungle.

The principal Mesoamerican culture encountered by Cortez and the conquistadors in the sixteenth century was the Aztec, whose empire dominated much of southern Mexico. The Aztecs also used lime plasters, but there is no evidence that they ever created concrete. Even though the Aztecs controlled richer agricultural lands than did the Mayans, the fuel requirements for making lime likely restricted the use of plaster to just their principal buildings and temples.

Perhaps one reason why the Mesoamericans never realized the full potential of concrete was related to the difficulty they would have encountered in making forms for the material, for they had never developed metal tools. The carpentry tools used by the Romans were little different from those used in the early nineteenth century: saws, chisels, awls, planes, and so on. Just look at the saw. With a saw, one can not only bring down a tree—though an axe might also be employed—but also make long planks with a very flat surface, similar to the ones employed by the Romans for making concrete walls. The other basic carpentry tools would be needed to create something truly amazing, like the wooden form used for making the Pantheon's dome. However, this still doesn't explain why another concrete application has not been discovered at Mesoamerican sites: artificial stone floors, a common feature of Neolithic temples and dwellings in the Middle East. Perhaps stonemasons were held in high regard for their talent at a craft that took years to master, while concrete was something almost anyone could do. Without the discovery of an inscription that may shed light on the matter, it will remain a subject of speculation.

At the time the Spanish had begun their conquest of the New World, many people back in Europe were excited by the news that a scholar-priest had introduced the definitive edition of an ancient book discovered a few decades earlier in an obscure monastery. Because of its highly technical language, the book had frustrated scholars attempting to decipher certain portions of the text. The book was Vitruvius's *On Architecture*, and in it were instructions on how to make Roman concrete.

VITRUVIUS REDISCOVERED

It would not be until the fourteenth and fifteenth centuries that the Renaissance began to lift Western civilization from the mire in which it had been stuck for almost a millennium. The Roman Catholic Church, which had for centuries been chained to the counterlogical and antihumanist traditions of the early church fathers, gradually embraced the teachings of St. Thomas Aquinas, a thirteenth-century proponent of Aristotelian logic who believed that the mind of God was both just and rational. It was a crack in the orthodox pavement through which a tree would grow. Scholars began rummaging around old monasteries looking for the surviving texts of pagan authors. The literacy rate slowly grew, though it was still largely restricted to the clergy, some merchants, and a few among the nobility.

In the 1300s, fragments of Vitruvius's book began to appear, and handwritten copies of these began circulating throughout Europe. For almost a century, it was assumed that the complete version of the book would never be found (many ancient works that did survive were nevertheless incomplete, like Tacitus's histories). Then, in 1414, a complete edition of *On Architecture* was discovered in a Swiss monastery. The discovery naturally caused a sensation, for here was the only complete work on architecture to survive from antiquity, and people were eager to know what great knowledge would be revealed. Initially, it would be little. The text was corrupt in many places—mostly transcription errors—and even the most erudite Latin scholars were often baffled by Vitruvius's technical language and his frequent use of Greek terminology (*emplecton*? Huh?). Ancient Greek, forgotten in the West for over a thousand years, was being studied once more, but the number of people who were fluent in the ancient tongue was still small, and their reading was mostly restricted to the literary classics, not technical treatises. People would puzzle over Vitruvius's book for decades. Finally, a Franciscan monk with a very unique résumé would come to the rescue.

FRA GIOCONDO'S BRIDGE

One of France's better kings was Louis XII. He reigned from 1498 to 1515 and was known to his subjects as the "Father of the People" (*Pére du Peuple*). He curbed corruption, reformed the legal system, reduced taxes, and kept the meddlesome and occasionally rebellious nobility in check. These actions and policies finally allowed the French to enjoy a stable and efficient government, as well as a period of domestic—though not for-eign—tranquility. Louis XII was also a decent general who fought a suc-cessful war against the Piedmontese warlord Ludovico Sforza, which won back Milan for France. (After twelve years, the unfortunate city was "lost" again, returning to its traditional role as a treasured football to be kicked back and forth by French kings, Italian princes, and German kaisers.)

Shortly after coming to power, Louis XII invited a remarkable man to his court: Fra Giovanni Giocondo, a Franciscan monk considered one of the lights of the already well-lit Renaissance. Fra Giocondo's title was royal adviser, an appropriately general job description for a man of such wide-ranging knowledge and interests. Fra Giocondo was one of the great clas-sical scholars of the Renaissance (he had discovered and published Pliny the Younger's correspondence to Emperor Trajan) and was as comfortable with Greek as he was with Latin. If this were not enough, he was also an accom-plished and highly regarded architect.[2] King Louis wanted his new royal adviser to build a magnificent bridge across the Seine River to the island where Notre Dame Cathedral stood. The monk was certainly up to the task, but for the bridge's construction, Fra Giocondo wanted to try some-thing novel: Roman concrete.

By the time of his bridge commission, the Italian monk had already been studying Vitruvius's book for years, and it is likely that no one then living could so easily grasp the Roman architect's language. Besides being an architect, Fra Giocondo was a pioneering archaeologist who examined ancient ruins with a civil engineer's eye, carefully noting the dozens of different building methods and materials employed by the Romans.

Fra Giocondo wanted to use the waterproof mortar Vitruvius described in his book to build the piers for the new Pont Notre-Dame—the first time in many centuries that Roman concrete would be employed in a construction project. But would it work? Fra Giocondo would be venturing in territory that had been unexplored for a thousand years. Essentially, he was putting his faith in the hands of his ancient predecessor and hoping that the strange formula of lime and pozzolanic soil would perform as described.

It did, and Louis XII was pleased with the magnificent new bridge.[3] Upon completion of this project, Fra Giocondo returned to Italy, where he designed and built several structures, including the famed Fondaco dei Tedeschi in Venice, in which Titian and Giorgione would paint some of their most famous murals. In 1511, Fra Giocondo published Vitruvius's *On Architecture*, a thoroughly annotated and richly illustrated edition that would be the basis of many later translations and that remained the most definitive Latin version for several centuries. Fra Giocondo died a few years later, leaving an architectural and literary legacy that is still treasured today.

Oddly, even though Fra Giocondo had solved the technical mysteries of Vitruvius's book, centuries would pass before concrete was used again. The Italian monk's courage in using the material to build the Pont Notre-Dame would not be emulated by his more conservative contemporaries. Concrete's comeback would have to wait. Its return would be gradual, but unlike before, this time it would refuse to disappear once more in the mists of time and ignorance.

Chapter 5

THE DEVELOPMENT OF MODERN CONCRETE

few miles north of Koblenz, Germany, along a picturesque stretch of the Rhine River that meanders through the low rolling hills of the Neuwied Basin is the town of Andernach. This innocuous municipality, chiefly known today for its "cold-water" geyser and the modest remains of a medieval fortification, is actually one of the oldest European settlements north of Italy. When the Romans founded the town of Antunnacum here in 17 BCE, it was already a Celtic settlement called Antunnuac, meaning "Antunnos's Village." The identity of this Antunnos has been lost over time and through shifting cultural allegiances and languages.

For hundreds of thousands of years, the land around Andernach was volcanically active, and many of the surrounding hills are actually the eroded remains of cinder cones. The Romans found the local igneous stone ideally suited for building purposes and, when pulverized, a perfect pozzolana for making concrete. The locals would later call the stone *trass*, probably a word that has its origins in *terra*, the Latin word for *earth*. The special properties of trass were lost in the upheavals that followed the fall of the Roman Empire.

Sometime in the sixteenth century, the people of Andernach discovered that trass was ideal for carving millstones. It was soft enough to easily chisel but hard enough to grind grain without incurring very much wear. It wasn't long before millstones became one of Andernach's chief exports. The leftover chips and powder from manufacturing the millstones were probably dumped into a pit—or the Rhine—before someone figured out a useful purpose for them. Some bricklayer evidently tried substituting powdered trass for sand in his lime mortar and discovered that once it had set, it was harder and more durable than regular mortar. It also set and held up well underwater. This forgotten experimenter had rediscovered the secret of Roman concrete. Sometime in the seventeenth century, word about the remarkable properties of trass reached the ears of the Dutch.

Dutch traders had plied the Rhine for many centuries. The late seventeenth century was the "Golden Age" of the Dutch Republic: the Dutch had recently thrown off Spanish rule and founded a republic that many economists regard as the first fully capitalistic society. Their excellent navy, manned by first-rate seamen, allowed the Dutch to grab colonies around the world, including the lucrative Spice Islands of Indonesia and its archipelago. One reason for the success of the Dutch Republic was its citizens' fiercely mercantile nature, for they were always on the lookout for interesting business opportunities. Trass looked interesting.

The Dutch tried mixing the trass with lime, and saw that it worked. Because they had been building dykes, canals, and levies for many years, they immediately recognized the commercial potential of this hydraulic building material. The once worthless leftovers from carving millstones were quickly scooped up by the Lowland traders.[1] Once this supply was gone, they probably accepted uncut rocks too small to carve into millstones. My guess is that they actually *preferred* the raw stones. Rocks were the most common form of ballast at that time, and by putting rocks in the holds of their ships, they could sneak it past customs, something you couldn't do with a barrel of powder. Once back in Holland, the traders would then pulverize the trass.

Almost as soon as the Dutch satisfied the domestic demand for trass, they began selling it to the French and British. They sold some trass, but not much: it was viewed as a specialty product whose use was mainly restricted to marine masonry. British engineer John Grundy and his son, John Grundy Jr., used trass for their sluice works on the River Witham in the mid-eighteenth century.[2] French engineer Bernard Forest de Bélidor mentions it as well in his book *Architecture hydraulique*, published in 1748.[3] If the Dutch had had more marketing savvy, they would have called the product "Roman cement" instead of "trass," which sounds too much like the English word *trash* or the French word *travers*, which means "failure." The Dutch would later change the name to "terras," which had a better ring to it—and was coincidently cognate with its etymological origins—but by that time, the French and British were already trying to create their own versions. The eventual result would be the discovery of natural cement and, later, something even better: Portland cement.

JOHN SMEATON'S DISCOVERY OF NATURAL CEMENT

In the second half of eighteenth century, one British engineer stood in preeminence over all others: John Smeaton. He was born in 1724 in Austhorpe, a small town that is now a parish of Leeds, England, and from early childhood he showed a precocious interest in mechanical devices and architecture. Despite these early signs of his natural proclivities, Smeaton dutifully followed his father's desire that he study law instead (the elder Smeaton's profession). Smeaton spent a couple years at his father's law firm but could not endure work for which he had no native aptitude. He left the firm with his father's grudging blessings and became a maker of scientific instruments instead. His improved marine compass and other advanced instruments were noted by the Royal Society, which made him a member. Smeaton went on to improve the efficiency of steam engines and introduced the term *horsepower* to calculate their relative workload capacity, an innovation often wrongly attributed to James Watt.[4] (We now know that Watt, inventor of the greatly improved steam engine utilizing a separate condenser, was overoptimistic about horsepower force, while Smeaton's estimate is closer to its true value.) Smeaton also improved the efficiency of watermills and windmills. For the latter, he developed a formula that addressed the effect that air pressure had on the velocity of objects, specifically the foils of a windmill vane. This formula was later refined and called Smeaton's Coefficient and was used by the Wright brothers in the construction of their early airplane. Smeaton is also known for having first coined the term *civil engineer*, which he applied to himself to distinguish his profession from that of military engineers.

Smeaton was similar in many respects to his ancient predecessor Vitruvius. Both men were fascinated by architectural and mechanical engineering,
and both were honest, conservative, and cautious individuals who first famil-
iarized themselves with the details of a particular field of study before
attempting to practice it or introduce any improvements. Smeaton was definitely the more innovative of the two, and there is hardly a subject that captured his attention to which he did not make some valuable contribution.

However, it was Smeaton's many civil engineering projects that firmly
established his fame. Smeaton does not seem to have had much of a life outside his work. He married at age twenty-two and had two daughters, but his
family probably did not enjoy his presence as much as they would have
liked. By his thirties, he was designing and overseeing the construction of a
number of major engineering projects in Scotland and North England,
some concurrently. At one point, Smeaton was simultaneously overseeing
the building of the Perth and Coldstream Bridges, the Ripon Canal
(designed by William Jessop), the Forth and Clyde Canal, and the canal-
ization of the River Lee (the Lee Navigation). When Smeaton had a quiet
moment outside a jangling carriage or away from work, he wrote scientific
articles or corresponded with his employers or friends.[5] One wonders
whether Smeaton, like Thomas Edison, required only several hours of sleep
each night, for the list of his achievements is quite long. Most of his major
architectural endeavors are still with us, as he built with the same solidity of
the Romans, especially his arched bridges, which would have been constructed in much the same way in ancient times. However, it is his lighthouse for which Smeaton is mostly remembered today. It is also a milestone
in the chronicle of concrete, but not for the reason usually given: its use of
hydraulic mortar. Smeaton had discovered something about concrete that
would eventually change the world.

THE EDDYSTONE LIGHTHOUSE

Lying at the mouth of the entrance to Plymouth Harbor in southwest England are the Eddystone Rocks. The rocks are treacherous stony outcrops

rising precipitously from the sea that have destroyed ships and claimed the lives of their seamen ever since the Bronze Age. Obviously, a lighthouse was needed. The first was designed in 1695 by Henry Winstanley, and he began building it the following year.[6] During construction of the lighthouse, a French privateer captured Winstanley during one of the many conflicts between the two kingdoms. The Sun King, Louis XIV, ordered Winstanley's release in a noble gesture that was reiterated by his famous statement "France is at war with England, not humanity." Once Winstanley was freed, he immediately went back to completing his lighthouse, which was first lit on November 14, 1698. It was an octagonal (eight-sided) wood structure that barely survived its first winter, so Winstanley rebuilt it. The result was a dodecagonal (twelve-sided) stone edifice constructed on a wooden frame. The new tower held up much better against storms, and Winstanley bragged that he would not mind being in the structure during the "greatest storm." On November 27, 1703, a few days after Winstanley toasted the lighthouse's fifth anniversary, a veritable hurricane called the Great Storm of 1703 smashed into the southeastern coast of Britain. Winstanley would get his wish, for he was at the lighthouse with five construction workers making repairs when the gale hit. The lighthouse held up to the storm's fury during the day, but by evening the stone blocks began to shift—they were bound by standard, non-hydraulic mortar—and this stressed the internal wood substructure. Finally, the whole lighthouse collapsed, killing Winstanley and several workers.[7]

Another lighthouse was built a few years later and lit in 1709. It was a firmly built wooden structure that seemed immune to the most ferocious storms. Nevertheless, like all lighthouses back then, a fire had to be continually maintained for its beacon, and since fire has a strong affinity for wood, the inevitable happened. A blaze broke out near the lamp in 1755 and quickly consumed the structure, killing one of the three lighthouse keepers.

Just weeks after the destruction of the second lighthouse, Smeaton was commissioned by the Royal Society to build its replacement. Evidently the society felt that if any man could build a permanent lighthouse on the Eddystone Rocks, that man was John Smeaton.

Smeaton, who was given considerable latitude in the design, was deter-

mined that his lighthouse be the best-built in the world. Before beginning
work on the structure, he hunkered down and began conducting experiments on stones and mortars. He quickly came to the conclusion that
granite, the most durable of building stones, would be perfect. And to make
sure that no storm would ever move the granite blocks, Smeaton specified
that they have dovetail joints at their ends to secure them in place, a provision that must have annoyed the stonecutters, who probably had to re-
sharpen their chisels several times a day. As a further precaution against the
elements wreaking havoc on the granite masonry, Smeaton began testing
the properties of different hydraulic mortars, and it was with these experiments that he would help usher in a new era in construction technology.

Since at least Roman times, the quality of limestone was judged by its
hardness and the purity of its whiteness (a strong indicator of a high calcium carbonate content). Smeaton decided to ignore accepted knowledge
and perform his own experiments. He tested lime taken from a variety of
different limestone outcrops in England. Smeaton began by rolling up balls
of lime and other materials (like trass or plaster) two inches (*ca.* 51 mm)
thick and then allowing them to dry before dropping the balls in boiling
water to test their hydraulic properties. He found, not surprisingly, that
pure lime, while quite hard, slowly dissolved in water. By adding trass to the
lime, Smeaton confirmed that it created a fine hydraulic mortar. However,
Smeaton also found that lime produced from limestone quarried near the
small town of Aberthaw on the south coast of Wales appeared to have very
good hydraulic properties, even without the addition of trass. To discover
its constituent parts, he submersed pieces of the Aberthaw limestone in a
water and nitric acid solution called *aqua fortis* (Latin for "strong water")
used to separate minerals in the mining industry. It showed that approximately 11 percent of the Aberthaw limestone was clay. Smeaton had rediscovered natural cement, which had been used off and on since the Neolithic
period, most notably by the Mayans. The Romans recognized that kiln clay,
in the form of pulverized pottery shards or bricks, provided *caementis* with
hydraulic properties, but they preferred using pure limestone because it
allowed them to control the admixture of pozzolana, whether it was
ceramic dust or volcanic soil. Thanks to Smeaton's discoveries, the primary

concrete cement and mortar that would be used over the next century would be natural cement sourced from limestone adulterated with clay.

Smeaton added trass to the natural cement and discovered that its hydraulic properties were further enhanced. He then substituted Italian pozzolana for the trass, and found that this combination was slightly better. Since Italian pozzolana was harder to obtain, Smeaton probably would have settled on using the trass instead but for a fortuitous coincidence. A few years earlier, a British merchant had ordered a large consignment of this same Italian pozzolana on speculation, hoping to sell it to the people who were about to build the Westminster Bridge. But the bridge builders showed no interest in using the material, and the merchant was stuck with the pozzolana. Smeaton bought the merchant's pozzolana—which no doubt made the merchant happy—and soon began construction of the new Eddystone Lighthouse.[8]

The granite stones were cut at the small town of Millway, near Plymouth, and then transported by boat to the Eddystone Rocks, where they were assembled and mortared into place. The lighthouse was completed in October 1759. Thanks to hydraulic mortar and the dovetail joints of the granite blocks, the lighthouse can be ranked as one of the most solid works of stone masonry in existence. It is justly regarded as one of the jewels of eighteenth-century British engineering. Smeaton would go on to build bridges and canals throughout Britain, most of which still remain with us today, carrying boats, people, cars, buses, and trucks to their appointed destinations.

By the late 1780s, Smeaton began feeling the increasing fatigue and decreasing acuity that comes to many of us upon entering our seventh decade. His wife, Anne, had recently died, his daughters had married, and perhaps he thought it was time to settle down. He retired to his home in Austhorpe, where he began collecting and editing his vast volume of papers and articles for publication. One pleasant September day in 1792, while strolling in his garden, Smeaton suffered a stroke and was carried into his house by several servants. Though mostly paralyzed, he was said to have "still retained his faculties." He died six weeks later, on October 28, 1792.

A story published after Smeaton's death serves as perhaps the engineer's best epithet. A man overheard several young boys debating on which historical figure they would choose to be. One young man said that he would like to have been Julius Caesar, while another opted for Alexander the Great. The third boy, displaying a wisdom belying his age, said that he would like to have been John Smeaton, explaining that Smeaton had improved peoples' lives, while the others had sought glory at the cost of lives. It is a fitting tribute to a man who never took out a patent on his discoveries, choosing instead to share his knowledge with the world.[9]

Smeaton's lighthouse remained in use until 1877, by which time the rock beneath the lighthouse had eroded so much from wave action that the lighthouse shook during stormy weather. In effect, the lighthouse was far stronger than the stone on which it was built. When word spread that Smeaton's lighthouse was to be demolished, a public outcry arose, demanding that it be saved. The lighthouse was laboriously disassembled and reconstructed on a square in Plymouth overlooking the sea. Today, it is as beautiful and solid as it was over a quarter of a millennium ago. However, Smeaton's lighthouse could have remained where it was, had it not been for a false notion about the effects of seawater and concrete, one that Smeaton himself shared. Smeaton, noticing that lime stucco mixed using seawater was not as strong as stucco mixed using freshwater, reasonably assumed that the same was true with hydraulic mortars. However, the Romans had discovered that concrete mixed with seawater to make monolithic structures worked well, as demonstrated by the harbor emplacements they built in Caesarea and elsewhere in the Mediterranean. If a cofferdam had been constructed around the eroded rock and filled with concrete and aggregate, Smeaton's lighthouse could have remained on the Eddystone Rocks to this day. The difficulty lay not only in attitudes about seawater and concrete but also in the fact that monolithic concrete construction was still at an experimental stage at this time. Despite a few notable exceptions, concrete was still primarily used as a hydraulic mortar or stucco during the first three quarters of the nineteenth century. Nevertheless, the discoveries made by Smeaton in his experiments on lime mortars marked a turning point in construction history, and men throughout Britain would continue to experiment with different mixtures and manufacturing techniques to create cement of better quality.

ROMAN CEMENT

The revolution begun by Smeaton did not take place immediately after the construction of the Eddystone Lighthouse but several decades later. Smeaton did not publish an article about hydraulic mortars until 1775, but it was a short piece, barely one and a half pages long, and it did not mention his detailed experiments that led to the discovery of natural cement. Among the papers he did see go to press just before his death was his "A Narrative of the Building and a Description of the Construction of the Eddystone Lighthouse with Stone," published in 1791.[10] In this work, Smeaton provides a thorough account of his experiments with various limestones and mortars. Not long after the publication of this paper, a patent would be filed for something that would soon be called "Roman cement."

First, let's take a quick look at an earlier patent filed in 1779 by the Irish-born chemist Bryan Higgins, who published a booklet the following year titled *Experiments and Observations Made with the View of Improving the Art of Composing and Applying Calcareous Cements, and of Preparing Quicklime. Theory of These, and Specification of the Author's Cheap and Durable Cement for Building, Incrustation, or Stuccoing, and Artificial Stone.*[11] Because Higgins was an otherwise esteemed chemist, early chroniclers of concrete's history have included his patent and book in the material's story, but his product was unstable and did not have the long-term endurance of good hydraulic mortars or concrete.

Of more interest are the two patents filed by James Parker, reportedly an English clergyman and civil engineer. (A later historian who conducted a search of many ecclesiastical and academic records was unable to discover an engineering, architectural, divinity, or any other degree awarded to Parker.) Parker's first patent, filed in 1791,[12] speaks of a hydraulic material using "bricks and tiles" and of calcinating the mixture "with a material not previously used for the purpose." The latter was simply peat, and as for the ceramic ingredients, people had already begun mixing pulverized bricks or tiles with lime after an English translation of Vitruvius's *On Architecture* by William Newton had been published in 1771.

Parker's second patent (1796),[13] filed five years after Smeaton's book was published, demands especially close scrutiny. It is for natural cement. An anonymous article published in 1830, based on an account "written down by him [the anonymous author] many years earlier" explains how Parker made his discovery of "Roman cement":

> It was first discovered by the Rev. Dr Parker in the year 1796 and like many other of our most useful acquisitions, was purely accidental. When on a visit to the Isle of Sheppey, he was strolling along under its high cliffs on the northern side and was struck by the singular uniformity of character of the stones upon the beach and which were also observable sticking in the cliffs here and there. On the beach, however, the accumulation of ages, they lay very thick. He took home with him two or three in his pocket and without any precise object in view, threw one on to the parlour fire from which in the course of the day it rolled out thoroughly calcined. In the evening he was please to recognise his old friend upon the hearth, and the result of some unpremeditated experiments with it has been the introduction to this country of strong, durable and valuable cement.[14]

Of course, this limestone had enough clay admixture to make natural cement.

This account of Parker's discovery raises many suspicions. Parker brings home some stones "without any precise object in view" and throws one of them in the fireplace where it—thoroughly "calcined" after just a few hours—rolls down onto the floor. He then conducts some "unpremeditated experiments" and discovers natural cement, independently of Smeaton and—at least by suggestion, if not by word—without having read the master's work on the subject.

Let's go back in time several years. In 1792, the year following the publication of Smeaton's book about the building of the Eddystone Lighthouse and his discovery of natural cement, Parker leased some land at Sheppey.[15] His kiln was already in operation at Northfleet, not too far from Sheppey, and he had established a London office by 1773.[16] And the story of Parker's subsequent discovery has as many holes as a block of Swiss cheese. A man who had patented a mortar five years earlier strolls around the land he had already leased—for what purpose?—picks up some limestone rocks and later calcinates one of them (apparently after only a few hours!) "without any object in view" and then performs "unpremeditated" experiments on them? This story is obviously based on Parker's own account, which apparently no one closely examined. The most rational conclusion is that Parker used Smeaton's research to file a patent that, by all rights, should have been denied but was not. In a stroke of marketing genius not uncommon among borderline shysters, Parker would call his product "Roman Cement" (something the Dutch should have done with trass a century earlier). Of course, it was nothing of the sort, but the name was an excellent choice. As with Higgins's work, chroniclers of concrete's history have listed Parker's "Roman Cement" as some sort of milestone, when, in fact, the achievement was Smeaton's. Parker's ill-gained patent would also delay by a few years the widespread use of an important mineral freely provided by Nature.

The year 1796 was a busy one for Mr. Parker. Sometime in February or March, Parker approached the British Society for Extending the Fisheries and Improving the Seacoasts of This Kingdom to interest them in his hydraulic mortar and stucco. On March 17, the society directed one of their engineers, Thomas Telford, to examine Parker's product. Parker then convinced Telford—after two meetings at the former's house—to write a glowing testimonial about his cement (dated April 12, 1796), which Parker immediately printed and distributed as an "impartial and disinterested report" from "an eminent Engineer," even though Telford's language was hardly objective. (Telford concluded his remarks by writing that "I was glad to embrace this opportunity of doing justice to a discovery which may become of considerable importance to the Public, and which appears to merit its attention."[17])

On July 27, 1796, Parker was granted his second and more famous patent for what he called "Parker's cement," which he soon re-dubbed "Roman cement." To disguise the fact that it was limestone adulterated with clay, he described it in the patent as clay containing "calcareous matter." Now armed with a patent and a glowing report by an "impartial" engineer, Parker sold his cement works, patent rights, and leased land to

Samuel and Charles Wyatt, and then sailed for America. There is no evidence that Parker produced a large quantity of Roman cement, but enough had been used to show Telford its qualities. Although Parker had leased the land on Sheppey, he did not own the mineral rights, a legal issue that he left the Wyatts to sort out at great expense.[18] Parker probably just took enough stone to make small batches of his cement, while being careful not to prompt the suspicions of the locals. It is not surprising that Parker went to America: he no doubt knew that large-scale extraction of the Sheppey stone would eventually lead to a court battle. He was also probably worried that someone would draw attention to Smeaton's work and contest his patent rights, perhaps one reason why he delayed applying for the patent until after everything else was in place. Fortunately for the Wyatts, and unfortunately for the world, no one did contest the patent. As for James Parker, he died shortly after arriving in America.[19]

THE WYATTS

Samuel Wyatt was an architect and builder, and Charles (Samuel's cousin) was a tinned copper sheet and pipe manufacturer. Both were respected businessmen who had good ties to the construction industry. The Wyatts renamed the London cement works Parker and Wyatt Cement and Stucco Manufacturers. Parker's name was preserved to publicize the fact that the Wyatts now owned the patent rights. Samuel soon withdrew from any active participation in the cement business, and Charles seems to have been in full control after 1800. Once the legal wrangling was over—a compensation package settled the issue—the stone from the Sheppey quarry continued once more to supply the company's needs. When limestone with similar properties was found on the Essex coast, the company used this for their product as well. The firm of Parker and Wyatt would be so successful that all natural cement would thereafter come to be known as Roman cement. (This would have no doubt pleased the late Mr. Parker.) Thomas Telford, whom we met earlier, used it for building bridges, harbor works, and the beautiful Chirk aqueduct in North Wales—an arched masonry structure resembling the aqueducts of ancient Rome. Telford's enthusiasm for the product was apparently real.

Thanks to Smeaton's discoveries, Parker's dubious patent, and Charles Wyatt's sound business practices, the primary concrete cement that would be used over the next sixty years in Britain would be natural cement. Although it was called Roman cement, some of it was actually better, since the hydraulic element—clay—was kilned together with the limestone, creating a stronger molecular bond. Hydraulic mortars and stuccos gained popularity in Britain, and they were soon being used beyond marine applications. The weather in Northern Europe is damp, and conventional lime mortars and stuccos eventually wear away and need to be reapplied. Stuccos were especially valued, as most people found exposed brickwork unsightly. The prevailing taste held that covering a brick building with stucco, which was then indented to give the appearance of stone mortar joints, imparted a sense of dignity to the structure. However, after a decade or two of freeze-and-thaw cycles, the stuccoed building's dignity would be considerably diminished. On the other hand, hydraulic stuccos were found to be hardier and, while not completely immune to environmental factors, they lasted much longer. The one disadvantage of Roman cement was its color, which was light brown. For this reason, stucco made from Roman cement often had to be whitewashed after drying.

A profusion of new—and allegedly better—hydraulic mortars and stuccos were patented and peddled in early nineteenth-century Britain. Most manufacturers claimed that the "superiority" of their product was due to novel or unique ingredients or an improved manufacturing process. A few of these products were indeed better, but most were not. One of the latter was "oil stucco." Since oil repels water, it was thought that the addition of linseed oil would make the hydraulic stucco even better. It seemed to work, and after drying, the surface could be smoothed to resemble polished stone. But what was gained in appearance and water resistance by the linseed oil was balanced by a lack of adhesion. Not many years passed before oil stucco began peeling off buildings throughout England, especially in London, and most of the firms manufacturing this particular product quietly closed their doors.[20]

In Britain, patents were—as they are now—valid for fourteen years. As the 1810 expiration date of Parker's patent approached, people around Britain and on the Continent made preparations to manufacture and market Roman cement, a brand name that had become a generic term for natural cement (trademarks often expired with patents at that time).

Among those in the best position to take advantage of the expiration of Parker's patent was Charles Francis of Vauxhall, London. Francis was an architect who primarily made his living as a brick, marble, terras (trass), and cement wholesaler. He also had a wharf on the Thames that served his extensive business. Francis's background as an architect gave him an intimate knowledge of the building industry and its raw material needs. By 1808, though only thirty years old, he was managing a thriving business and had established a host of valuable commercial contacts throughout Britain. With his intimate knowledge of lime kilning and cements, Francis believed that he could produce Roman cement of equal if not better quality to Wyatt's. By 1808, he was already making plans to be among the first to benefit from the expiration of Parker's patent two years later.[21]

Francis designed more efficient kilns and wanted to secure firm quarrying rights to the appropriate stone (he had no desire to be involved in a legal imbroglio like the Wyatts had experienced). The amount of financing required for an operation of this scale could not be met by his own resources, so he sought a partner. In late 1808, a mutual friend introduced Francis to John Bazley White, a former banker who was then an executive with a firm importing goods from the East Indies. White was intrigued by Francis's plans. He also appreciated the fact that Francis had proven himself a good businessman, knew the industry well, had a strong base of loyal customers, and possessed numerous business contacts. A legal partnership was formed between the two men in July 1809, and the firm Francis & White was born. It would go on to become one of the most successful manufacturers of Roman cement in Britain and eclipse the fortunes of Parker and Wyatt.[22]

Francis immediately began traveling around Britain, looking for the appropriate clay limestone. He found that most of the promising outcrops were, like those in Sheppey, along the east and south coasts of England. Once the rights were secured for the stone, Francis built a large kiln that had elements both new and old. English forests had been shrinking for centuries, and the huge demand for lumber to build the vast fleet of ships used by the Royal Navy in the Napoleonic Wars (two thousand trees had to be harvested for each of the larger vessels) wiped out many of the few remaining British woodlands. Fortunately, the new energy demands of the Industrial Age coincided with England's expansion of her coal mining and gas industries. Satisfying the vast demands for fuel to power steam engines, foundries, and limekilns was no longer a problem. The kilns that Francis built seemed, at first glance, no different from Cato's. The ancient Burgundy bottle shape remained, but the kilns were far larger, and the way they were fueled and operated was also different. Coal—and later coke—allowed higher temperatures and greater efficiency than wood. Boys were hired for a few pence a day to break down the limestone with hammers to a size suitable for kilning (no piece could be more than a few inches in circumference), a monotonous task that kept the youngsters busy from dawn to nightfall. A layer of coal one foot thick (*ca.* 30 cm), was laid down, upon which a layer of limestone rocks of the same thickness was laid. This pattern of alternating layers of limestone and coal was repeated until the top of the hearth was reached, just below the chimney. The kiln was then lit from beneath the bottom iron rack that held the first layer of coal. These first layers of coal and limestone were allowed to burn for a couple days and nights before being pushed out with the rack to allow the next layer of limestone and coal (the latter having ignited by this time) to cook. The amount of limestone that could be burned at one time in such a kiln was far higher than in Cato's time—reportedly over a couple hundred tons in the larger bottle ovens. The kilned chunks of lime and clay, which were then pulverized, were said to be "as light as cork." As demand grew, more bottle kilns were built, often physically adjoining one another to save masonry and space. Bricks, being fireproof, were the preferred building material. The mortar used for the masonry often contained a mixture of brick dust, which helped against the heat, but the masonry seams of the kiln's interior still had to be repointed regularly because of exfoliation. Eventually, larger square kilns replaced the bottle kilns, their shape being more conducive to layer racking.[23]

For some reason, the cement produced by Francis & White became the preferred product of its day. The company often used stone from the same quarries as Parker and Wyatt, but the cement produced by the former was judged to be better. Perhaps Francis discovered that by weighing the rock, he could better judge its relative proportion of clay and limestone. (The company offered different grades of cement, though this was not uncommon at the time.)

Another reason for the company's success could be the degree of cooperation and professional advice they offered their clients, which was considerable. Most notable in this respect was a major engineering project that, upon its completion, would be called the "Ninth Wonder of the World." The builder of this wonder was Marc Brunel, whose life and career is worth reviewing, for, as the old saying goes, "You can't make this stuff up."

MARC BRUNEL

The Chinese curse of having to live through "interesting times" could certainly be applied to Marc Isambard Brunel. Brunel was born to a prosperous French farmer and his wife in the Normandy village of Hacqueville in 1769.[24] Brunel, like Smeaton, exhibited a childhood interest in building things, and he enjoyed peering into clockworks to better understand their operation. Both men also had fathers who initially preferred that they take up professions to which they were temperamentally and intellectually unsuited. Unlike Smeaton, Brunel faced extraordinary difficulties throughout his life. Since his family was devoutly Roman Catholic, they followed the tradition of consigning the inheritance of the estate to their first-born son, while the second was pledged to the Church. In other words, Marc was destined—at least in his parents' eyes—to become a priest. The priesthood required a firm background in the classics, but Brunel showed no interest in learning Greek or Latin. He did, however, display a remarkable aptitude in mathematics, drafting, and music (the boy—then eight years old—could also make his own musical instruments).[25] Despite these early signs of a proclivity toward engineering, Brunel's father instead pushed him to do better in ancient languages. Brunel finally rebelled. At age eleven, Brunel firmly announced to his father that he had absolutely no desire to be a clergyman and instead wanted to be an engineer. Nothing could sway him from this determination, just as nothing could sway his father's insistence that he become a priest, so the father packed the boy off to the Seminary of Sainte Nicaise in Rouen, hoping that Marc would forget his silly notions and learn to serve God instead. Fortunately for young Brunel, the seminary's superior was an open-minded individual who believed that God-given talents should be expressed, not suppressed. Noting that the boy enjoyed designing and building things, he allowed him to learn carpentry, a craft that the youngster was soon practicing with the skill of a master cabinetmaker. The boy was also given paper and charcoal pencils and allowed to draw. Instead of landscapes, Brunel sketched the ships in Rouen's harbor, producing renderings of startling realism that clearly demonstrated the youth had a particularly fine eye for proportion and detail. Brunel's father, informed that the boy's gifts lay outside the Church, finally gave in. Brunel was sent to live with his cousin, Mme. Carpentier, whose husband, François Carpentier, was the American consul in Rouen. Carpentier, a retired sea captain, instructed the boy in naval matters, while another family friend, Vincent Dulague, who taught hydrology at the Royal College at Rouen, tutored Brunel in the sciences. The French Royal Navy was then undergoing progressive reforms instituted by Charles de La Croix (Marechal de Castries), the minister of marine affairs. De La Croix was assiduously recruiting young men with strong technical backgrounds and promising them good opportunities for exercising their gifts.

In 1786, Brunel became a midshipman on a French frigate that embarked on a six-year tour of duty in the West Indies. While onboard, Brunel used his spare time to learn English, bone up on his astronomy, and design a superior quadrant that he then constructed of brass and ivory. Brunel's ship was in the Indies when the French Revolution broke out in 1789, and when the frigate returned in 1792, the country was in chaos. Since the ship's captain was receiving conflicting orders and reports from Paris, he decided to pay off the crew and dismiss them from service. Royalist sentiment was very strong in Brittany and Normandy, and Brunel shared these views. One year after his dismissal from the navy, Brunel was in Paris personally observing the tumultuous events that were reverberating across France and Europe. King Louis XVI was then being tried for treason, and a prominent revolutionary leader, Maximilien de Robespierre, was vehemently arguing that the former monarch be sentenced to death. According to one account, Brunel was at a café when word came that the king had been found guilty and would be beheaded. Brunel, perhaps forgetting that he was in Paris and not Rouen, cursed Robespierre and predicted that he would one day suffer the same fate for his cruelty. Several revolutionaries at the café immediately arose to defend Robespierre and began questioning Brunel's loyalty to the republic. A melee broke out, and Brunel barely escaped. Another account relates that Brunel, who had brought his dog with him to Paris, continually addressed the canine as *Citoyen* (citizen) in public, the common greeting among republicans. In any event, Brunel was forced to leave Paris to avoid arrest and probable execution. The fate of his dog is unknown.

Brunel found refuge with the Carpentiers in Rouen, who enjoyed diplomatic immunity. While there, Brunel fell in love with their English governess, Sophia Kingdom, and proposed to her. Unfortunately, the Revolution turned even uglier, and "traitors" were now hunted down with a ruthlessness that often ignored diplomatic privileges and foreign nationalities. Brunel had no choice but to escape France and leave Sophia behind. With the help of a friend, he obtained a passport that allowed him to travel to America for the purpose of obtaining wheat for the army. He boarded the American ship *Liberty* and sailed for New York. Brunel probably felt that he was now out of danger, but it was not to be. A French warship approached the *Liberty* and ordered her to heave to so that she could be searched for political refugees. Brunel could not find his passport. Stifling his panic, he took a pen, some paper, and scissors, and secreted himself in a remote hold of the ship. Brunel's drafting skills came to his rescue. It took two hours for the French officers to find him, but by then he had counterfeited a passport good enough to pass inspection. Even then, it had been a close call, since many officers in the French Navy knew the young man. The ship continued on its voyage to New York, and Brunel made sure not to lose his new "passport."

Arriving in New York, Brunel saw to his horror that a squadron of French naval ships was docked in the harbor and the city's streets were crawling with French officers and seamen. Since the American government was then on friendly terms with France, an extradition request might have been honored. Or he just might be knocked on the head, taken onboard one of the French vessels, and put in chains to await the ship's return to France. Thus, Brunel had no choice but to flee New York. He went to Albany, hoping to find a friend, Pierre Pharoux (they were fellow passengers aboard the *Liberty*), who was then surveying the Black River Valley in upstate New York. He met up with Pharoux and became a member of his surveying party. Since Brunel's surveying skills were superior to Pharoux's, the latter cheerfully allowed him to take charge of these duties. This portion of the state was still largely unexplored, and the team depended on the goodwill of the local tribes to perform their work. Apparently Brunel made a good impression on the local Native Americans; fifty years later, Oneida tribesmen were still talking about a wonderful white man called "Bruné."

At one point, John Thurman, a wealthy merchant with strong political connections, joined the surveying party. Thurman was interested in developing this remote part of New York and was naturally eager to know what areas held the best prospects for farming, lumbering, and road building. After spending some time with Brunel, he quickly recognized that the young man's surveying, architectural, and mechanical abilities were uniquely suited to the needs of the growing country. After the survey party had completed its assignment, Thurman used his influence to obtain work for the Frenchman, and it was not long before Brunel was receiving more commissions than he could accept. Relations between France and America were also souring, so he felt safe in applying for American citizenship, which he was quickly awarded. Brunel formed friendships with many prominent figures in the United States, including Alexander Hamilton and Pierre L'Enfant, the planner of the nation's new capital, Washington City. At L'Enfant's suggestion, Brunel submitted a design for the first Capitol building, which, while admired by the judges, lost out to another's plan. Through Hamilton's influence, Brunel was awarded the post of New York's chief engineer. While serving in this position, Brunel oversaw the construction of a number of buildings, built a cannon foundry, and supervised the fortifications of the Narrows at New York Harbor.

One day, Hamilton invited Brunel to his house for dinner. There, Hamilton introduced him to Pierre Delabigarre, another French political refugee who had come to America and become a citizen. Delabigarre told Brunel that the British Navy was having problems obtaining enough ship's pulley blocks, since each one had to be laboriously carved by hand. A single warship used up to fifteen hundred blocks, and the British Navy purchased over a hundred thousand blocks each year for new ships or to replace broken ones.

Brunel saw his opportunity. He was certain that he could design a machine that could mass-produce the blocks. Besides, he wanted very much to go to England. Britain was the heart of the new Industrial Revolution, where innumerable opportunities awaited talented and inventive individuals. America, on the other hand, was still a largely rural society and would remain so for a couple of generations to come.

Brunel had another reason for wanting to go to Britain. After his escape from France, Sophia had been arrested. Britain joined the European coalition fighting against the revolutionary government, which prompted the latter in October 1793 to pass a degree ordering the arrest of all British citizens residing in France. Sophia was also under suspicion for her relationship with a royalist "enemy of the State." She languished in prison, existing on bread mixed with straw, and on several occasions barely avoided being sent to the guillotine. After Robespierre's overthrow and execution, a more moderate faction had come to power, and most political prisoners and British citizens, including Sophia, were released. Sophia left France for England, and Brunel wanted to join her in London.

Taking with him what money he had saved and letters of introduction from his prominent friends, Brunel sailed for England in 1799. Thankfully, it was a smooth voyage without incident. Arriving in London, Brunel was reunited with Sophia, and they soon married. Brunel's marriage would be the happy bedrock in his oft star-crossed life, for he would continue to live in "interesting times."

His first two months in England were extraordinarily busy. Besides marrying Sophia, he worked on the designs for his block-making machine, took out a patent on a writing and duplicating device (possibly inspired by Thomas Jefferson's copying press), and invented a machine for twisting thread. By 1801, he had built a working model of his block-making machine with the assistance of the famed toolmaker and inventor Henry Maudslay. Brunel approached the firm of Fox and Taylor, which supplied the blocks to the British Navy, with his machine model. The company showed no interest, telling Brunel that it had spent many years perfecting its method of manufacturing blocks and saw no possibility that it could be improved upon. Brunel then took his plans to Lord Althorp (George Spencer), First Lord of the Admiralty, and Sir Samuel Bentham, the noted inventor and naval architect. The latter had been working on the same problem, but his equipment could produce only a rough block that needed to be finished by hand, while Brunel's device could perform all the manufacturing steps. At Bentham's recommendation, Brunel won the contract to provide blocks for the British Navy. Maudslay built the complicated machines, and the navy was soon receiving blocks at low prices and in large quantities. What had once required sixty men now required just six, and their output was incomparably higher. The factory was one of the earliest examples of mass production.

Things went smoothly for a time, but Brunel had difficulty obtaining payment from the admiralty for the blocks he was delivering to them. He had invested £2,000 of his money in the venture, yet he was seeing nothing in return. He finally received £1,000 "on account," but six years would pass before he was sent a more substantial payment—£17,000—though it was still less than what was owed. (The British government's heartless and myopic penny-pinching ways would become legendary.) In the meantime, Brunel had patented some improvements to sawmill machinery. There was a big demand for lumber by the British Navy, which was now fighting Napoleon on the high seas. The war kept Brunel busy, but this time he just built the sawmills and let others deal with the government. His most notable effort was the steam-powered sawmill at Chatham, near its docks, which increased lumber output while reducing yearly manpower costs from £14,000 to £2,000.

Learning that the British Army needed thousands of boots, which, like the pulley blocks earlier, were made slowly by hand, Brunel designed and built machinery that could perform much of the laborious work—the earliest example of the mechanized mass production of shoe wear. Brunel obtained a contract for fifteen thousand boots of various sizes. He had just received his £17,000 for the blocks and had recently become a British subject, and so felt more comfortable about filling the government's order. Of course, after the boots were made, peace broke out, and the government informed him that they didn't need the boots after all. Not only was Brunel stuck with the boots, but he also needed to pay the suppliers who had provided him with the tons of leather and hobnails to make them. Granted, he had also made some unwise investments, but it was the affair with the boots that pushed him over the edge to insolvency. He petitioned the government for redress while holding off his creditors as best he could—for years—but it was all for naught, and he was sent to a debtors prison in 1821. Brunel, who had saved the government of his adopted country God knows how many pounds—although the Exchequer probably knew—now found himself behind bars.

Seeing no prospect for release, he started corresponding with the Russian tsar, Alexander I, who had earlier offered him a well-paid position at court overseeing various engineering projects. After describing his many "vexations," Brunel suggested building a tunnel under the Neva River in St. Petersburg. As a young man in Rouen, Brunel had picked up a floating piece of wood and, noticing the telltale holes of a shipworm, took it apart to see how the creature could make its way through the hard cellulose fibers. The shipworm—actually a mollusk related to clams—possessed a small, two-piece tubular shell near its head that protected it while it bored through the wood. This gave Brunel the idea of a shield that could be used for tunneling purposes. This inspiration would later have important ramifications.

Brunel was universally well liked, and while he languished in prison, his many friends in high places were lobbying for his release. When the British government learned that it might lose one of its brightest subjects to the Russians, it cut a deal with Brunel. Essentially, it offered to pay off his debts if he agreed not to work for the tsar or for any other foreign government. It was a wise move, for Brunel was about to embark on a project that would bring fame to Britain and earn him knighthood. However, Brunel being Brunel, would not accomplish his goal without encountering some difficulties. These would remain interesting times for him.

THE THAMES TUNNEL

The largest use of hydraulic cement since the days of ancient Rome was the construction of the Thames Tunnel in London. It was Ralph Dodd who first suggested building a tunnel underneath the Thames River in 1798. Dodd, an engineer from Northumberland who had observed coal miners excavating beneath the River Tyne without any apparent hazard, proposed a tunnel be built to connect the districts of Gravesend and Tilbury. One or two years later (some accounts give late 1799, others, early 1800), work began. A shaft was dug at the Gravesend side, but it continually filled with waterlogged sand. The project should have ceased as soon as it became clear that the ground was unsuitable for excavation, but Dodd believed he could work around the problems and persisted in his efforts until the funding eventually ran out two years later.

A new enterprise, the Thames Archway project, was created in 1804 to build a tunnel linking Rotherhithe and Limehouse (now Wapping). Robert Vazie, another Northumberland engineer, was chosen to supervise the effort. Though the ground was not as bad as that at Gravesend, Vazie also encountered flooding problems that continually put the tunnel project behind schedule. Frustrated with the slow progress, the directors hired Cornish engineer Richard Trevithik to replace Vazie. Trevithik was a gifted mechanical engineer who had made a number of improvements to steam-engine technology. (He was one of the earliest advocates of steam-powered transportation and even built working examples to demonstrate the concept, but no one was interested in the scheme.) At the time he was offered the tunnel project, Trevithik was enjoying some popularity for the success of his steam-powered dredger, which was being used to keep the Thames waterway clear for shipping. The Archway Tunnel directors apparently thought that he was the right man to deal with mud. Trevithik hired a group of Cornish miners to build the tunnel, but the men, more used to dealing with rock than river slime, had difficulty adapting to the new conditions. Still, adapt they did, and work progressed steadily. In 1808, with the tunnel over two-thirds completed, a break in its ceiling caused a catastrophic flood that nearly killed Trevithik and several coworkers. A story persists that the flooding was caused in the following manner: someone told Trevithik that the tunnel was out of line and bearing slightly off course. Trevithik then reportedly broke through the tunnel's ceiling to take a look, and so caused its flooding. The story sounds like a malicious fable spread by one of Trevithik's colleagues, with whom he was always quarreling. One cannot read of Trevithik's accomplishments and at the same time believe he could do such a witless thing.

Undaunted, the Thames Archway project announced a public competition for proposals to build the tunnel. Charles Wyatt—of Parker and Wyatt fame—won the competition. He suggested that the Thames's riverbed be dredged along the proposed course of the tunnel. Into this trench, prefabricated brick cylinders—built using Wyatt and Parker's hydraulic Roman cement, of course—would then be laid and joined underwater. On paper, the scheme made sense, but when the company hired John Isaac Hawkins to test the viability of the proposal, the shifting riverbed and the difficulty of connecting the cylinders proved insurmountable. An engineering report later concluded that building a tunnel under the Thames was "impracticable," though a close reading of the report suggests that the term "virtually impossible" was a better description. The Thames Archway Tunnel project came to an end.

Nevertheless, the idea of constructing a tunnel under the Thames stubbornly persisted. To talented and visionary individuals, nothing is more tempting than accepting a challenge to perform the impossible. And Brunel wanted to do just that. Perhaps no one since John Smeaton was as capable—and stubborn enough—to tackle such a formidable project.

Brunel approached investors with his tunnel-shield idea for excavating under the Thames. The previous two tunnels were cramped, three-feet-wide (*ca.* 91 cm) pedestrian passageways that barely allowed two people to pass each other. Brunel's tunnel would be more massive: twenty feet high and thirty-five feet wide, consisting of two parallel arched corridors open to each other and large enough to permit both foot traffic and carriages. Brunel provided detailed drawings of his proposed "tunnel shield" that would protect the workers and secure the excavation effort as it progressed. He described the shield as an "ambulatory coffer-dam." The investors were interested enough to pay for a series of test borings of the Thames's river bottom, and the best results came from the borings performed between the banks of Rotherhithe and Wapping, less than a mile from Trevithik's tunnel. Here, the borings pulled up firm "blue clay." Convinced that Brunel's plan was practical, the Thames Tunnel Company was incorporated in June 1824, and 2,128 shares were issued at £50 each. Once the tunnel was completed, the shareholders believed that a small toll would eventually cover the costs of construction and thereafter provide a steady income.

Work began in Rotherhithe on March 2, 1825, a little over a hundred feet from the banks of the Thames. A vertical shaft had to be excavated down to the desired depth before work could proceed on the tunnel proper. Brunel handled this in an ingenious fashion: he had a fifty-foot-wide, flat iron ring assembled on the spot where the shaft was to be excavated. Upon the ring, he constructed a circular brick tower using Roman cement. The brick walls were built several wythes (layers) thick, and the tower's interior was strongly braced by wooden timbers and iron tie rods. As the walls of the tower rose, the workmen dug out the earth within. As the excavation progressed, so did the tower's height and weight, and the structure slowly began sinking into the ground. Once the masonry was finished, a steam engine was assembled on top of the tower. The engine was used to pump out the water that sometimes flooded the hole and operate a conveyor that carried up the endless chain of buckets filled with excavated earth that needed to be dumped (the men had previously been handling that task by climbing ladders, buckets in hand). Gradually, despite minor setbacks, the tower sank to the desired depth of 65 ft (*ca.* 20 m). The vertical shaft was completed in November 1825. Now it was time for the horizontal excavation work to begin using Brunel's shield.

Or maybe not. The chairman of the tunnel project, William Smith, seemed intent on undermining Brunel at every step. As the tower began sinking and Brunel was finishing up the final details of the tunnel shield's design, Smith got cold feet. He told Brunel that the shield was an unnecessary luxury and that the tunnel could be constructed using more traditional methods. Brunel pointed out the failure of these conventional approaches in the previous tunneling attempts. Smith held to his position, while Brunel went ahead and ordered the shield's construction (it would be built by his old friend, Henry Maudslay). Since a shield was in the original plans—and its maker, Maudslay, was now famous and politically well connected—Smith gave in. Smith was a Member of Parliament and astute at recognizing the limits of how far he could obstruct Brunel—not that he ever gave up trying.

Maudslay delivered the several hundred cast iron and wrought iron components that made up the tunnel shield to the worksite at Rotherhithe. Each was lowered by crane to the bottom of the shaft to be assembled. Once the shield was put together, it was rightly considered an engineering wonder. The eighty-ton behemoth consisted of twelve frames, aligned vertically to one another, and each contained three compartments where the men could excavate the earth ahead of the masonry work. Every frame held dozens of poling boards—five hundred in all—that penetrated the face of the tunnel, called the "drift," several feet in depth. These poles served to both hold the frame in place and help loosen the clay soil ahead. Each frame, supported at its base by a broad iron shoe connected by a ball joint, could be moved forward by powerful screw jacks that abutted the brickwork to better follow the progress of the various labor teams. Of course, this inspired a lively competition, so no one frame advanced much beyond any of its companions. The shield began its slow journey north to Wapping on November 25, 1825.

As work started on the horizontal shaft, Smith and Brunel began arguing over which Roman cement to use. Brunel wanted to use Francis & White's Roman cement, while Smith insisted that a cheaper alternative be used, one produced by a friend, Matthew Wilkes, an immensely wealthy businessman. Wilkes had also seized the opportunity to open a cement works after the expiration of Parker and Wyatt's patent and was aggressively peddling his product. There is no evidence of Wilkes having bribed Smith,

but it would not have been surprising, since Wilkes had an unsavory reputation (he had made his fortune from pirating and the slave trade). Brunel
had not liked using Wilkes's inferior cement on the tower, but now that
excavation was beginning under the Thames, he insisted on the one manufactured by Francis & White, pointing to tests he had conducted that
showed it to have better hydraulic properties. The chairman would not
budge until Brunel's incessant complaints and letters made the issue too
tiresome to contest, so Smith finally allowed the engineer to use his favorite
cement. (Smith and Wilkes must have enjoyed some satisfaction when,
later, a barge carrying a large quantity of Francis & White's cement to the
worksite sank in the Thames, a misfortune ascribed to an accident.)

Still, Brunel realized that just one batch of defective cement could jeopardize the entire enterprise, and he required that a sample be taken from
each cask and tested—a practice that many years later would become standard in the construction industry. Four hundred samples were tested each
week, which means that hundreds of tons of cement—probably representing close to half of Francis & White's total production—were being
used each month to mortar the five-wythe-thick brick walls of the tunnel.

Water was more of a problem than originally expected. The cheery
results of the borings proved deceptive, for while the soil was mostly clay, it
was also veined with water passages and sometimes pocketed by huge cavities
filled with water-soaked mud. Both mud and clay were inundated by the
detritus of centuries, and the tunnelers occasionally plucked the odd artifact
out of the slime. However, it was the more recent rubbish that proved more
troublesome, and the tunnel often stank of the still-decaying garbage that
dropped down from the ceiling or popped out from the drift. This slimy,
coagulated waste generated methane that made the men dizzy or detonated
the occasional "flashes" that singed hair or eyelashes. Fortunately, the quantity of gas produced was never high enough to cause a major explosion.

It was not only the workmen who were suffering from the poor air.
Brunel and the other engineers were also laid low by the fumes. One of the
assistant engineers, a man named Riley, fainted and was carried out of the
tunnel. He became feverish, then delirious, and a week later he slipped into a
coma and died. It is not certain whether the man had died from the gas or

FIGURE 19. Portrait of John Smeaton, famed engineer and architect. Smeaton discovered "natural cement" and used it to build the famed Eddystone Lighthouse (shown in portrait's background).

simply succumbed to one of the innumerable diseases that Victorian medicine was powerless to challenge, but most men at the worksite attributed Riley's death to the "bad air." (The only other fatality up to this time had been a drunken worker who had fallen from the shield and landed on his skull.) Soon after Riley's death, Brunel was afflicted as well, writing in his diary that

FIGURE 20. William Aspdin, serial swindler and fabulist. Aspdin is the probable inventor of modern Portland cement, a discovery he falsely attributed to his father for marketing purposes.

a "peculiar and indescribable sensation came over me—a haze rose before my eyes, and, in the course of half an hour, I had lost the sight of my left eye."[26] He was forced to spend several weeks recuperating, but he grew restless and returned to the tunnel works, even though his sight remained weak. When the resident engineer William Armstrong fell ill and seemed reluctant to return to work, Brunel appointed his young son, Isambard Kingdom Brunel, to replace him. Isambard, who had been working as an assistant engineer on the tunnel project since its beginnings, had just turned twenty. Despite the nepotism involved, it was a good choice. Isambard was every bit as brilliant as his father and would one day eclipse the fame of Brunel *père*.

Based on the borings, Brunel had estimated that the tunneling work would proceed at three feet each day, but with the ground proving at times to be more liquid than solid, he was lucky to move just one. He asked that a

FIGURE 21. The remarkable French-born British engineer Marc Isambard Brunel. Brunel used hydraulic concrete mortar to build the Thames Tunnel.

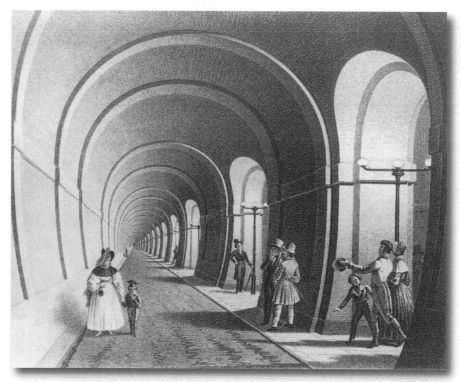

FIGURE 22. The Thames Tunnel, the first tunnel constructed under a navigable river. The eighteen-year-long effort claimed a number of lives and seriously injured Brunel's health.

spillway be constructed to evacuate the water, but Smith and the directors said that it would be too expensive to build, so men were forced to work hand pumps and carry water buckets from the flooded gap between the brickwork and the shield (steam-powered pumping was now impractical with the tunnel so deep). Although it was a decision the directors would soon regret, Brunel had grown tired of fighting Smith, and so the spillway was not built. Nevertheless, the prospect of catastrophic flooding, though often unspoken, remained lodged in everyone's mind. On January 4, 1827, Brunel wrote in his diary, "Every morning I say, 'Another day of danger is over!'"[27]

By March, the amount of water coming in between the shield and the brickwork steadily increased, and so did the appearance of sundry items in the mud that were clearly of more recent origin, such as broken pieces of contemporary porcelain, barely corroded nails, and shipping tackle. These bits of newer rubbish gave Brunel some concern, but all the 1825 borings had shown that a thick level of gravel overlaid the mud and clay of the river bottom, and earlier measurements had indicated that the tunnel was at least twenty feet below the latter. If they were close to the Thames's bottom, gravel would have washed in with the water. All the engineers agreed that the appearance of gravel was a dangerous sign, but so far it had not been in evidence.

The volume of water seepage continued to grow. Now forty men were working the pumps full-time and carrying buckets back to the vertical shaft. The additional manpower expense easily offset the cost of Brunel's proposed spillway. The water also put everyone's nerves on edge. The brick foremen, who took naps near the shield, began shouting in his sleep that the tunnel was flooding. A brief panic followed until one of his coworkers realized what was happening and woke him up.

Still, both the men and the engineers continued to have forebodings. One day, while Isambard Brunel was having breakfast, a workman ran from the shaft, shouting, "It's all over, it's all over, the river's in and they're all drowned except one." Isambard and his assistant, William Gravatt, dashed down the shaft and ran to the drift-work. They found only a wet lump of clay that had fallen from the ceiling of the shield. The high-strung tunneler was fired. Nevertheless, the threat of flooding, however real, was always at the drift and not from the brick walls, which, thanks to Francis & White's cement, remained watertight.

By April 1827, the amount of water coming into the works increased, and though the leakage was still manageable, it troubled Brunel. After much pestering, Brunel was able to convince the directors to hire a diving bell to explore the river bottom. Isambard and Gravatt volunteered for this hazardous task, but both men had to steel themselves before venturing into the bell, which was then hoisted upward by a heavy crane bolted to a ship especially designed for managing the device. The bell rocked back and forth, jostling the men as the crane positioned it over the water before lowering it into the murky depths of the Thames. It was probably the earliest instance of a diving bell—a recently invented device previously restricted to salvage operations—being used in a major civil engineering project. Isambard and Gravatt sat on a small seat positioned between two thick glass windows on each side of the bell, which was more tubular than spherical. The bell's bottom was open, and the air within was held in place by its own pressure. Of course, if the bell hit an obstruction that tipped it over, its occupants would be in trouble. As the bell descended and approached the river bottom, the men could see why so much water was now entering the tunnel. Trevithik's steam-powered dredgers, which had been employed earlier to clear the Thames's waterway for shipping, were now being used to harvest gravel. A deep depression had been scooped out of the river bottom directly above the tunnel. Probing the base of the depression with an iron rod, Isambard soon struck the top of the shield. Isambard realized that something needed to be done quickly to protect the tunnel.

A hard lining called a "steening" was prepared, by which the exposed earthen walls between the brickwork and the shield were reinforced with a layer of Roman cement concrete. The seeping water activated the fast-setting properties of the cement, and the seal seemed to hold. The leakage around the gap dropped somewhat, but the water, following the path of least resistance, began permeating through the small gaps of the shield itself, especially where the poling boards penetrated the drift. The men in the lower chambers of the frames often found themselves working knee-deep in water or fetid muck.

By early 1827, the directors—and Brunel—were getting nervous, as the funds allocated for the project were running low. Since there was much curiosity about the tunnel, the directors decided to raise money by charging people one shilling apiece for a tour of the works. Commoners and aristocrats rubbed shoulders to watch the progress, and all agreed that it was a remarkable endeavor that, once completed, would be an engineering triumph for Britain. Another possible motive for the tour was that the directors wanted to cultivate potential investors to refill the kitty. The shield was now so far from the Rotherhithe end as to look like but a dot in the distance. Evidently, the directors hoped that visitors would instead take notice of the tunnel's magnificent archways and not the mess the men were working through at its terminus. Higher-class visitors, who were viewed as potential investors, were given a complete tour of the works, including Brunel's tunnel shield. Unfortunately, the visitors proved to be more trouble than they were worth. The laborers had to work around them, being especially careful not to splash mud or water on the guests' clothes as they carried buckets back to the vertical shaft. No new investors were recruited. As for the tour fees, less than a hundred pounds were realized from them.[28]

By mid-May 1827, 540 feet of the tunnel had been completed, almost half its projected length of 1,296 feet. Brunel's shield, aided by the concrete lining, was now steadily advancing a foot or more each day. While this was far less than his original estimates, it was better than it had been for many past months. On May 13, Brunel wrote in his diary, "So far the shield has triumphed over immense obstacles, *and it will carry the tunnel through.*" In an entry Brunel scribbled later the same day—perhaps after remembering his checkered fortunes—he expressed worries: "[A] disaster may still occur. *May it not be when the arch is full of visitors! It is too awful to think of it!*"[29]

A few days later, as Brunel was giving a tour to Lady Sophia Raffles—wife of Sir Thomas Stamford Raffles, founder of Singapore—and a group of her friends, water began leaking from frame No. 11, not an unusual occurrence, but still troubling. Brunel would write later that he was "most uneasy all the while, as if I had a presentiment." After Brunel escorted Lady Raffles and her entourage out of the tunnel, one of the assistant engineers, Richard Beamish, noticed the leak and tried to staunch the flow. The water quickly began pouring out at an alarming rate and then become a torrent. A tunneler grabbed Beamish's arm and shouted, "Come away, sir, come away; 'tis no use, the water's rising fast." The other workmen, who had already begun running toward the staircase at the end of the tunnel, now found themselves propelled as much by the roaring water behind them as by their feet. A wooden office that was positioned one hundred feet from the shield was picked up bodily by a wave and, accompanied by hundreds of empty cement casks, formed a treacherous flotsam. The flood doused all the gas lamps in the tunnel, and men now struggled in the dark to reach the exit before it was too late.[30]

Aboveground, near the shaft, Isambard was going over some paperwork with Gravatt when they heard noise coming from the tunnel and saw men pouring out. Gravatt wrote that young Brunel immediately ran toward the works, and he quickly followed, but they could not get down the stairs because of the press of retreating workers. Gravatt encountered one tunneler who said that it was "all over." Gravatt, remembering the earlier false panic, called the man a coward but then recognized the true extent of the calamity when he saw water begin rising up the vertical shaft. The laborers climbing the staircase inexplicably stopped to look at the rising water, as if mesmerized by it; one later described it as "splendid beyond description."[31] Isambard shouted orders at the men to keep moving with all speed. This seemed to snap them out of their spell, and they began scrambling upward again. Gravatt recounted that they saw a man "in the water like a rat" and "quite spent" clutching the stairway's handrail, unable to go further. "I was looking how to get down, when I saw Brunel [Isambard] descending by rope to his assistance. I got hold of one of the iron ties, and slid into the water hand over hand with a small rope, and tried to make it fast round his middle, whilst Brunel was doing the same. Having done it he called out, 'Haul up.' The man was hauled up. I swam about to see where to land. The shaft was full of casks. Brunel was swimming too."[32] Isambard and Gravatt were finally able to make it to the top of the shaft and were the last ones out of the flooded tunnel works.

By some miracle, no one was killed.

Two days later, Isambard hired the diving bell again to inspect the river bottom. He soon saw that the concave depression in the river bottom had deepened. This time it was not due to the dredgers; the intense tidal flows in the Thames were now scouring out a depression. Brunel did not need to probe with a rod to find the shield, for it was now partially sticking out. At one point, he took off his shoe, pulled up this pants sleeve, immersed one leg into the water, and could actually feel the cold, hard iron of the shield with his foot. He and his father had already discussed what course to take, and the previous twenty-four hours had seen the uninjured members of the tunneling crew filling thousands of cloth bags with clay, which were then packed on a flat barge that was now anchored above the shield. Back aboard the diving bell's ship, Isambard ordered that all the bags be dumped in the depression. The dumping continued for weeks, during which time Isambard made repeated trips with the diving bell to supervise the effort. It took approximately twenty thousand cubic feet of clay to fill in the depression and stop the leak. Now the pumping began. By this time, Brunel, who, like his son, had been working throughout the crisis with little sleep, turned over his duties to Isambard. The details are sketchy, but it seems that the elder Brunel may have had a mild nervous breakdown compounded by exhaustion or vice versa.[33]

After the pumping had lowered the water level in the tunnel by several feet, Isambard and his assisting engineers boarded a small dingy and rowed down the flooded corridor to inspect the damage. Because of the huge volume of mud that had been deposited by the flooding, the water became too shallow for rowing after several hundred feet. The men had to abandon their oars and began pressing their hands against the tunnel's ceiling to propel themselves forward. When the boat could no longer be moved, they used their lamps to look around. Just ahead they saw that the mud had risen above the waterline but seemed to level off a couple of feet beneath the tunnel ceiling. Gravatt took a lamp and left the boat to test the consistency of the mud. It seemed just solid enough to support the weight of one man. Gravatt crawled one hundred and twenty feet across the mud before reaching the shield. Here he could see the tops of several frames, pushed back from the steening and, just above, the bags of clay that had been used to stop the breach. After the men had returned to the vertical shaft and Brunel made his report, two of the company's directors expressed their wish to also survey the damage. Gravatt and several men escorted them down the tunnel in the dingy. At some point, the boat was capsized when one of the directors stood up to change positions with another. One of the directors, who could not swim, drowned.

Once the water had been pumped out, the mud carried away, and the shield put back into place, the tunneling work resumed. A celebratory dinner was held in the tunnel on November 10, 1827. In one passageway, the directors and tunnel investors dined to music played by the military band of the Royal Coldstream Guards, while in the adjoining passageway, the common laborers feasted on less exalted fare and drank grog. Spirits were high, and most of the directors were optimistic about their chances of obtaining the additional funding to complete the project. Most London newspapers had favorably covered the pumping and excavation operations, pronouncing it a triumph of British ingenuity. Even the sniping reporters had mostly stopped calling Marc Brunel "Monsieur Brunel" and begun referring to him instead as "Mister Brunel." This was a personal and psychological victory for the engineer, whose loyalty to Britain—and competency in English—surpassed most of his compatriots. As he was still recovering from exhaustion and spent nerves, the elder Brunel did not attend the dinner. Nevertheless, he was buoyed by the resumption of tunneling operations and the positive press coverage it was given. His spirits began to revive, and soon he was back at the tunnel overseeing its operations at his son's side. The future looked rosy.

Progress slowed again, for the frames of the shield kept going out of alignment or sinking into the mud. The frames had to be jacked up, while men rammed in gravel and oakum (loosely twisted fiber) underneath them. The "bad air" also returned. When the ventilator failed, men collapsed and once again had to be carried out. Isambard and the engineers also suffered from the "vapors," since they spent as much time at the tunnel works as any of the workers. Still, there was room for optimism: improvements were made to the ventilator, and, as the men grew used to making adjustments to the frames, the work took less time, and it appeared as if the excavation might resume its old pace.

Then Nature intervened with a vengeance. Unusually high rainfall that winter caused the Thames to rise, and by January 1828, the river surpassed its previous high-water mark by three feet and flooded portions of London. The high water levels also increased the weight of the river by millions of tons, exerting tremendous pressure on its bed below and on the tunnel works beneath it. On January 12, 1828, the water of the Thames broke through to the construction site. The water smashed its way in between the brickwork and No. 1 frame, ejecting a massive high-speed torrent like a colossal fire hose. The gaslights were instantly doused, and in the darkness the men could feel the water rising around them at a faster pace than had the previous irruption. The panicked men made another mad rush for the vertical shaft, and this time, Isambard Brunel was among them. The water rose so fast that the men at the back of the crowded mass making for the exits were forced to swim. Isambard was trying to help two of his assistants, Ball and Collins, when a huge wave knocked him down and carried him forward. When Isambard came up for air, he found that the swell had brought him down to the vertical shaft, next to the stairway. He grabbed onto it and shouted down the flooded tunnel, "Ball! Ball! Collins! Collins!" The water was rising fast. Tunnelers pulled young Brunel up and prodded him up the stairs, but he could hardly walk, having thrown the ball-joint of his knee, and so, hopping on one foot, he was half-carried to the top of the shaft. Aboveground, a blanket was thrown over the shivering Isambard, who sat down and, in a daze, continually repeated the men's names. Besides Ball and Collins, four other men drowned in the disaster, and many were injured. When a doctor examined Isambard, internal injuries were discovered as well. The young man would be confined to bed for the next several months. Marc Brunel was once again in charge.[34]

Once more, thousands of bags of clay were dumped into the Thames to cover the breach. The tunnel was pumped out a second time, and another mass of muck was again shoveled into buckets that were carried out by hand. This time, however, there was no resumption of work. Funds for the project had been exhausted, and so, in August 1828, the far end of the tunnel was bricked up. Sealed behind the masonry was the enormous iron shield. The pioneering device, once justifiably touted as an engineering marvel, was left to corrode.

To more than a few people, the project seemed over. The enormous construction effort that had been said to be impossible apparently proved to be just that. Yet, it would not go away. There was a collective sense of injured pride. It seemed out of character for the British to simply surrender to the elements and say, "Well, we gave it a good shot, but it was just too much for us." In short, a large number of people, including many prominent lords and MPs, felt that it was shameful to give up on the Thames Tunnel project, and they began appealing for the nation to step in and finance its completion. Resolutions were passed, and it was not long before negotiations began between the project directors and the government for a loan. Of course, it's best not to hold one's breath while negotiating with a government. Talks dragged on, broke off, and resumed again in the same sluggish manner. William Smith was finally removed as president, a development that must have given Brunel some joy. Years passed. Finally, in

December 1834, a deal was struck. The government agreed to a loan of £247,000. This, combined with money raised from other sources, allowed the construction work to begin anew.

Work began in August 1835 with the dismantling of the old shield, no easy task, since some of its constituent parts had now fused together from rust. Once the shield was removed, Brunel's new and improved tunnel shield was installed, and excavation work resumed in March 1836. Five months later, on August 23, the tunnel flooded again. The breach was stopped, the tunnel was pumped out and cleared, and work continued. Brunel's poor health impeded his supervision of the tunnel project's work, and after the August flooding, his nerves were shot as well. He resigned and turned the work over to his assistant engineer, Mr. Gordon. (Isambard was now engaged in other engineering projects.) The air in the tunnel affected Gordon's health as well, and he soon turned the supervisory work to a Mr. Page. Shortly after the tunneling work had recommenced, another flood stopped operations on November 3, 1837. After this disaster had been cleaned up, work continued before another subfluvian deluge on March 20, 1838, stopped operations once more.

The tunnel project was becoming a joke. Numerous doggerels were repeated, including this one:

Good Monsieur Brunel
Let misanthropy tell
That your work, half complete, is begun ill;
Heed them not, bore away
Through gravel and clay,
Nor doubt the success of your Tunnel.

That very mishap,
When the Thames forced a gap,
And made it fit haunt for an otter,
Has proved that your scheme
Is no catchpenny dream;—
They can't say "twill never hold water."[35]

The great engineer usually addressed as "Mister Brunel," was now being referred to as "Monsieur Brunel" once again, as if to emphasize his foreign origins, the implication being that a British-born engineer would have performed better—though three British-born engineers had already tried and failed miserably to construct a tunnel under the Thames.

After the damage of the March flood was cleaned up, work progressed for two years without another flooding incident, but only very slowly. Apparently, there was more refuse dumping on the Wapping side of the Thames, for the methane leakages increased. Particularly dreaded was the "black mud," an organic waste in its latter stages that always brought with it more gas than usual. The ventilator helped, but when it failed one day, men dropped like flies and had to be carried out of the tunnel. The directors finally agreed to restrict the men's hours in the tunnel to allow their lungs to recover.

On April 3, 1840, when the tunnel had almost reached the point that marked the far bank of the Thames, the water rushed in once more. Thanks to improved safety measures and the new shield—which slowed the flooding enough to give the men a chance to escape—only one laborer was killed in the flooding. The flood of April 1840 was the last, for the tunnelers were soon beyond the far bank of the river, and the mercurial Thames was no longer above them.

While the work continued at a slow pace—pockets of methane still caused fainting among the men or erupted in flashes—everyone was now confident that the worst of it was over. The sniping satires had stopped, and people began once more calling the tunnel's architect and engineer *Mister* Brunel. After March 24, 1841, even this form of address would not suffice, for on that day a young Queen Victoria knighted Brunel at the Rotherhithe construction site next to the tunnel's entrance. He was now Sir Marc. It was the crowning moment of a long, difficult, star-crossed career.

A few months later, Brunel suffered a stroke that paralyzed the right side of his body. He had to spend the following year convalescing and attempting to exercise his stiff limbs, but he continued to receive reports on the tunnel's progress and offer his advice. The tunnel reached the vertical shaft at Wapping on August 1, 1842, but another eighteen months were spent outfitting it with gas lines, lamps, and building the elaborate public staircases in the shafts at each end. The Thames Tunnel was finally opened to the public on March 25, 1843, to wide acclaim and celebration. Despite his ill health—the right side of his body was still partially paralyzed—Brunel attended the opening ceremonies. In the next fifteen months, over one million people would visit the first tunnel constructed under a navigable river.

Because of Brunel's age and the effects of the stroke, the Thames Tunnel would be his last project. He still consulted on engineering matters and especially enjoyed giving advice to Isambard, who was now one of the most noted engineers in Britain. Brunel's "interesting times" were over at last. He spent the last few years of his life with Sophia, content with his past achievements and enjoying the autumnal glow of a glory that took several difficult decades to achieve. On December 12, 1849, Sir Marc died at his home in London. He was eighty years old. He was buried at Kensal Green Cemetery, where Sophia would later join him, and, later still, his son, the now-famous engineer Isambard Kingdom Brunel. It is supremely satisfying to recount the life and career of a remarkable man and to see it conclude with a "happy ending."

The original plans for the Thames Tunnel called for a gentle slope at each end to allow carriages to pass through, but there was not enough funding left to purchase the additional real estate and build the roadway. It remained an overlarge pedestrian corridor, where people bought trifling souvenirs and were entertained by organ grinders, fortune-tellers, and—in the evening—the occasional prostitute. In 1865, the East London Railway Company purchased the tunnel to serve as an underground conduit linking rail service between Rotherhithe and Wapping. It now forms a small part of the London Underground network.

The Thames Tunnel would have been inconceivable without the use of concrete cement, and during the eighteen years of construction, the material had undergone some changes and refinements. The quality of most Roman cements was variable—one reason why Brunel tested every batch. Even if one commercial source was more trusted than another, slight differences in the setting periods or hydraulic properties could cause problems on the worksite. The search for a better and more stable material would eventually bring an end to the use of Roman cement and see the rise of its replacement: Portland cement.

JOSEPH ASPDIN

On October 21, 1824, a struggling forty-four-year-old bricklayer in Leeds named Joseph Aspdin was granted a patent, BP 5022, for a hydraulic mortar/stucco he called "Portland Cement." Exactly one hundred years later, representatives of the British Cement Makers Federation and the American Portland Cement Association unveiled a plaque in Leeds commemorating the event. Speeches were given that recounted the enormous contribution this "humble bricklayer" had made to human progress and how he made possible the benefits we enjoy today. In virtually every chronicle of concrete's development, Joseph Aspdin is portrayed as having played a leading role in the material's advance. In Kendall F. Haven's book *100 Greatest Science Inventions*,[36] Joseph Aspdin's invention of Portland cement is ranked among those revered one hundred (his contribution is listed between Archimedes's compound pulley and Charles Babbage's analog computer). In truth, it is highly unlikely that Aspdin personally contributed much of anything to the development of modern concrete. It is probable that those august officials from the British and American cement industries had dedicated the plaque to the wrong person for the wrong reasons. Few discoveries in the Industrial Age are shrouded in so much mystery—or obscured by so many deliberate fabrications—as Joseph Aspdin's "invention" of Portland cement. However, enough information has been uncovered to tease out some details. And the leading candidate for the invention of Portland cement is not Joseph Aspdin.

The details of Joseph Aspdin's life are very spotty. He was born in Leeds sometime in late 1788 (the exact date has been lost) to bricklayer Thomas Aspdin and his wife (whose name has also been lost to us). Since he was baptized on Christmas day of that year, it is assumed that he was born earlier that month. Joseph was the firstborn of the Aspdins' six children. He grew up in a crowded and struggling household, but his father was kept busy at his craft: Leeds was undergoing significant growth in the late eighteenth century. It had long since become the nexus of Britain's wool industry, and the recent introduction of the mechanized spinning and looming devices made Leeds the textile capital of Britain. Since virtually all the massive factories that were then popping up in Leeds were built of brick, a bricklayer could expect steady, if meagerly paid, employment. As was common in those days, the sons took up the father's profession, and Joseph became a bricklayer.

On May 21, 1811, at the relatively ripe age of thirty-one, Joseph married Mary Fotherby. The marriage certificate states his occupation as "bricklayer," a profession that he had probably been practicing since at least his fourteenth birthday. By late 1816, he had his own business and address, for the 1817 Leeds directory lists him as "Joseph Aspden. Bricklayer. Ship-In Yard, Back of Shambles."

The misspelling of his name was probably a typo, and not due to Aspdin's ignorance: many bricklayers were either illiterate or barely literate, but Aspdin seems to have had enough schooling to read and write decently. By this time, Aspdin was the father of several children, one of whom, William, would play an important role in concrete's story. There is no question that Joseph Aspdin possessed some ambition and curiosity, for he was obviously conducting experiments with different cement formulas for several years prior to his patent application. These experiments were probably hard for Aspdin to conduct, since there was no local source of limestone, and rail transport had yet to be introduced. Adding to this difficulty was another problem: bricklayers employed in the construction of a residence or factory were provided with a fixed amount of mortar, the use of which was overseen by a sharp-eyed foreman. It was not uncommon for bricklayers to mix too much sand in the mortar so they could squirrel away a portion of the hydrated lime for their own use. Thus, there was only one place where a poor—or tightfisted—person in Leeds could obtain limestone: paved roads. Aspdin was twice fined for pilfering limestone from the highways of West Yorkshire. For every time he was caught, Aspdin had no doubt made a dozen successful plunderings, so he could probably afford the penalties.

Aspdin's experiments did produce cement that he felt was worth patenting. The relevant portions of the patent are provided below.

My method of making a cement or artificial stone for stuccoing buildings, waterworks, cisterns, or any other purpose to which it may be
applicable (and which I call Portland cement) is as follows:—I take a
specific quantity of limestone, such as that generally used for making or
repairing roads, and I take it from the roads after it is reduced to a
puddle or powder; but if I cannot procure a sufficient quantity of the
above from the roads, I obtain the limestone itself, and I cause the
puddle or powder, or the limestone, as the case may be, to be calcined.
I then take a specific quantity of argillaceous earth or clay, and mix
them with water to a state approaching impalpability, either by manual
labour or machinery. After this proceeding I put the above mixture
into a slip pan for evaporation, either by heat of the sun or by submitting it to the action of fire or steam conveyed in flues or pipe under or
near the pan till the water is entirely evaporated. Then I brake the said
mixture into suitable lumps and calcine them in a furnace similar to a
lime kiln till the carbonic acid is entirely expelled. The mixture so calcined is to be ground, beat, or rolled to a fine powder, and is then in a
fit state for making cement or artificial stone. This powder is to be
mixed with a sufficient quantity of water to bring it into the consistency of mortar, and thus applied to the purposes wanted.[37]

The guileless, semiconfessional part about obtaining the limestone is a hoot,
but let us move on to examine exactly what Aspdin was describing. He was
taking good, pure limestone and grinding it to a powder form that he then
kilned. On the other hand, he may have lost the explicatory thread of his tor-
tuous description and mentioned the powder part prematurely. Perhaps he
meant to say that he broke the limestone into pieces small enough to kiln
easily, after which he reduced them to a powder. (The other way would have
been far more difficult and might have fused the fine particles during the
firing.) There may have been another reason for this strange description. It
was not uncommon for inventors back then to deliberately falsify an ingredient or aspect of the manufacturing process to protect themselves from imi-
tators. Aspdin goes on to relate how he takes clay and mixes it with water and
then dries it until the "water is entirely evaporated." Further obscuring both

his intentions—and chemical logic—Aspdin goes on to kiln the "said mixture" until it reaches a state in which it can be "ground, beat, or rolled" into a powder and used for cement. Again, Aspdin has either lost the thread of his description or is being deliberately vague, for *nowhere does he mention mixing the lime with "dried" clay*. One assumes he has, since he kilns the mixture until the "carbonic acid is entirely expelled," which points to limestone being involved. (References to carbonic acid are common in nineteenth-century mortar patents, and simply point to the then-imperfect understanding of how calcium carbonate was transformed to calcium oxide.) But this also makes no sense, since he has already calcined the powdered (or chunks of) limestone, and so completed the removal of the "carbonic acid." Most likely he thought that the second kilning removed even more of the "acid," but, of course, there was none. As we have already seen, the patent clerks of the early nineteenth century were either overworked or slow on the uptake when it came to basic chemical processes, or perhaps both. Lime-based cement patents were being filed left and right during this period. The clerk who approved Aspdin's application probably thought, "Oh, no! Not another patent for a cement mortar and stucco!" and probably did not read the document carefully before registering it, which was common at the time.

The importance given to Aspdin's patent is principally due to several reasons: reading more into the text than what was there, ignorance of the work being done by others at the same time, false information promulgated later about Aspdin's product by another party, and, of course, the designation "Portland cement," which later became the name for standard modern concrete cement.

Later authorities, looking at the patent after many years of technological progress in cement manufacturing—and flawed accounts of Aspdin's methods—interpreted it in such a way as to construe a step in the production process that was not really in the text. This requires a brief explanation. After Roman cement became popular as mortar and stucco, many people tried to make it better in some way, or they pretended that *their* version was better—oil stucco being one example of many failed attempts. However, two discoveries in the first half of the nineteenth century did improve the quality of cement. One enhanced the material marginally, while the other represents a significant advancement in cement technology. The first was a process we will call "slurry mixing," the ancestor of today's "wet process" of cement mixing. Slurry mixing was *probably* developed previously—much is murky during this period—by two other Englishmen, James Frost[38] and Edgar Dobbs.[39] To control the right proportions of clay and limestone, and at the same time strengthen the bond between them, the powdered lime-stone—still unkilned—was thoroughly mixed with clay and water, and the whole was allowed to dry until it assumed a paste-like form. This paste was cut into portions small enough to kiln. After kilning, the cooked pieces were then pulverized to make cement. The second critical discovery—dependent on the first—was something called "clinkering." Cement makers using the slurry-mixing process were careful not to "overcook" the mixture in the kiln. If the mix was kilned too long, the resulting material was completely vitrified. In other words, it would become a hard, rock-like substance, called "clinker," which was very difficult to pulverize. Over-burned and blackened bricks that could not be sold were also called clinkers, and this is probably where it got its name. Cement clinkers were also deemed useless and tossed away, since the cement manufacturers wanted an easily pulverized product. If they had taken the trouble to grind up the clinkers—and admittedly this would have cut down the life span of millstones significantly—they would have discovered the finest cement the world had yet seen. While slurry mixing might be read into Aspdin's patent, nothing suggests that he discovered the wonders of clinkering. If he had, it definitely would have been noticed by his competitors and others, for it would have stood head-and-shoulders above anyone else's cement.

As for the name Portland cement, it was hardly novel. Years earlier, John Smeaton remarked that his hydraulic mortar set as hard as "Portland stone,"[40] the famed limestone used in the construction of many prominent buildings. And at least one other cement manufacturer, William Lock-wood, was manufacturing a fine product he called "Portland cement"[41] several years prior to Joseph Aspdin's patent being granted.

So why has so much attention been paid to this particular patent filed by an obscure bricklayer and cement maker? For that we have to thank Joseph Aspdin's son, William. It can be argued that no single individual has contributed so much to the history of concrete—or so corrupted that history—as William Aspdin. Sadly, corruption was an inextricable component of William Aspdin's character.

WILLIAM ASPDIN

William Aspdin was born on September 23, 1815, in Leeds. In 1825, he and his family moved to Wakefield, just south of Leeds. His father had formed a partnership with William Beverley, and the two had built a cement works there. Beverley, who owned a successful brass foundry in Leeds, capitalized the venture. Joseph Aspdin was allowed to manage the cement works in Wakefield, while Beverley remained in Leeds to serve as their commercial agent. Wakefield was probably chosen because land and labor were cheaper there than in burgeoning Leeds. Sometime around late 1829 or early 1830, William began work at the cement facility as an apprentice. Things did not go well at Wakefield. Assuming that Joseph Aspdin was slurry-mixing the cement—a more expensive procedure than simply burning the clay-adulterated limestone to produce Roman cement—he would have found it difficult to compete against rivals who offered a cheaper product. Slurry mixing might have made Aspdin's cement slightly better, but consumers probably balked at having to pay for this marginal improvement. Besides, Roman cement had proved itself, while the claims for "Portland cement" were based on the attestation of its maker and not on any independent tests. To remain competitive, Joseph Aspdin almost certainly provided both Roman cement and his Portland cement.

In any event, Beverley was probably anxious about the venture by the mid-1830s. His foundry had prospered—he now had an iron works as well—but the same could not be said of his cement division. When the newly formed Manchester & Leeds Railway presented a plan in 1837 that showed a route running through the Wakefield cement factory, he decided to dissolve his partnership with Joseph Aspdin. Aspdin was forced to disassemble his kiln and relocate it to a nearby patch of land then being used as a market garden. Cement production probably did not begin again until

1840, for there is no published evidence of his firm until the 1841 Wakefield directory. The entry reads: "Joseph Aspdin. Ornamental Chimney Pipe & Roman cement Manufacturer, Kirkgate, Wakefield." There is no longer any mention of Portland cement.

In August of 1841, something strange transpired. On August 3, Aspdin drew up a deed transferring a 50 percent share of the business to his oldest son, James, and not William, who had been working at his side for some fifteen years or more. What makes this strange is that James was not in the construction trades. He had decided not to follow in his father's footsteps and had instead studied accounting and become a bookkeeper. Several days later, on August 6, Joseph Aspdin published the following notice, dated a few days earlier, in the *Wakefield Journal & West Riding Herald*:

> TO BUILDERS AND OTHERS
>
> I, Joseph Aspdin of Wakefield, cement maker, take this opportunity of returning my best thanks to my friends and the public, for the numerous favors I have received at their hands for many years past; and beg to inform them that I have just taken my son, James Aspdin into Partnership with me, and that we shall hereafter carry on business under the firm of "JOSEPH ASPDIN & SON." I think it right at the same time to give notice that my late agent, William Aspdin is not now in my employment, and that he is not authorized to receive any money, nor contract any debts on my behalf or on behalf of the new firm.
>
> Cement Works, Wakefield Joseph Aspdin
> 2nd August, 1841[42]

We can only speculate why this rupture occurred, though William's later life seemed to suggest a host of reasons, as we shall see. All we know for sure is that William Aspdin had already left his father's firm the previous month (July) and moved to London. He briefly returned the following year to marry Jane Leadman, the daughter of a butcher in Barnsley, a village several miles south of Wakefield. (The marriage certificate shows that no member of the Aspdin family witnessed the nuptials.) William Aspdin then returned with his wife to London, where he would seek his fortune.

PORTLAND CEMENT

William Aspdin's schooling prior to his apprenticeship seemed to have been a bit better than his father's, since he was loquacious and enjoyed writing, and he employed both skills for sales bombast or for relating quite colorful, and thoroughly fictitious, tales. Like all good con men, William could be rather convincing, for he always had the ability to attract investors. He might have gone far in the cement industry and made a large fortune; that he did not do so can be attributed to a character defect: William was an incorrigible liar and swindler.

William Aspdin's departure to the "Big City" was hardly an unusual move for an ambitious person in Britain. London was, and remains, the cultural, political, trade, and communications hub of the United Kingdom. Still, setting off for London and taking up residence there was not something most people could afford to do. William must have saved his money or, more likely, stealthily embezzled funds from his father's firm to finance the move. This would explain the sudden falling out with his family and the disinheritance. Nevertheless, something must have given William the confidence to cut all his familial and material ties to Wakefield and strike out on his own. He had apparently discovered a process that radically improved cement, no doubt stumbling across it after a batch of slurry-mixed cement was overcooked. He then decided to experiment with the vitrified stone by pulverizing it. William evidently kept the secret of the clinkering process to himself, for there is no evidence that his father made clinkered cement after his son's departure to London, let alone anytime before the family fissured.

His first documented appearance in London was recorded in the 1842 directory "William Aspdin. Cement Manufactory. Church Passage, Rotherhithe." William was not there long, for the 1843 directory shows that he had moved to "Upper Ordnance Wharf & 342 Upper Rotherhithe St." He was not far from the Thames Tunnel and might have attended the opening ceremonies that year with his wife. Certainly, the tunnel captured his imagination, for he would later use it in one of his most outlandish fabrications. For the present, he was making cement and seeking backers. Although Roman cement overwhelmingly dominated the market, the slurry-mixing method had become more common. William was probably on the lookout for the clinker—overcooked rejects to everyone else—that were so valuable to his process. Who knows what excuse he used for taking these remnants off the hands of the other cement manufacturers, but they were probably happy to get rid of them. It was really ingenious: William did not even have to build a kiln. He only had to hammer the clinker into a powder; certainly not an easy task, but it was the only work he had to perform aside from packaging it. No limestone to purchase, no kilning, no employees. His only overhead were the casks and, of course, his residence/office at the Ordnance Wharf.

It was not long before William Aspdin found partners: John Milthorpe Maude and his son Edmund. John Maude had also come from Leeds as a young man to make his fortune in London. He was a shipping broker who had enjoyed considerable success and was now looking to invest in something promising that he could eventually turn over to his son to manage (he was then sixty-five). Neither he nor his son knew very much about cement. Still, he probably had William's cement tested to confirm the latter's claims for it. The company was formed in the late summer or early autumn of 1843 under the name J. M. Maude, Son & Co. The sole purpose of the firm was to manufacture and sell Portland cement. William ran much of the operation and kept aspects of his manufacturing process cloaked in secrecy.

It is clear that William Aspdin had learned from his father's experience that introducing a new product from a new company was a difficult undertaking. Instead, he presented the cement to his partners as a product long established in the Leeds area and manufactured according to a secret process known only to him and his father. (In truth, he was the only Aspdin with a trade secret.) He probably acknowledged the falling-out with his father, but he no doubt gave a very different and self-serving reason for it. Shortly after the company was formed, a circular was sent out. It is an excellent example of William's chicanery in action. It reads:

PATENT PORTLAND CEMENT

The manufacturer of this cement has for many years been carried on by Mr. Aspdin at Wakefield in which neighbourhood and throughout the northern counties of England it has been successfully and extensively used; owing to the heavy charges attending its conveyance to the London market its consumption there has necessarily been limited and although its superiority over other cements has never been contested by those who have been induced to give it a trial, the high price at which alone it could be supplied has hitherto proved a serious impediment to its more general introduction into the metropolis. Messrs J. M. Maude, Son & Co. have now the satisfaction of announcing to the public that they have made arrangements with the son of the patentee for the purpose of carrying on the manufacture of this valuable cement at their extensive premises at Rotherhithe, and whilst they will be enable to supply it at a considerably reduced price, they have also the satisfaction of stating that in consequence of improvements introduced in the manufacture, it will be found for the following reasons infinitely superior to any cement that has hitherto been offered to the public:—

(1) Its colour so closely resembles that of the stone from which it derived its name as scarcely to be distinguishable from it.

(2) It requires neither painting nor colouring, is not subject to atmospheric influences, and will not like other cements, vegetate, oxydate, or turn green but will retain its original colour of Portland stone in all seasons and climates.

(3) It is stronger in its cementative qualities, harder, more desirable, and will take more sand than any other cement now used.

It was very clever to present the cement as long established in Yorkshire and "throughout the northern counties," and to say that it was only the high cost of transport that prevented it from gaining a foothold in London. In those days, one could not simply pick up a phone and make a call to ask, "Is this stuff really legit?" Leeds was far, far away, and only with the introduction of the locomotive and the interconnection of long rail lines would it become less than a four-day journey from London, the approximate time it now takes a car traveling from New York to San Francisco.

Skipping over the circular's colorful—if misleading—introduction, one cannot find fault with any of Aspdin's claims. All the qualities ascribed to Portland cement were true in every respect. It was by a wide margin the finest concrete cement in the world. Still, it would hardly be the first or last time that a major discovery was made by a duplicitous scoundrel.

At roughly the same time they released this circular, Maude, Son & Co. engaged the highly respected London building firm Grissell & Peto to perform independent tests on Portland cement and several of its prominent rivals. The tests showed that Portland cement was almost twice as strong as the best Roman cements. Records for the firm are scanty for the next couple of years, but we do know that Maude, Son & Co. purchased Parker & Wyatt's old cement works in Northfleet in 1846, so sales must have been good. That same year, John Maude retired and a new partnership formed that included Edmund, Maude's other son, George, a certain William Henry Jones, and, of course, William Aspdin. Oddly, less than a year later, the firm went into bankruptcy.[43] The reasons are not clear, but later events might provide a clue.

William quickly found new partners: William Robins and his son-in-law, George Goodwin. Aspdin was able to retain—or he and his new partners were able to *obtain*—the cement plant at Northfleet. As before, the new partners had no experience in the cement industry, and they left management of the company in Aspdin's hands. It was at this time that William published an advertisement in several issues of the trade periodical the *Builder* that was as widely believed as it was unquestionably false. In it, he claims that his father's patented cement was used in 1828 to stop the breach that caused the first flooding of the Thames Tunnel while it was under construction:

> It was not until Portland [cement] had been manufactured seven or eight years that its value became apparent and its superiority over all other cements manifest. Then it particularly arrested the notice of Sir

[Marc] Isambard Brunel, the eminent engineer and constructor of the Thames Tunnel who tested it with "Roman cement" until he was thoroughly convinced of the great superiority of the Portland, by finding it three times stronger than any other cement then known to the public. Although at that time it cost 20s to 22s per cask, besides the carriage to London, yet Sir [Marc] Isambard Brunel determined (notwithstanding his ability to procure "Roman" at 12/- per cask, delivered on the spot) to adopt it chiefly for his purposes, as its merits required no other recommendation than an impartial trial. When the Thames broke through the Tunnel in 1828, and filled it with water, a large quantity of this Cement was thrown into the river which effectively stopped up the cavity and enabled the contractors to pump out the water; and soon afterwards the work resumed its wonted appearance subsequently obstructed for want of funds.[44]

This fabrication is so breathtaking in its scope that it would have embarrassed that master of prevaricators, Baron Munchausen. (Most of William's inventions involving famous individuals—he later claimed to have personally convinced Sir Robert Peel not to push for a tax on cement—were published after the individuals in question were dead.) According to both Isambard Brunel's and Richard Beamish's journals, clay was used to stop the flooding, not cement. The only time during the construction of the Thames Tunnel that cement was not used as mortar was for the steening built after the first flooding, and that undoubtedly came from Francis & White, whose firm produced the cement that Marc Brunel always insisted be used for the project. Also incredible is the claim that his father's Portland cement was "three times stronger" than Francis & White's product. This was quited far-fetched, for William's own product, which was quite good, could be claimed to be only twice as strong—a modern engineer estimates 1.8 times[45]—but beyond that point we find ourselves in the fairyland realm of sales blarney.

Also untrue is William's assertion that this took place "seven or eight years" after his father began producing Portland cement. This would mean that Joseph Aspdin began producing quantities of Portland cement around

1821, at a time when he was still a poor experimentalist covertly stealing limestone from the Yorkshire roads. Yet this fantastic boast was quoted without comment by a historian in the late nineteenth century and later regurgitated as *fact* in dozens of books, pamphlets, and, of course, Internet sites. Repeat a falsehood often enough, and it assumes the appearance—though not the substance—of truth.

Sales were brisk for William and his third group of partners. The firm also won a prize for their cement at the famous 1851 Great Exhibition in London. The future looked bright, but no company with William Aspdin on its board of directors could ever enjoy prosperity for long. Shortly after the firm's triumph at the Great Exhibition, William obtained £300 from the firm to purchase a steam engine for the cement works. It was later discovered that the engine had cost only £80 and the receipt for it had been forged. This prompted the directors to launch an investigation into William's other dealings at the company. They discovered that money allocated for rent payments found their way into Aspdin's pocket instead, as did wages paid to fictitious employees. As the investigation probed deeper, things got worse. It seemed that there was hardly any major transaction that was not, in whole or part, confiscated by William. When his partners confronted William with the evidence they had gathered, William cursed them roundly and left the premises. The partnership was legally dissolved on November 7, 1851.

William Aspdin quickly found a new partner, Augustus William Ord, a well-to-do retired army officer. Ord possessed the three qualifications Aspdin deemed necessary for a business partner: money, ignorance of the cement industry, and enough faith in William to allow him full reign in running the operation. The new company, Aspdin, Ord & Co., was formed in 1852 and was producing cement by the end of that year.

Despite William's many outlandish frauds and deceits, most people in the cement industry recognized that he indeed made a superior product. For several years, they could only offer lower prices but never compete with him at the quality level. Now, one of them could: a man working for one of William's competitors, John Bazley White—of Francis & White fame—had rediscovered the secret of clinkering.

Francis & White had amicably dissolved their partnership in 1836, the former forming a family company called Charles Francis & Sons, and the latter doing the same with his new firm, J. B. White & Sons. Isaac Johnson, who had worked since his teens in the earlier partnership, stayed with John Bazley White and soon rose to become the manager of his cement works. White recognized the superiority of Aspdin's Portland cement and offered to sublicense its manufacturing on generous terms. Aspdin refused for obvious reasons: the secret process for making Portland cement was the only thing of value he possessed. After Aspdin's refusal, Isaac Johnson stepped forward and told White that he could probably discover Aspdin's secret.

Isaac Charles Johnson lived to a very old age—he died exactly two months shy of his one-hundred-and-first birthday in 1911. He personally witnessed the growth of the cement industry from a small cottage industry to an international manufacturing behemoth. When Johnson was born, even the use of hydraulic mortar was rare; by the time he died, concrete was being used to build everything from streets to skyscrapers. His personal recollections are valuable but sometimes suspect. Johnson is a perfect example of "the last man standing at a gunfight"—the sole survivor who tells everyone else what happened. His most engaging story is about his espionage efforts to discover William Aspdin's secret manufacturing process. Johnson writes that Aspdin had built a twenty-foot-high wall around the factory, and, as a further precaution, the only entrance was through the offices, so that all those coming in could be screened. Johnson tried to find out from his competitor's employees what was involved in the process. Perhaps he bought them beers at the local tavern or slipped them some money. Johnson does not enlighten us, so that aspect remains a mystery.

Yet William Aspdin was no fool. He used various stratagems to distract attention from the clinkering process and to make it appear as if something else was responsible for the quality of his product. He had a large tray of various chemical powders—apparently labeled for all to see—from which he would take small handfuls to toss into the mix at various stages.[46] Although he does not say so, Johnson probably tested these—one was copper sulfate— but he could not obtain any good results. Johnson then acquired a sample of Aspdin's cement, perhaps covertly provided by William through one of his employees. Johnson had it analyzed by a chemist, who found that 50 percent of it consisted of calcium phosphate (then called "phosphate of lime"). Johnson felt that he had now discovered Aspdin's secret, for calcium phosphate is the most common chemical component of bones.[47] Johnson went to the local butcher shops and bought as many pig and cattle bones as he could, and then began his experiments to replicate Aspdin's secret formula. He spent considerable time testing different proportions of bone dust before realizing that he had been fooled once again. Johnson then tried slurry mixing a formula of two parts ground chalk (a limestone rich in calcium carbonate) to one of clay. This attempt also looked like it would be unsuccessful, since he still could not produce a cement as good as Aspdin's. One day he overcooked the paste, producing hard, "useless" clinker instead. He was about to throw it away when curiosity prompted him to pulverize the clinker and try using the resulting powder as cement. It worked. Until the day he died, Johnson claimed that it was he who had discovered *true* Portland cement. In fact, it was the same accidental breakthrough that William Aspdin had stumbled upon years earlier.[48] Johnson also claimed that "the Portland cement of Aspdin was no more like the cement that is made today as chalk is like cheese!"[49] This is nonsense, for Aspdin's cement was held in high regard, and the independent tests conducted by Grissell & Peto certainly confirmed its superiority. While improvements would be made over the coming years to various brands of Portland cement, Johnson's early formula for White's product was probably no more "cheese" than Aspdin's.

Whatever the case may be, the upshot of all this was that William Aspdin now had serious competition from J. B. White & Sons, and it would not be long before other firms also discovered clinkering. The process was no longer a secret held by one man.

As is so often the case when an established brand finds generic versions of its product being offered at the same or lower prices, Aspdin placed advertisements warning the public to be suspicious of such cements and hinted at skullduggery by asserting that his competitors employed manufacturing methods "bought or borrowed." He also published "tests results" comparing his Portland cement to others. Unlike before, when the tests were conducted by a respected third party, Aspdin oversaw the testing—

that is, if we assume any testing was actually conducted. Of course, the results were impossibly skewed in his favor. Aspdin must have realized that the end was near for *his* Portland cement, as it was no longer the only one bearing that name or offering such quality. He scrambled to find another cement-manufacturing process that would again give his product an edge. In December 1851, he was granted a patent for "the manufacture of Portland and other cements from alkaline wastes."[50] In other words, William was claiming he had found a way to reprocess the waste products from soap manufacturing to create cement. By 1854, he was producing such cement in the Newcastle area to supplement his Portland product.

We cannot gauge the success of this new cement, as the same irrepressible and self-destructive duplicity in William's character broke to the surface again. Aspdin refused to pay rent for his manufacturing premises or make payments for his leased equipment, claiming them to be "unsatisfactory." The more likely reason for his refusal is that he absconded with the money set aside for these purposes. In February 1855, William Aspdin was arrested for unpaid debts. Although he was able to borrow enough money to obtain his release, William would be involved in one lawsuit after another for the next couple of years. Since he had burned too many people in Britain to ever again do business there, William decided to leave his homeland.

TAWDRY LAST YEARS

The third and final act of William Aspdin's life finds him in Hamburg. Thanks to the meticulous record keeping of the Germans, it is easy to track Aspdin's actions there. A residency form filed with the local authorities shows that he arrived alone on May 22, 1857. He wisely stated his profession as "Buyer" (*Kaufmann*—all governments back then preferred their foreign visitors to be buyers rather than sellers).

Hamburg and London had been major trading partners since the days of the Hanseatic League, and William made sure that the advertisements he placed in the British construction periodicals were translated and printed in

German trade journals as well. William Aspdin's various companies had sold much Portland cement to dealers in the major cities of northern Germany, and his name and brand were well known there.

The political situation in the area around Hamburg was complicated at the time. Hamburg was a self-governing "free city" (*Freistadt*). Across the Elbe River to the south was Prussia, and to the west, north, and east was Holstein, a German-speaking region of Denmark that nevertheless belonged to the German Customs Union (*Zollverein*). William's activities would take him to three distinct regions in this one, rather small area—a particular advantage for someone wishing to flee debt collectors or contractual lawsuits. It is as if he had planned an exit strategy years in advance.

We do not know how fluent William's German was, but even at a basic conversational level it would have somewhat curtailed his natural persuasiveness. Perhaps it was for this reason that Aspdin sought out those of his countrymen who had already established themselves in the area. He found an expat named Robert Fawcus, who had come to Hamburg from Hartlepool, England, a few years earlier. Fawcus, along with Alfred Buschbaum, had purchased Klueudgen & Co., a small firm they quickly transformed into a successful coal-importing business. Fawcus and Buschbaum had contracted Aspdin to build their company a Portland cement plant, and it was the fee from this arrangement that financed Aspdin's move to Hamburg.

Either Aspdin had moved fast, or he was already in correspondence with Fawcus before coming to Hamburg. The site for the cement plant, located at Bill Horner Kanalstrasse 10, was purchased exactly one week after William's arrival (May 30, 1857), and the city's planning commission granted their construction application less than three weeks later. The completed plant was inspected on October 4 and was allowed to begin operations. By this time, William had brought his wife and six children over from England and installed them in his new residence in Bille Waerder (now Billwerder), an island on the Elbe River in southwest Hamburg, close to the new factory.

In the rush to build the cement plant, either Aspdin or his partners had clearly not done their homework. Instead of locating the factory near limestone or clay deposits, he decided to import both from England (were kickbacks involved?), even though good sources for these minerals could be found in nearby Prussia and Holstein. Also, his plant was upstream on the Elbe, south of Hamburg's main shipping harbor. According to the regulations established by that city and the *Zollverein*, this required his company to declare the imported goods twice: the first time, when they arrived at the plant; the second time, when the manufactured cement was "exported" back to Hamburg. The company eventually filed for and was granted a remission from the double duties, but by then they had already produced 89,000 pounds of Portland cement for which nonrefundable import duties had been paid. Five years would pass before the firm would see a profit, but by that time Aspdin was long gone. He had agreed to build the plant and oversee its initial production runs, but nothing more.

In April 1860, a prominent acquaintance of Aspdin's, Adolph Tesdorpf, a member of the Hamburg senate, arranged a meeting between the Englishman and Carl Ferdinand Heyn, a businessman in Lüneburg, a town 45 km (30 miles) southeast of Hamburg in Prussia. Heyn and his brother owned a successful sugar refinery there, as well as nearby land that contained a substantial outcrop of high-quality chalk, the perfect stone for cement production. An agreement was soon reached that called for Aspdin to build a cement plant in Lüneburg for the Heyn brothers. He would also serve as the plant's manager. We do not know if his role as the factory superintendent was to be a permanent position, but Aspdin, then forty-five years old, must have thought about ending his peripatetic existence and settling down.

The Gebrüder Heyn Portland cement plant was up and operating by early 1861. Initially, things seemed to go well, and the factory produced eighteen thousand barrels of quality cement in its first year. However, as was always the case with William Aspdin, the same dark, self-destructive impulses emerged. Despite his success—or to spite his success—Aspdin began drinking heavily and, as a result, quality control at the factory slipped dramatically. Barrels of cement began bursting in the firm's warehouse, and the company had to dispose of their entire stock by dumping it into a pit. Either Aspdin was buying inferior clay, or the problems were due to a secret invention Aspdin developed to test the readiness of the kilned cement. According to coworker Carl Heintzel, no one but Aspdin was allowed to inspect the device. Now that most people in the industry knew the process of making Portland cement, Aspdin apparently felt that he had to have some kind of confidential and proprietary process to call his own and use as leverage. Instead, it caused the Heyn brothers to doubt his competency. Carl Heyn later wrote that, despite his manager's boasting about having been in the business since childhood, "he really doesn't know what he's doing."[51] After eighteen months at the Lüneburg facility, Aspdin either resigned or was fired.

Aspdin moved with his family from Lüneburg to Altona, now a suburb of Hamburg but then a town in Holstein. One report has it that he changed his son's name from "William Aspdin" to "William Altona Aspdin." He may have written down such a name in a residency form to curry local favor, but Danish and German officials were (are) hesitant to legally alter birth names, so the change was probably unofficial.

While in Altona, Aspdin looked around for other business opportunities. He evidently thought that Holstein was a good place to scout out prospects, since his reputation in Hamburg and Prussia was no longer held in high regard.

William soon found another potential partner, Edward Fewer, an Englishman of Irish extraction. Fewer insisted on contractual protections and an equal say in the running of the business. That William agreed to these terms indicates how desperate he must have been at the time. In early 1862, a contract was drawn up and signed by the two men in Altona. The cement company Edward Fewer & Co. began operations in 1862 at a plant in Lägerdorf next to a virtually inexhaustible outcrop of high-quality Holstein chalk.

Of course, any company with which William Aspdin was involved would experience problems, but this time, the outcome would be very different. After just six months, Aspdin was kicked out of the partnership and given a small compensation check for his trouble. William immediately moved to the small town of Itzehoe, near Lägerdorf, where he published a notice in the local newspaper, the *Itzehoer Nachtrichten*. This is an English translation of the German text:

After the dissolution of the business relationship in Lägerdorf between Mr. Edward Fewer and myself, I can no longer bear any responsibility for the quality and worthiness of the cement that will henceforth be produced by Edward Fewer. Nor in the smallest degree may he associate my name with his brand. Itzehoe, July 9, 1863.[52]

It is almost comically ironic that "henceforth" (*fortan*) Edward Fewer's cement business would thrive.[53]

Aspdin, already a heavy drinker, drank more. One spring day in 1864, while walking down a street in Itzehoe—or perhaps stumbling along in a drunken stupor—he fell and most likely struck his head on a paving stone. He died soon thereafter, on April 11, 1864. He was forty-eight years old. William Aspdin lies buried in the town's Protestant cemetery.

Had he been an honest individual, Aspdin might have used his superior product to become the dominant player in the cement industry, enriching himself, his partners, and their shareholders. Instead, his serial swindling left nothing in its wake but ruined fortunes, estranged family members, and no one whom he might rightly call "friend" (there is no evidence that he ever owned a dog). One cannot but feel acute compassion for his wife and children, who disappeared from the public records in Germany and presumably moved back to England.

Upon his retirement in 1889, Edward Fewer sold his large and prosperous cement plant to the Alsen'sche Portland Cementfabrik. This would later become the international cement company Alsen, which is now part of the even larger Swiss firm Holcim, one of the largest concrete cement manufacturers in the world. Its success serves as an object lesson for "what might have been." Had a virtuous version of William Aspdin existed in a parallel universe, he would have achieved more. And had this alternative Aspdin lived as long as Isaac Johnson, and then suffered the same mishap, he might have been killed by his own product.

So, did William Aspdin discover clinkering and, thus, *true* Portland cement? The research compiled by the esteemed British engineering historian Major A. C. Francis seems to suggest that credit should go—however grudgingly—to Joseph Aspdin's wayward son. I would tend to agree, but a curious discovery made several years ago presents us with a puzzle. In April 2008, archaeologists working at the dockyards in Bristol, England, uncovered the concrete floor of a factory building designed by Isambard Kingdom Brunel to manufacture the engines used for his ship, the SS *Great Britain*, the world's first propeller-driven, oceangoing vessel. After the Thames Tunnel was bricked up in 1828, Sir Marc Brunel's son shifted much of his attention to solving mechanical engineering challenges in the growing rail and shipping industries. (His contributions in these fields were substantial and later earned him a second-place position—Winston Churchill was first—in the BBC's program *100 Greatest Britons*, which polled the UK public to determine the greatest people in British history.[54] The massive concrete floor using heavy aggregate, measures 20 m wide by 50 m long (*ca.* 66 ft by 164 ft) and 400 millimeters thick (*ca.* 1.3 ft). The metric measurements appear like something Brunel *fil* would use—he studied engineering in France and saw the utility and common sense of the new system. However, the concrete seems a product of another era, for, according to Professor Geoff Allen of the Interface Analysis Centre at the University of Bristol, *Portland cement* was used in its construction. Where did it come from? It certainly did not originate from William Aspdin or his son, since both were then incapable of producing and/or delivering it, especially in such quantities. If William Aspdin had been involved, he certainly would have told the world about it, as he had about so many other things he did and did not do. The quality of Roman cement varied greatly, and some was quite good. Chemical analysis of early Portland cement—like the kind made by William Aspdin and Isaac Johnson—would show the presence of lime, aluminosilicates, and so on, but so would Roman cement made with limestone adulterated with the right amount of clay. Whatever the case may be, the old factory floor at the Great Western Dockyard in Bristol presents us with a mystery requiring more investigation.

OTHER PIONEERS

At the same time John Smeaton was investigating the hydraulic properties of concrete mortars, some unknown Briton had already rediscovered Roman *caementis* or, rather, lime concrete. It was a simple mixture of gravel combined with lime that a Neolithic builder or Cato would have immediately recognized. Called "grouted gravel," it was being used as a foundation material in Britain by the end of the eighteenth century. Although Portland cement concrete would eventually replace the decidedly non-hydraulic grouted gravel, use of the latter for foundation work persisted until the dawn of the twentieth century.

If one were to accept that Joseph Aspdin's 1824 patent for Portland cement describes the process of slurry mixing, he would still be far from the first person in Britain to have used this method. Englishman James Frost was certainly employing the process around the same time Joseph Aspdin filed his patent, and it's possible that he was using such methods earlier, though they are not described in a patent. James Frost began experimenting with hydraulic cements as early as 1810, but he was not satisfied with the results. He traveled to France in 1821 to study under Louis Vicat, the French engineer who was the first person since Smeaton to conduct tests of various limestones to gauge their hydraulic properties. Vicat recognized that clay played an active role in the special characteristics exhibited by natural cements, and he conducted a series of tests to determine the ideal amount of clay to limestone for producing the best results. France has an abundance of clay limestone deposits from which good Roman cement can be produced. The deposits of such limestone in Britain—almost all of it near the coasts of England and Wales—were limited and close to exhaustion by the middle of the nineteenth century. Indeed, it was common in the 1850s to see hundreds of boats employed at dredging clay limestone up from the seafloor just off south England's shores, the land portions of the same outcrops having already been removed to make Roman cement. This scarcity drove up the price of Roman cement. Since limestone rich in calcium carbonate—including chalk—and clays with high aluminosilicate content are quite common on the island, clinkered Portland cement probably saved the

British cement industry from disaster. This also explains why Roman cement remained popular in France for a longer time than it did in Britain. It was not only cheaper to manufacture, but its source materials were—are— locally abundant. Quality production of Roman cement is still carried out in France by Vicat S.A., founded in 1853 by Louis Vicat's son, Joseph.

Louis Vicat's experiments, more extensive and detailed than Smeaton's, would have tremendous influence on the French cement industry, which he, in part, fathered. Vicat also discovered that the proper admixture of clay to limestone was between 15 to 20 percent, while the English often employed 30 percent or more in their Roman, and some early, Portland cements. He also began early experiments with *artificial cements*, by which the amounts of lime and clay are controlled; a method employed by Frost, Dobbs, and Joseph Aspdin. Vicat's architectural masterpiece was the Souillac Bridge in southeastern France, the world's first concrete bridge. Vicat had more faith in the strength of concrete than most of his British contemporaries. Vicat's work inspired many of his countrymen to explore new formulas and applications for the material.

Under Vicat's mentoring, Frost realized all the mistakes he had been making in his earlier experiments and returned to England the following year to set up a cement plant. Strangely, he filed a patent in 1822 that makes even less sense than Joseph Aspdin's. In it, he describes a cement that can be made "without alumino (sic)," something that is impossible, since alumina is a critical ingredient.[55] One's first reaction to this patent is to dismiss Frost as someone who was hopelessly behind in his basic chemistry. However, if one takes into account his knowledge gained in France, and the report by a later eyewitness of his manufacturing process, it becomes clear that the patent was simply a device to throw off his competitors, many of whom *did* have a limited understanding of chemistry. Frost knew exactly how to make good cement, but he was also aware that industrial rivals would subject an Englishman who had studied under the renowned Louis Vicat to extra scrutiny. In tribute to his mentor in France, Frost would name his product "British cement." (Nationalism and marketing considerations usually trump gratitude.)

In 1828, Charles William Pasley visited Frost's cement factory, by which time it had been in operation several years. Pasley was a highly respected military engineer who was investigating the hydraulic qualities of the various cements then being produced by firms throughout Britain (his notes make no mention of Joseph Aspdin's product). For some reason—perhaps the high regard he held for Pasley—Frost opened up to the engineer and showed him his process for making cement. Pasley's notes show that Frost was practicing slurry mixing, though there is no mention of kilning the paste until vitrification to make clinker. Pasley describes Frost grinding the chalk into a powder and mixing it with water and clay and "by opening a small sluice, [the mixture] was allowed to flow into a . . . reservoir where it usually remained some months and acquired the consistency of a stiff paste. In this state the material was cut out of the back in lumps and laid on open shelves to dry. When dried, the lumps were broken into smaller pieces and burnt in a kiln of the common inverted cone-like form and in the same manner as lime with alternate layers of fuel and cement."[56]

Pasley's description is the earliest of slurry mixing *observed in practice.* James Frost would later introduce another product that he called, appropriately enough, "British marble," which was created by crushing chalk and flints to make cement that was quite white when it dried. However, Frost was not happy with the slim profits he was realizing in his cement business. He sold his plant to Francis & White and immigrated to the United States, where he enjoyed a successful career as a civil engineer.

Edgar Dobbs is another figure who evidently used mixing. Whether it was the wet or dry process, we do not know for certain. However, the patent he filed in 1811, while more coherent in its description than many other cement patents of the day, also includes a few red herrings to throw off competitors. He specifies mixing the lime with one or more of the following ingredients, most quite fishy: clay, shale, road dirt (?), mud (?), sandstone, earths (?), and so on. The only component that makes any sense is clay and shale (if the latter was ground). The components were then mixed and kilned. Dobbs seems to have been in business for only a few years, for his company disappeared from the local directory by 1817.[57]

Another person mentioned in the chronicles and timelines of concrete's progress is William Jessop, whose West India Docks in London were reportedly built of hydraulic cement. However, Jessop used standard lime mortar to build the docks. Apparently, some historian simply assumed that he had used the then-available hydraulic mixes, either Roman cement or trass. In fact, Jessop ordered six thousand tons of Dorking limestone to make his *non-hydraulic* mortar for the brickwork at the West India Dock.[58]

The key milestones in the first half of the nineteenth century were the discovery of the properties of natural cement, followed by wet (slurry) mixing, and then clinkering. The other advances are tied to the recognition that hydraulic concrete cement could be used in more ways than simply serving as a waterproof mortar.

NO LONGER JUST A MORTAR

By the early nineteenth century, clay and plaster of paris (the latter made from gypsum) had long been used with molds to cast decorative fixtures or busts of famous individuals. Unfortunately, neither material holds up well to the elements, especially plaster of paris, so such products were restricted to indoor displays. It was not long before people began experimenting with a mixture of cement and sand to create concrete castings that could hold up well outdoors. The first person to do this was James Pulham, a talented artist who worked for cement dealer William Lockwood. (Lockwood would later produce his own cement under the brand name "Portland cement" several years before Joseph Aspdin filed his patent.[59]) By 1802, Pulham was casting concrete vases, coats-of-arms, pilasters, friezes, architraves, cornices, and sculptures.[60] A few years later, Lockwood used Pulham's gifts to construct a house resembling a Gothic castle that incorporated large portions of cast concrete.[61] Pulham's talent contributed immeasurably to Lockwood's success in the cement business. More than anyone else at the time, they demonstrated that concrete could do more than just bind masonry blocks. Sadly, both James Pulham and William Lockwood are rarely mentioned in modern histories chronicling the material's progress, although their influence on the early nineteenth-century building industry was substantial. Not only had their house demonstrated that concrete could be used as a monolithic building material, but their cast outdoor ornaments were direct precursors of the concrete objects d'art—and kitsch—that now grace millions of gardens around the world.

Although largely ignored by most people during much of the nineteenth century, the idea of using concrete to cast walls and floors to make houses was an appealing challenge for a few brave souls active in the cement industry. Besides Lockwood, William Aspdin attempted to build a mansion using his Portland cement, but only a third of the structure was completed before financial problems with his last English firm forced him to stop construction. Perhaps a dozen concrete houses were built in England in the 1850s, and a few still remain.

Only with the introduction and wide-scale use of iron, and then steel, reinforcement would monolithic concrete construction finally take wing. By that time, Britain, home to so many innovations associated with the "new" building material, would be left behind by other countries.

Chapter 6

REFINEMENTS, REINFORCEMENT, AND PROLIFERATION

At the same time that William Aspdin was setting up the Portland cement plant outside Hamburg, another cement factory a little over 325 km (*ca.* 220 miles) to the east, in the Prussian city of Stettin (now Szczecin, Poland), had already been in operation for over two years. Thanks to the efforts of chemist and entrepreneur Hermann Bleibtreu, Stettin would be home to some of the largest and most successful Portland cement companies in the world. By century's end, German Portland cement would be considered the finest made, and the nation's production of this building material would dwarf that of its birth country, Britain.

Hermann Bleibtreu was born in 1821 in the Rhineland village of Pützchen bei Bonn, now a suburb of Bonn. His father, Leopold, was a successful businessman who owned a lignite mine and factory. (Lignite is a soft form of coal—somewhere between peat and standard coal—that was widely used as an industrial fuel in Germany.) Leopold encouraged his son to study chemistry, a field that would obviously be useful in the coal industry. Unlike John Smeaton and Marc Brunel, Hermann Bleibtreu would find his father's designated career for him in keeping with his own interests. Hermann attended the universities of Bonn and Giessen, then matriculated at the Royal College of Chemistry in London. Although Bleibtreu made significant contributions to both the lignite and cement industries, it is his work in the latter for which he is most remembered. While in England, Bleibtreu was impressed by the growth of the cement industries there. After returning to Prussia, Bleibtreu began scouting for an area that had good sources of limestone and clay and was close to major transportation routes. Outside Stettin on the Oder River, he found clay rich in aluminosilicates. Just north of Stettin, the Oder empties into the Stettin Lagoon (*Stettiner Haff*), where the large island of Wollin holds huge deposits of chalk. The chalk could be sent by barge to Stettin, where it would be mixed with the local clay and kilned to make Portland cement. The finished product could then be shipped north to the Baltic Sea, and thence to the Scandinavian or Russian ports, or sent south to the city of Küstrin on the Oder, where it could be loaded on boxcars and transported by rail to nearby Berlin. In 1852, Bleibtreu, with the assistance of Consul Paul Gutike, formed a partnership with a local firm, Gruben und Fabrikanlagen, to build an experimental factory near Stettin to serve as a test bed for the project. The results were excellent. Bleibtreu and his partners quickly built a full-size factory in 1855 that produced twenty-five thousand barrels of Portland cement in its first year. Bleibtreu would go on to found several more cement plants throughout Prussia, and, in 1862, he entered his product at the International Industrial Exhibition in London, where it won first prize for its quality. The German cement companies, especially those in Stettin, would continue to win gold medals at industrial exhibitions and shows around the world.

The judges at the 1862 London Great Exhibition could not have realized that this award represented a symbolic moment in the history of Portland cement. The Germans, once their attention was concentrated on a particular industry, quickly dominated the field. Beginning with the work of Hermann Bleibtreu, the Germans invested tremendous time and research exploring the chemical interactions of the ingredients comprising Portland cement and steadily improved the mix. Unlike their British counterparts, virtually every major German cement company would have at least one chemist on its staff. Experiments were made with various mixtures, and tests were then conducted on their respective setting times and resulting compressive strengths. Detailed analysis was made of the various clays, especially their alumina and silicate content. Various other minerals, such as iron oxide and gypsum, were added to the cement mix in assorted quantities to see their effect on the quality and curing properties of the resulting concrete. Meanwhile, a torpid complacency had settled in among the directors of the various British cement companies, which slowed technical innovation to a leisurely crawl.

Not a few warned of the consequences of this complacency. A major industry leader, Gilbert R. Redgrave, wrote in 1895:

> I regard the present time as, in many respects, a very critical period in the history of the cement trade. Our own country, the original seat of manufacture, has been distanced in certain directions in consequence of the superior scientific skill and energy of foreign rivals. The supremacy we have long enjoyed has undoubtedly been to some extent wrested from us by the products of Continental industry and enterprise, and in the absence of some united action and intelligent leading, our manufacturers are threatened with a competition which they are not adequately armed to encounter.[1]

The larger British cement firms did take concerted action, but of a different kind. Of more interest to them were the low prices offered by smaller British cement companies whose owners were desperate to remain in business. If the larger British cement producers had, like the Germans, concentrated on improving the quality of their product and greater manufacturing efficiency, the cheaper competition would have fallen by the wayside. Instead, a round of mergers began, and the larger firms that came out of this consolidation soon formed a commercial trust: the Associated Portland Cement Manufacturers (APCM) in 1900. Initially representing firms that comprised some 35 percent of all cement production in Britain, the APCM quickly grew, and its members would soon increase that proportion to 75 percent of total production in the United Kingdom.[2]

By this time, the APCM's share of the world market had dropped precipitously, and the once rich overseas markets, like the United States, had all but stopped buying British Portland cement. Besides Germany, Belgium and France had become major players as well. Germany's neighbors were paying closer attention to her success in the cement industry, and so they also began investing heavily in research and development. German technical journals were translated to French and avidly read.

The British consulates scattered throughout the world were required to file commercial reports on the imports and exports of their host countries. These reports were collected and published each year, and they chart the slow decline of the British exports of Portland cement in the last quarter of the nineteenth century. By 1905, one British consul in the United States wrote that British cement in that nation had all but disappeared, and, though much of it had been supplanted by domestic production, the Americans were still importing huge quantities of Portland cement from Belgium and Germany. This was true in most of the world. If one wanted the best Portland cement, the German product was preferred. If both price and quality were important, the Belgian cement, while not quite up to German standards, would suffice. The domestically produced cement could handle the rest.

That the British companies were slow to take notice of the situation could be attributed to the fact that the use of concrete in the building industry had increased substantially during the same period. More cement was being manufactured and used in Britain, and the country continued to ship the surplus to those of its colonies where protectionist trade policies were enforced, like India. Elsewhere, it was a different story. Many Commonwealth nations, such as Australia, began producing their own cement and supplementing the rest of their needs with product imported from abroad. When World War I broke out in 1914, Australia severed its commercial ties with Germany, from which it was importing almost 40 percent of its Portland cement. To the annoyance of the mother country, the Australians switched to importing cement from Denmark, and not from Britain. Officials of the British government, suspecting that the cement destined for Australia was actually being produced in Germany and simply rerouted through Denmark, launched an investigation. They discovered that the cement being shipped to the folks "Down Under" was indeed of Danish origin. Apparently, almost every industrialized country, including tiny Denmark, was manufacturing cement of a higher quality than the British could produce.

What saved the British cement industry from collapse were the two World Wars, which effectively removed their primary competitors from the world market. After 1945, a long period of peace settled over Western

Europe, and competition in the cement industry gradually returned. Today, the vast majority of Britain's domestic cement production comes from plants owned by foreign-based companies, mostly French or German. The one remaining British-owned cement company of any consequence, Tarmac, now produces less of its product in the United Kingdom than its Mexican-owned rival, CEMEX.[3] The story of the British cement industry is an instructive example of how industrial preeminence can be lost through complacency and an obsession with increasing profit margins at the expense of research and development.

PRODUCTION INNOVATIONS

Commercial declines, however, are rarely apparent at their onset, especially when a country enjoyed such a spectacular predominance as the British cement industry did in the mid-nineteenth century. And while innovations had slowed, they did not cease in the Britain.

British cement companies understood that simply increasing the size of their static bottle kilns, while producing better yields, was still inefficient, especially in terms of fuel expenditures and manpower. Rotary kilns seemed to offer a solution. They had been used in the alkali industry since the 1850s to separate salt from brine. The problem was that the alkali rotary kilns operated at far lower temperatures than those needed to vitrify the clay-limestone mix to produce cement. The first rotary kilns developed for the British cement industry were more robust, produced higher temperatures, and, while there were technical differences among them, they did share common traits. The kilns consisted of a long, thick iron tube raised at one end and resting on iron rollers. Intermeshed gears, usually spur or worm gears driven by a steam engine, slowly rotated the kiln.[4] The raw material to be kilned was dumped into the raised end of the tube and slowly made its way to the other end of the kiln, where it dropped into a cooling pit before grinding. Heat was supplied by an oil or gas flame projecting from an iron pipe inserted into the middle of the opening at the lower end. Sometimes, coal dust was injected into the kiln, which was then ignited by the flame.

The first rotary kiln developed for the cement industry was patented in 1877 by Thomas Crampton, a noted British engineer now mostly remembered as the "father" of the submarine telegraph cable. Crampton's rotary kiln was too complicated and inefficient, and it never went beyond the experimental stage. A few years later, another Englishman, Frederick Ransome, introduced a slightly better rotary kiln. Ransome's kiln, patented in 1885, consisted of an iron cylinder 21 ft (*ca.* 6.4 m) long and 3.5 ft (*ca.* 1 m) in diameter.[5] The problem with Ransome's patent was that it called for unmixed limestone and clay powder to be kilned. This was not a deliberate ruse to protect his patent: Ransome was ignorant of the need for thoroughly mixing the two minerals before kilning. In short, he did not comprehend at this late date—the 1880s—the importance of clinkering. Not surprisingly, Ransome encountered great problems in his attempts to make his kiln work properly. After a few years of effort, Ransome recognized his error, but he still could not get his kiln to achieve the operating efficiency needed to induce cement factories to adopt the technology. Further attempts by other British inventors to produce a practical cement rotary kiln over the next two decades also failed. Though much was learned by these failed attempts, the first commercially viable cement rotary kiln would be invented in another country.

Perhaps discouraged by the poor results obtained with rotary kilns in Britain, the Germans tried another approach. Their answer was the vertical shaft kiln. Small lumps of mixed clay/limestone were dumped in or near the top of the circular shaft, where the mixture was then heated from below. Various versions of the shaft kiln were used. Some had the flames projected from jets at the side of the shaft, while others mixed fuel with the lumps to accelerate the kilning process. Compartments or choke points within the shaft were used to control the amount of material kilned at one time. After kilning, the cement clinker would then be dropped to a cooling chamber at the shaft's base, or into a pit beneath it for later grinding. The shaft kiln reduced the amount of fuel and manpower needed to produce cement and helped give Germany another competitive edge in the industry. Shaft kilns are still occasionally used today for small-scale cement production and are sometimes seen at major construction sites.

STEEL REINFORCEMENT

The use of concrete as a monolithic building material, and its reinforcement through iron or steel, was a gradual and complex development. This may seem surprising to us, but some advances are obvious only in hindsight. Masonry construction had been used for thousands of years, and to the people of the nineteenth century, this building technique was both tried and proven. It was also venerated, for were not all the great buildings, from the pyramids to St. Paul's Cathedral, constructed in this manner? The longevity of this building technique, and the respect given it, imposed a kind of subjective blindness. By the eighth decade of the nineteenth century, the few buildings constructed of concrete were there for all to see, but most people perceived them as only novelties—if they took any notice of them at all. We suffer from the same cognitive disorder, for no doubt future generations will shake their heads in wonder about how we built things in our day.

Until very recently, the various histories of concrete identified Frenchman Jacques Monier as the inventor of reinforced concrete. A gardener and landscape artist, Monier tried building vases and planting tubs using concrete reinforced by iron wires.[6] Monier patented his technique in 1867 and later sold the rights to his invention to two Germans, Wayss and Bauchinger. Wayss published a book in 1887 called *Das System Monier* that focused much attention on Monier's "discovery" and the possibilities offered by iron reinforcement.[7] Actually, Wayss and several Austrian and German colleagues had greatly elaborated and improved upon Monier's primitive ideas concerning concrete reinforcement, but the name stuck: for the next three decades, the concept of reinforced concrete was often referred to in Europe and the United States as the Monier system or Monier construction.

However, Monier was hardly a pioneer in the field of reinforced concrete. Recent research has uncovered earlier precedents for the technology. In 1861, another Frenchman, François Coignet,[8] published a treatise on concrete reinforced by iron bars, which he had patented six years earlier. Predating Coignet's paper by over a decade was the work of Jean-Louis Lambot, a gentleman farmer in southern France. In 1848, Lambot constructed a concrete rowboat reinforced with iron bars and mesh. He frequently used it to row across the pond on his estate just outside Miraval in Provence. The boat was 3.6 m (*ca.* 11.8 ft) long, with a beam of 1.35 m (*ca.* 4.4 ft).[9] The hull was 30–40 mm (*ca.* 1.2–1.5 in) thick. The boat sprang a leak one day, and Lambot was forced to swim back to shore. Protected by anaerobic mud at the bottom of the pond, Lambot's reasonably well-preserved boat was recovered over a century later. Unless some earlier inventor comes to light, Lambot may rightly be called the "father of reinforced concrete construction."

For the earliest example of reinforced concrete construction in the building industry, credit goes to an Englishman in Newcastle, William Boutland Wilkinson. Wilkinson was a plasterer who went on to found a firm that did thriving business by using Portland cement to cast concrete paving stones. Wilkinson was granted a patent in 1854 for "improvements in the construction of fire-proof dwellings, ware-houses, and other buildings."[10] The patent provides a clear description how to use a network of flat iron bands or disused iron cables to reinforce concrete walls or ceilings. He built an attractive cottage of reinforced concrete in 1865 to demonstrate the technology. Wilkinson's patent shows that he had given much thought to the processes involved, as well as to the structural stress issues. Still, Wilkinson could interest no one in reinforced concrete construction, and so he redirected his attention back to making cast paving stones, of which he sold many thousands.

While these events were transpiring in Europe, a number of American architects and engineers were also experimenting with iron-reinforced concrete. The large-scale use of concrete in the United States began decades earlier, shortly after the groundbreaking work of the British and French pioneers. The most notable example was the construction of the Erie Canal (1817–1825). The scale of such a construction project was beyond the ability of most Americans, so New York governor DeWitt Clinton sent American engineer and fellow New Yorker Canvass White to Britain to study the canals, aqueducts, and culverts of northern England and southern Scotland.[11] White did his homework well and returned to the United States with hundreds of pages of notes and drawings. White recognized that the hydraulic properties of natural cement would be key to ensuring the canal's strength and durability, so he began searching New York State for outcrops of limestone adulterated with clay. He found them in abundance in Madison County. His advocacy of natural cement met initial resistance, but these doubts were swept away when White demonstrated the material's hardness and hydraulicity. White patented his cement in 1820 and would use it to build America's greatest engineering endeavor of the early nineteenth century.

The first experiments with iron reinforcement of concrete in the United States took place fifty years later. In 1871, only six years after Wilkinson built his cottage in England, William Ward built a large house for himself on Comly Avenue in Port Chester, New York. It still exists and is popularly known as "Ward's Castle." Not only is Ward's Castle the earliest example of reinforced concrete construction in the United States, but it was also the largest reinforced concrete structure built up to that time in the world. Ward's home exerted a tremendous influence on American and European architects and engineers.

One American encouraged by Ward's work was Thaddeus Hyatt. Hyatt decided to conduct thorough tests of reinforced concrete's strength. Since the equipment needed for conducting such tests was not available in the United States, Hyatt traveled to London in 1877 to collaborate with David Kirkaldy, a pioneer in the development of industrial test machinery. Hyatt's work was critical in formalizing the relative strengths of reinforced concrete slabs, beams, and columns;[12] Hyatt also discovered that the thermal expansion attributes of concrete and iron, as well as their elongation properties under a particular load, were virtually the same for both materials. This conclusively demonstrated the suitability of reinforced concrete for construction purposes. Another important contribution at determining the strength characteristics of reinforced concrete was the work done by François Hennebique, a Frenchman who had been conducting experiments for some years on the material, independent of Monier and the Germans. In 1879, Hennebique demonstrated the efficacy of using iron bands to overcome the weak tensile strength of concrete.

Despite all this work that proved the utility of reinforced concrete construction, Europeans and Americans remained skeptical of the building material. This kept the adoption of reinforced concrete for construction purposes restricted to those few adventurous individuals, like James Ward and François Hennebique, who had the resources to either build or perform major experiments with the material. Apparently, most contractors and architects did not yet trust reinforced concrete. A key person in changing those perceptions and in establishing respect and acceptance for reinforced concrete construction in the United States and the rest of the world was Ernest L. Ransome, a contractor in San Francisco, California.

ERNEST RANSOME

Ernest Ransome was born in 1852 in Ipswich, England. He was the son of Frederick Ransome, the man who had such trouble trying to get his cement rotary kiln to work properly. Frederick Ransome owned a company in Ipswich that manufactured agricultural implements. The firm enjoyed enough success to allow Frederick to conduct expensive experiments that, like the rotary kiln, would not bring financial disaster if they happened to fail. Ernest began his apprenticeship at his father's company in 1859 at the strikingly young age of seven years old. Besides the rotary kiln, another of Frederick's experiments resulted in the discovery of a particular cement made of powdered limestone mixed with a small quantity of silicate of soda that was then briefly submerged in a solution of calcium chloride. According to Ernest, the product was sold "in all parts of the world." A curious comment then followed: "In America, the new process was introduced in 1870 by the Pacific Stone Company of San Francisco, of which company I was the superintendent for four years."[13] In this reminiscence of his early years, Ransome makes no mention of his birth date, for it would have highlighted how young he was when he traveled to the United States. Did he have his father's blessing for bringing this manufacturing process to the New World? Or was he a runaway? The trip from Ipswich to San Francisco would have taken some months, and fare for America's transcontinental railroad was steep. Did Ransome find work as a common seaman

on a ship sailing to California? The reasons and circumstances of this precocious move are shrouded in mystery. While this abrupt relocation and
other aspects of his subsequent career have certain parallels with William
Aspdin's life, Ernest Ransome was incontestably the more honest, imaginative, and industrious of the two. Ransome's partners would not lose money;
in fact, one in particular would do very well indeed. Ransome was grateful
to those who assisted him in his career and was always ready to point out his
indebtedness to them.

Ransome's decision to come to San Francisco, and not New York, seems
odd at first. Perhaps it was a romantic impulse justified by practical considerations. San Francisco was growing fast. It doubled its population every few
years, and the construction trades were doing quite well there. The Gold
Rush had come and gone, but the 1870s saw greater fortunes being made just
over the state line at the silver mines in Nevada. Although the silver was in
Nevada, the riches flowed into San Francisco where most of the mining magnets, like James Flood, John McKay, James Fair, and George Hearst (father
of newspaper publisher William Randolph Hearst), lived and invested most
of their profits. San Francisco was also home to several railroad barons, as
well as others who made fortunes by cornering prime industries (Claus
Spreckels's sugar monopoly or Francis Marion Smith's domination of the
borax market) or by making astute real estate investments (James Phelan,
William Ralston), or through outlandish swindling (Henry Meiggs, Philip
Arnold, and others). The unofficial capital city of the "Wild West" also had
a unique and extraordinary allure. Oscar Wilde wrote of the city, "It is an
odd thing, but every one who disappears is said to be seen at San Francisco.
It must be a delightful city, and possess all the attractions of the next
world."[14] Rudyard Kipling observed that "San Francisco is a mad city—
inhabited for the most part by perfectly insane people whose women are of
remarkable beauty."[15] Ransome's older American contemporary, Hinton
Helper, wrote, "I have seen purer liquors, better segars, finer tobacco, truer
guns and pistols, larger dirks and bowie knives, and prettier courtesans here
in San Francisco than in any other place I have ever visited; and it is my unbi-
ased opinion that California can and does furnish the best bad things that
are available in America."[16] That Ernest left England and abandoned a posi-

tion at his father's successful firm suggests an adventurous spirit and a longing for independence. Thus, the reasons for his move to California are hardly strange: for what young man possessing such qualities in the late nineteenth century would not want to go to San Francisco?

As noted, Ransome quickly found a position at the Pacific Stone Company. The firm was located on Greenwich Street, between Gough and Octavia Streets (now in the Pacific Heights district), where it supplied cast concrete paving stones, vases, and architectural adornments. Ransome convinced them to switch to his father's calcium chloride and silicate soda-based cement. It did not require kilning and could be mixed onsite with ground limestone from the nearby Santa Cruz Mountains. The Ransome mix also enjoyed another advantage over Portland cement: the latter—like most heavy goods—was still being shipped around Cape Horn from the East Coast of the United States or from Britain. Thus, a barrel of Portland cement cost $8 (approximately $160 in today's dollars) in San Francisco, several times its price on the Eastern Seaboard.

In 1875, Ransome left the Pacific Stone Company to found his own firm bearing the eponymous name Ernest L. Ransome. It was located at 10 Bush Street, near Battery Street, just south of the waterfront. At that time, this was an industrial area, and Ransome's business was surrounded by furniture manufacturers, iron foundries, and textile factories. He did a little bit of everything he knew how to do. He made vases and statuary and sold sodium chloride and silicate of soda, the critical components used for making his cement. To maximize his exposure, he used the city directory to post as many separate listings as possible, each highlighting a particular product or service. Ransome's "business hours" were from noon to 2:00 p.m. It is likely that he spent most of the day working for customers or pounding the pavement, hustling up business and leaving his card with prospective clients. He would then probably return to the shop to meet with people or quickly wolf down his lunch. He had no residential listing, so he probably slept in a backroom of the premises. Ransome was probably too busy to do much research during the first few years of his business, but he seemed to have been doing well enough by the early 1880s to spare himself enough money and time to begin experimenting with concrete.

These were productive years for Ransome. While concrete construction was still rare in California, he noticed that when it was used, cracking usually occurred afterward. This was due to slight shrinkage of the concrete as it set. The previous solution to this phenomenon was to simply slop more concrete into and over the crack, hardly an elegant solution. He solved the problem by patenting a process by which expansion joints were incorporated into the design[17]—now an almost universal practice.

Around 1882, Ransome began switching from his father's cement to the Portland product. By this time, the cost of Portland cement had plummeted, as it was now being produced locally. Very locally: one of the state's largest producers, the California Portland Cement Company, was located on Beale Street, just a few blocks away from Ransome's business. However, this cement company had been in business several years before Ransome made the switch, so why did he wait so long? Although Ransome does not enlighten us, it is probably because his father's formula did not work well with metal reinforcement, since calcium chloride also accelerates iron corrosion. And this would not do, for one of Ernest Ransome's many contributions to the concrete industry was the invention of the modern reinforcement bar: the "rebar."

Even by the 1880s, reinforced concrete buildings were exceptionally uncommon, and the few isolated examples used iron mesh or bands. The latter, often called "barrel bands," were produced by the millions of feet each year for binding the countless wooden casks that were then used for shipping and storing everything from nails to wine. In short, iron barrel bands were used because they were cheap and readily available. Ransome felt that something better was needed, for the thin barrel bands bent easily under a load, and their flat surfaces were hardly ideal for securing concrete. He began doing experiments using two-inch-thick square rods, no doubt obtained from the Pacific Rolling Mill Company, a nearby ironworks that produced iron rods and cables.

During this time, most major cities in California were changing their wooden sidewalks over to more durable ones made of either mortared stone or concrete. In the beginning, most of these were being privately replaced by owners of the homes or businesses adjoining them. In 1883, word of Ransome's reinforced paving panels made the rounds, and the prominent architect George W. Percy of the firm Percy and Hamilton approached Ransome to install a sidewalk using his panels in front of the Masonic Hall he was building in Stockton, a city 80 miles (*ca.* 128 km) east of San Francisco. It was during this commission that Ransome began twisting the iron bars to better grip the concrete on all sides and throughout its length. To twist the rod, he attached it in some manner to his steam-powered cement mixer (the details are sketchy), which he then turned on for a few moments. He discovered that this "cold twisting" of the bar also gave it additional tensile strength. In 1884, he patented the "Ransome system" of concrete construction using his reinforcement bar. His particular design would be widely used in reinforced concrete construction for the next thirty years, and it can be found in many ruins of old concrete buildings dating from this period.[18]

Ransome showed Percy the superiority of his system by applying heavy loads to a beam made of concrete using his reinforcement bars. Percy had long been intrigued by the potential of reinforced concrete construction, especially a method employed by a friend, Peter Jackson. Jackson, who had been inspired by Thaddeus Hyatt's work, had developed a reinforcement method that used parallel lengths of thin iron cables held in place by small tie beams.[19] The latter were simply barrel bands with holes drilled in them for the cables to pass through. Percy felt that the tie beams were a possible weak point that could lead to shearing action under load stress. On the other hand, Ransome's system using thick rebar was much simpler and demonstrably better. Ransome convinced Percy that his reinforced concrete could be used not only to make sidewalks but also floors, walls, ceilings, and almost anything then being done in wood, masonry, or iron. Percy became a convert to reinforced concrete, though initially only in regard to floor and ceiling work. It was a major break for Ransome, and the two men would soon collaborate on several projects.

Still, it would be an uphill battle, as Ransome explained in a book published twenty-eight years later:

> The introduction of the twisted iron was no easy matter, and when I
> presented my new invention to the technical society of California, I was

simply laughed down, the consensus of opinion being that I injured the iron. One gentleman kindly suggested that if I did not twist my iron so much I might not injure it so seriously . . .

But all this criticism led to exhaustive tests, and when the professors found that my samples stood up better than the plain bars, one even went so far as to suggest that I had doctored my samples. This led me to twist half of each test rod only, and the superior strength of the cold twisted iron was finally admitted, and in due time, when steel became common, even better results were had with cold twisted steel.[20]

The organization that Ransome referred to was the Technical Society of the Pacific Coast, of which George Percy was a founding member. Percy no doubt encouraged Ransome to ignore such criticisms and helped him organize testing procedures whose results would finally convince society members of the superiority of his reinforcement system over contending methods.

Ransome's first major project that fully utilized his reinforced concrete system was the "fireproof" warehouse for the Arctic Oil Company Works in San Francisco (1884). It was built to replace an older wood frame building. It was the first large commercial structure built of reinforced concrete and would help remove doubts about the viability of material in constructing major buildings. Two more buildings followed. One was the modest Alvord Lake Bridge (begun in 1886 and finished in 1887) in the city's Golden Gate Park. While it is usually cited as the first reinforced concrete bridge, it is actually an arched pedestrian tunnel under the park's main thoroughfare, Kezar Drive, now Dr. Martin Luther King Drive. (The first true reinforced concrete bridge was the one built in 1894 by the French concrete pioneer François Hennebique in Wiggen, Switzerland.) Nevertheless, the Alvord Lake Bridge is easily the world's oldest *surviving* reinforced concrete structure using modern rebar and has been designated a civil engineering landmark by the American Society of Civil Engineers. Ransome received the commissions for these two projects before his collaboration with Percy. The Arctic Oil Works may have been of his own design, while the design of the Alvord Lake Bridge is attributed to John Hays McLaren, the horticulturist and "father" of Golden Gate Park.

Three projects that Ransome and Percy did collaborate on were the Bourn and Wise winery building in St. Helena (1888), the California Academy of Sciences display hall and offices in San Francisco (1889), and the Sweeney Observatory in Golden Gate Park (1891). All were designed by Percy, with Ransome performing the concrete construction work. For the first two, Percy hired Ransome to build only the reinforced concrete floors used in both buildings. The third structure was more ambitious in its use of reinforced concrete: the remarkable Sweeney Observatory built on Golden Gate Park's Strawberry Hill. The observatory was a beautiful structure made entirely of reinforced concrete, the first such designed by Percy, who had previously used it mostly for floor and ceiling work. The observatory, more a viewing platform than an astronomical observatory, was built with money donated by the eccentric San Francisco millionaire Thomas Sweeney. A winding gravel path led pedestrians or horse-drawn carriages though a magnificent castellated entrance to a one-hundred-by-seventy-five-foot courtyard arranged like a horseshoe. Flanking the portal were two towers holding spiral staircases leading to the viewing platform above. From the platform, much of the park could be surveyed, as well as the breaking waves of the Pacific Ocean to the west. Beneath the portal was a reflecting pool. The observatory's portal, mirrored by the reflecting pool below, was a common subject for artists and photographers. Although built of reinforced concrete, the observatory was cast to resemble a sandstone masonry structure. To enhance the similitude, the concrete was lightly tinged with red (probably by adding iron oxide to the mix). The observatory became so popular that a second story was added the following year that featured large glass windows to shield visitors from the occasional strong winds blowing off the Pacific. The Sweeney Observatory was featured on many postcards and in tour guides of the city.[21]

Another collaboration between the two men was the Girls' Dormitory (1891) at Stanford University near Palo Alto, California. Since the dormitory needed to be built quickly, Percy suggested that it be constructed of reinforced concrete. The suggestion was accepted. Ransome went to work, closely following his friend's blueprints. The large dormitory, Roble Hall, was completed in seven months. Three years later, the two men collaborated again on another project at the university: the Leland Stanford Junior Museum of Art (now the Iris & B. Gerald Cantor Center for Visual Arts).

By the late 1880s, Ransome had become a very busy man, but he needed capital to finance a nationwide expansion of his construction company and also to cover the tooling and manufacturing costs to produce the cement mixer he had designed and recently patented. In 1889, he formed a partnership with Francis Marion Smith. Smith was a very wealthy man who had cornered the borax market a few years earlier. (Smith's brand, "Twenty-Mule Team Borax," pitched by the actor Ronald Reagan when he hosted a 1950s television show, is still sold today.) It was a wise move on both men's part: Ransome would use Smith's seed money to make the new company the nation's leading concrete construction firm, as well as to produce his patented concrete mixers that would soon dominate the industry; while Smith, whose borax company would later be taken over by creditors when he became financially overextended, could still die a wealthy man, thanks in large part to the success of the Ransome and Smith Company.

One of Ransome and Smith's early efforts was, appropriately enough, building the Pacific Coast Borax Company's refinery in Alameda, California (1893). It was the second major commercial structure to be built mostly of reinforced concrete by Ransome. Smith assigned Ransome to build another borax refinery in Bayonne, New Jersey (1897). The latter building received much publicity in the industry press and was frequently pointed to as proof that reinforced concrete could tackle almost any construction task.

The use of reinforced concrete in building construction grew steadily over the next few years. Enumerating all of the many projects undertaken by Ransome and Smith goes beyond the scope of this book. However, one building incorporating Ransome's patented methods, mixers, and rebar would demonstrate more than any other structure in the world that a new era had dawned in the construction industry: the world's first reinforced concrete skyscraper.

THE INGALLS BUILDING

The first skyscraper was the Monadnock Building in Chicago, Illinois. It was built in 1891 by the architectural firm Burnham & Root and, upon its completion, could boast a number of "firsts": the tallest brick masonry structure in the world, the tallest commercial building in existence at the time, and the first to use aluminum (for its staircases) on a large scale. It had 17 stories and rose 214 feet high. It dazzled everyone who saw it. The term *skyscraper*, the name for the tallest sail on clipper ships (also called a *moonraker*), would soon become the noun denoting any especially tall office building. The Monadnock ushered in an era of competition among architects and building companies for designing and constructing tall, taller, and *tallest* buildings that continues to this day.

The Monadnock Building was both pioneering and archaic: despite its impressive dimensions, it was also one of the last great office buildings to be built of brick masonry; the final flowering of a species doomed by the introduction of cheap steel. The improved efficiencies in steel production introduced by Andrew Carnegie, plus the recent discovery of the vast iron ore deposits at the Mesabi Range in Minnesota, brought the price of steel down to such a point that it now became economically practical to construct buildings using steel frames. And, because steel's strength was far greater than that of masonry—the latter's weight-bearing abilities were restricted to compressive loads—one could build even *taller* buildings than the Monadnock skyscraper.

Another important consequence of the steep drop in the price of steel was that stronger rebar could now be manufactured with the tougher alloy. Thus the strength of reinforced concrete increased substantially as well. Ransome quickly began making his rebar of steel, as did everyone else who had patented steel-reinforcement schemes (there were competing systems, but none survived for very long). The cost of Portland cement was also going down, thanks in large part to the Keystone Portland Cement Company in Coplay, Pennsylvania, which built the first American rotary kiln for this purpose in 1892.

New converts to reinforced concrete construction were being made

FIGURE 23. Two early pioneers in reinforced concrete construction pose together in an automobile, circa 1905: Thomas Edison (*left*) and Ernest Ransome (*right*). Edison once owned the world's largest Portland cement plant. Ransome invented modern rebar and refined the tools and methods of reinforced concrete construction.

almost every day. Especially attractive were its economic benefits. Despite the drop in the price of steel, reinforced concrete construction was still less expensive than the steel-frame method. However, most people in the concrete industry recognized that one important hurdle remained: the construction of a major edifice like the Monadnock Building to demonstrate the strength and cost benefits of the material. At the time that the Monadnock Building had been constructed, the tallest reinforced concrete structures were only four stories tall, all of them constructed by Ransome. A skyscraper made of reinforced concrete would also demonstrate that claims made by representatives of the trade unions, such as the allegation that the

FIGURE 24. The Ingalls Building, the world's first reinforced concrete skyscraper.

FIGURE 25. Hoover Dam in Arizona. So vast a quantity of concrete was used in its construction that some of the material is still curing—more than seventy-five years after the dam was dedicated in 1935!

material was not strong enough to support large loads and thus posed a public danger, would be proved both false and self-serving.

The opportunity for constructing a reinforced concrete skyscraper came in 1901, when railroad magnet Melville Ezra Ingalls decided to build an office building in Cincinnati, Ohio. He chose the architectural firm Anderson and Eisner to design the structure. W. P. Anderson took on the task and convinced Ingalls to construct the building using reinforced concrete instead of steel frame, the method then universally employed to put up such towering edifices. That was the easy part. The hard part was convincing the Cincinnati Planning Commission, whose members were no doubt being heavily lobbied by the trade unions to deny the building a permit. The process dragged on for months. In the meantime, Anderson and Ingalls went ahead with the project, securing the vast amount of Portland cement and rebar needed to build the skyscraper. Almost up to the day that construction work commenced on the building, the planning commission was still refusing to grant a permit. The reason they finally caved in at the last minute is not known, but it is likely because Ingalls's powerful political connections—and money—finally intervened to bring the matter to a satisfactory conclusion. Satisfactory to Ingalls, that is. The tens of thousands of men employed as bricklayers around the country must have realized that the concrete skyscraper represented a sea change in the building industry, and the only lucky ones among their ranks were those close to retirement age. The era of masonry construction of major buildings, which stretched back millennia, was slowly coming to an end.

Another beneficiary was the San Francisco firm of Ransome and Smith. Anderson had chosen the Ferro-Concrete Construction Company to serve as the contractors for the mammoth undertaking. The company had already licensed Ransome's patented reinforced concrete construction methods—including his rebar—and they also used Ransome cement mixers. It was the best kind of unpaid advertising imaginable—assuming the Ingalls Building didn't collapse, of course.

Reinforced concrete buildings were still rare. To put this in perspective, the amount of concrete cement used for the Ingalls Building represented one-half of 1 percent of all cement used in the United States. Considering that the thousands of homes and business buildings then being constructed were using concrete for their foundations, the amount leftover for monolithic concrete structures was meager indeed.

When the Ingalls Building was completed, it appeared no different from the steel-frame skyscrapers being constructed in most major cities of the United States. With sixteen stories it rose to 210 ft (*ca.* 54 m), just under the height of the Monadnock Building, and several stories fewer than the steel-frame skyscrapers being built. Nevertheless, it was more than twice the height of any reinforced concrete structure then in existence. (The previous record holder was the six-story Weaver Building in Swansea, Wales, built by François Hennebique in 1897.) The Ingalls Building measured 50 ft by 100 ft (*ca.* 14.25 m by 31.5 m), and, like most office buildings of its time, the first two stories were dedicated to business storefronts. Legend has it that so many people were certain it would fall, a local newspaper stationed a photographer near the building to capture its collapse on film. The story is probably apocryphal.

The Ingalls Building was featured in news stories around the globe and was, of course, given special attention in the construction and architectural journals. It is also one of the very few landmark reinforced concrete structures of the period that is still with us. Although Ernest Ransome did not build the skyscraper, his patented methods and equipment were used in its construction, and so vindicated his pioneering vision as nothing else could. Ransome continued designing and building reinforced concrete structures. The factory building he constructed for the United Shoe Company in Beverley, Massachusetts, in 1906, was considered the most advanced of its kind in the world and reportedly exerted a strong influence on the young German architect Walter Gropius. Ransome did more to win acceptance for reinforced concrete construction than any other single individual. When he died in 1917, tributes to his achievements poured in from around the world. Today, he is largely forgotten.

Although trade union guilds and unions continued to oppose reinforced concrete construction, their opposition largely vanished in the wake of the 1906 earthquake and fire in Northern California. Concrete advocates pointed to the resiliency of concrete structures to both the tremor and the inferno that arose in its wake.[22]

Ernest Ransome was also one of the last of the self-educated men who made significant contributions in the early days of reinforced concrete construction. Two of Ransome's contemporaries, also largely self-educated men, would play a role in concrete's story as well. One would use reinforced concrete with spectacular success; while the other probably should have stayed away from the building material and continue doing what he did best: creating electromechanical wonders.

Chapter 7

THE WIZARD AND THE ARCHITECT

THE WIZARD OF MENLO PARK

On a hot August evening in 1906, America's premier inventor, Thomas Edison, stood up to give a short, impromptu speech at a dinner reception held in his honor in New York City. Edison was asked what new marvel he would present to the world. After a moment's hesitation, he said, "Concrete houses." After all, he told the rapt audience, concrete was fireproof, termite-proof, immune to mildew and dry rot, and would stand up well to most natural disasters. The very recent San Francisco earthquake was still fresh in everyone's mind, and concrete structures had reportedly performed better during the disaster than those built of wood or masonry. Although houses had been constructed of reinforced concrete for several decades (e.g., Ward's Castle), they had usually been expensive, custom-built affairs. Now Edison was asserting that roomy concrete houses could be built on an industrial scale so that the cost of each would be "about one thousand dollars," less than the price for most modest homes in 1906.[1]

Edison's pronouncement made headlines the following day. Edison seemed to be earnest, for he was then in the final stages of constructing the world's largest concrete cement plant in Stewartsville, New Jersey.

Edison's involvement in the cement business came about accidentally. Long fascinated by ore-refining equipment, he had patented an iron-ore processor in 1881 that used powerful electromagnets to separate the higher-yield iron ores from the lower-grade variety. Rich deposits in Pennsylvania, New York, and New Jersey were being depleted toward the end of the nineteenth century, and the price of steel had risen as a result. Not unreasonably, Edison believed that his new process would make him a fortune. Shortly after filing his patent, Edison formed the Edison Ore Mining Company. The company purchased ore-bearing land in northern New Jersey, and a couple of years later it opened its processing plant near the mines and town of Ogdensburg. From the beginning, things did not go as planned. Edison's ore separator did not work as well in reality as it did on paper and frequently broke down. Undaunted by repeated failures and soaring costs, Edison continued to improve the processes and design new equipment. It wasn't long before the Ogdensburg operation had morphed into an investment pit that sucked in money like a black hole does stars. By the end of the century, Edison had fixed most of the bugs, and the ore separators were finally refining quality iron ore from poorer stock. Unfortunately, the fixes had come too late: vast iron-ore deposits had been discovered on the Mesabi Range in Minnesota that were so pure they required little or no processing; steam shovels just scooped up the ore and dumped it into open-top railroad freight wagons that transported it to the eastern steel furnaces. As a result, the price of quality iron ore dropped precipitously, and there was simply no way that Edison could provide the same product at equivalent prices. The Wizard—and the extraordinarily patient stockholders of the Edison Ore Mining Company—lost millions.

Nevertheless, Edison still had the ability to turn lemons into lemonade. As he was recovering from his ore-processing fiasco, he watched the meteoric rise of reinforced concrete and realized that his huge ore crushers were perfect for reducing limestone to the consistency needed to make cement. Coincidentally, there was a limestone quarry just a few miles from Ogdensburg, near Stewartsville. It was a perfect match. The Wizard bought the quarry and formed the Edison Portland Cement Company in 1902. The heavy ore equipment was disassembled, packed up, and moved to the new site. Since Edison never did anything by half measures, he also designed and built a one-hundred-fifty-foot-long rotary kiln—a colossus that was almost double the size of the largest rotary kilns then in existence. By August 1906, when Edison announced his intent to produce cheap concrete houses, his massive Stewartsville plant was close to completion.

To this day, we do not know whether Edison was really serious about his proposal in 1906 to manufacture concrete houses. He was certainly serious about producing Portland cement, but the banquet announcement was probably nothing more than a trial balloon sent aloft to judge public reaction to the idea. He certainly had not lifted a finger to advance such a project, nor have we been able to find any plans he might have made before his dinner speech. We do know that he *did* take the project seriously after thousands of letters poured in from excited would-be home buyers. Even Archduke Franz Ferdinand, heir to the Austro-Hungarian Empire, sent a letter to the Wizard eagerly requesting details. Edison, chastened by his previous failures, was reluctant to invest in anything that did not have an existing market. The positive response to his announcement signified to him that the market did exist. Whatever his thoughts may have been beforehand, Edison was now determined to manufacture inexpensive reinforced concrete houses—and lots of them.

Even as Edison continued to push concrete houses publicly, he was too busy with other inventions to devote any of his immediate time to the project. Initial work did not begin until 1908, when he constructed a model of his concrete house based on a design he had commissioned from the New York architectural firm Mann & MacNeille. Edison was determined that his homes not look plain and told the firm's architects to make the structure as attractive as possible. The firm came up with a two-story design described as being in the style of "Francis I." In reality, the architecture would be best described as a version of the "craftsman house" that was then so popular. The model home impressed no one, and Edison's employees jokingly referred to it as the "chicken coop."[2]

That same year, Edison began conducting his first serious experiments in reinforced concrete construction. Large wooden molds were created, and the pouring and setting properties of Portland cement were then examined. Apparently, Edison believed that concrete could be poured into molds designed for houses in much the same way as it was done for much larger structures, such as the Ingalls Building. However, molds for skyscrapers are obviously far larger than those for houses, and so pouring the concrete for larger structures is a relatively easy operation. In the case of the Ingalls

Building, the pour was performed slowly, and each layer of concrete was allowed to set before more was added. However, Edison wanted to create his two-story house, both frame and walls, in a single pour. This presented problems: the mold dimensions for his house were considerably smaller and thinner, making it difficult for the lumpy concrete to wrap around the interior rebar and flow into all the cavities. Modern cement also has a tendency to set rapidly, usually within an hour, unless a setting retardant (usually gypsum) is added to the mix. Since Edison estimated that he would need at least six hours to complete a pour for each house, this presented some difficulties. He also discovered another problem with the single-pour method: the heavier aggregate tended to drift toward the bottom of the molds if it flowed beyond a certain distance, creating an unstable mix. Edison was also not happy with the wooden forms that served as the molds: they tended to warp and left unsightly marks on the concrete's surface after they were removed.

If Edison had read the existing technical papers or talked to people in the concrete industry, he would have discovered that increasing the amount of gypsum in the cement would have prolonged the setting period and given him the extra hours he needed. He would have also learned how to fix the problem of aggregate settling in a long flow—gravity versus viscosity— by simply switching to two different flow points. Also, others had solved the wooden mold problems. To prevent warping, the current custom was to securely brace the wooden mold with crosshatched boards. Sanding and coating the surface where the concrete came into contact with the mold also eliminated the aesthetic imperfections. Yet, instead of following these established and field-proven construction practices, Edison decided to invent a new formula for making the cement and sought a better material for creating the molds.

If these problems all had ready solutions, why didn't the Wizard make use of them? Edison often refused to learn from others and frequently had to discover things for himself. If pressed, Edison might have said that he was always looking for a better solution, but in this case, his refusal to seek help from others was due to stubbornness. While stubbornness is often useful— as it was in Edison's exhaustive search to find the right material for his electric lightbulb filaments—it sometimes blinded him to relevant facts or possible remedies. For example, Edison was so absorbed in fixing the problems of his ore-refining equipment in Ogdensburg that he ignored the news coming out of Minnesota about the rich iron deposits discovered there and the implications this would have for his mining operation.

As the Wizard hunkered down to solve the already-solved problems of his concrete houses, it seemed to some like another fiasco in the making.

A couple of years passed, with more experiments, formulas, and processes being tried. And all the while, Edison continued to promote the concrete houses that he would "shortly" release to the world. He also added a humanitarian touch by proclaiming that his concrete houses would represent "the salvation of the slum dweller," and that tenement housing would soon be "a thing of the past." Edison also announced that he would make no profit in the venture; he would freely license the technology to anyone who agreed that the majority of the concrete houses so constructed be reserved for the working classes and that the profits realized not exceed 10 percent. By this time, Edison had patented a new concrete formula with bentonite clay that purportedly made the casting process for houses easier. (In truth, most of the improvements had probably been achieved when Edison reduced the size of the aggregate and used more gypsum to lengthen the setting time.) Licensing issues aside, if the Wizard could convince Americans to switch from wood to concrete houses, he would still stand to make a tidy profit with his newly patented cement.

One of Edison's neighbors, Frank D. Lambie, an expert in designing assembly line machinery, was so excited by the idea of concrete houses that he offered to fabricate the molds at his own expense. Lambie knew that many prominent industrialists were looking for decent yet inexpensive housing for their workers, and Edison's concrete homes seemed to provide the perfect solution. Unfortunately, the Wizard decided that the molds for his houses needed to be constructed of cast iron, not wood. Even though the switch to cast iron would drive up the mold's costs and weight considerably, Lambie evidently decided that the scheme would still be practical if he could obtain a large enough construction contract.

It took Lambie most of a year—and his savings—to create the huge cast-iron molds for the Mann & MacNeille–designed house. By the time he was finished, the mold set contained between 2,300 and 2,500 parts (the final number depending on the options exercised by the buyer) and weighed more than 450,000 pounds—this, at a time when heavy loads could be transported only by rail or large wagons drawn by teams of Clydesdale horses.

As a working exercise, Lambie built two of the houses in Montclair, New Jersey, near the Wizard's Menlo Park facility. The construction work did not go smoothly; the operation took weeks instead of days, and Lambie was forced to fill the molds one story at a time instead of through Edison's "single-pour-for-both-stories" procedure. After discussing his problems with Edison, the latter decided that the Mann & MacNeille design was too complicated, and he turned to his own company draftsmen to draw up plans for a smaller house with a simpler building plan.[3] We do not know what Lambie's reaction was to the sudden realization that his expensive and elaborately machined cast-iron molds were now probably worthless, but it could not have been a happy one.

Edison's new house design also proved disappointing. It was a very simple and unadorned two-story affair that was little more than an upright rectangular box with windows. Edison's houses would be plain after all. The first story consisted of a living room and kitchen, with a small cellar below. The second story included two bedrooms and the home's only bathroom. On the practical side, the new house would be easier to build, and its cast-iron molds would consist of just 500 parts and weigh *only* 250,000 pounds. (Edison still refused to reconsider employing wooden forms, though everyone else in the concrete industry was using them effectively and without problems.) Once the new cast-iron molds were ready, Lambie and his crew built a house of the new design in South Orange, New Jersey. Aside from a few minor hiccups, the concrete castings went well.

By 1911, the public was losing interest in Edison's concrete houses. Five years had passed, and, aside from a few model homes, the project had yet to be realized on an appreciable scale. Still, the Wizard remained an inventor as well as a major cement producer (his product was now being used throughout New York City), and he therefore felt that he had to come up with *something* soon that was made of concrete. He decided that it would be furniture, which could be produced relatively quickly and required far simpler molds, and using heavy aggregates could be skipped entirely. In December 1911, Edison unveiled a concrete phonograph cabinet before members of the American Society of Mechanical Engineers. After extolling the allegedly superior acoustics of concrete, he went on to describe his pre-IKEA Arcadia:

> I'm going to have concrete furniture on the market in the near future that will make it possible for the laboring man to put furniture in his home more artistic and more durable than is now to be found in the palatial residences in Paris or along the Rhine.
>
> And will it be cheap? Of course it will. If I couldn't put out my concrete furniture cheaper than the oak [furniture] that comes from Grand Rapids I wouldn't go into the business. If a newlywed, say, now starts out with $150 worth of furniture on the installment plan I feel confident that we can give him more artistic and more durable furniture for $200. I'll also be able to put out a whole bedroom set for five or six dollars.[4]

When quizzed about the heavy weight of concrete, Edison made the incredible claim that his concrete furniture would be only "33.3 percent heavier" than its oak counterparts, although he could probably reduce the difference to "25 percent."[5] Edison's claim that his concrete furniture would be only 25 percent heavier was rather amazing, considering that a cubic square foot (ca. .0283 cubic m) of concrete tips the scales at 150 pounds (ca. 68 kg), while the same volume of the oak is 59 lbs (ca. 26.7 kg). Although Edison told reporters that the furniture would be built using a proprietary "concrete foam," it could not have been radically lighter than the standard mix, since air entrainment only marginally reduces concrete's weight, usually by 3 to 9 percent.

To prove the durability of his concrete furniture, the Wizard packed up his phonograph cabinet and sent it on a round-trip journey that included stops at New Orleans and Chicago before finally returning home to the Big Apple, where the cabinet would be unveiled at a cement industry show. Edison affixed signs on the shipping crate that read "Please drop and abuse this package." However, after the crate had returned to New York, Edison canceled a scheduled press conference to show off the cabinet's resilience to abuse. We do not know what transpired, but it seems probable that the package had arrived in a damaged state—the transport handlers may have accepted Edison's dare with relish. Edison no longer spoke of the advantages of concrete furniture and stopped all efforts to produce it. A thousand people in the cement industry could have told Edison that dropping anything made of concrete was a bad idea, but Edison, as usual, had to first learn it for himself.

The furniture fiasco apparently also dampened Edison's enthusiasm for concrete houses, for he now divorced himself from the project, declaring that he had already "shown the way," and it was up to others to "fulfill the promise." Poor Frank Lambie was in a quandary. Having invested so much of his fortune in the concrete molds, he could hardly have had much leftover to fulfill the promise. (Lambie had already moved into one of the two original Mann & MacNeille houses to save money.)

Shaking off his despondency, Lambie became a salesman, with all the energy, aggression, and desperation emblematic of a struggling one-man company. He began approaching the titans of industry to promote his low-cost concrete housing for their employees. Lambie pressed Henry Ford especially hard when he learned in the summer of 1914 that the carmaker was considering building two thousand houses for his autoworkers. Lambie sent blueprints, photographs, and brochures of his houses to Ford, and in his accompanying letters he emphasized his friendship with Thomas Edison, whom he knew Ford admired. Ford considered Lambie's proposal and, as was typical of him, took his time. After months had passed with no answer from Ford, Lambie put forward an extremely attractive price for a single-family house: $525 each, "about the price of one of your automobiles." Fancier houses for middle management could be had for $1,025, and spacious villas for upper-echelon executives would be only a couple thousand more (Lambie had apparently not scrapped the molds for the Mann & MacNeille house).

The low prices caught the thrifty automaker's attention. Ford sent the blueprints for the $1,025 home to an architect friend, Albert Kahn, for comment. After perusing the blueprints, Kahn told Ford that such a house could not be built for less than $1,500. When Ford passed Kahn's remarks on to Lambie, the latter wired back that the Pittsburgh Crucible Steel Company of Midland, Pennsylvania, was pouring that very design for $200 less than the original quoted price of $1,025. By the early summer of 1915, Ford began negotiating with Lambie in earnest and even released a press bulletin announcing his intention to build two thousand concrete homes for his employees.[6]

Lambie, swept up by his success, overreached himself: he now attempted to convince Ford to build *one million* concrete homes. All Ford had to do was invest one hundred million dollars (roughly two billion in today's inflation-adjusted dollars) in the project, for which he would eventually reap *three hundred million dollars* in profits. No one knew more about the savings that could be accrued through mass production than Henry Ford, and he could usually perform a rough cost-benefit analysis in his head without resorting to a calculator or slide rule. Ford did not need to consult Kahn a second time to know that Lambie's revenue claims for his grandiose million-home project were nonsense. It was probably at this point that the carmaker also began to question Lambie's projected costs for the two thousand concrete homes planned for his employees. Ford withdrew from the housing deal just before the contract was to be signed, and Lambie was back to square one.[7]

Yet Lambie persevered, barely sustaining himself with the thin profits earned through small concrete construction projects. In late 1916, he was able to convince Charles Ingersoll, who had made millions with his reliable one-dollar pocket watch, to invest in the Lambie Concrete House Corporation. The two men decided on a pilot project in the growing community of Union, New Jersey. Forty houses were planned, the first eleven of which were built in late July 1917 on a street named Ingersoll Terrace, in honor of Lambie's new partner.

The eleven homes were constructed with little trouble. Lambie had learned much in the past several years and by now probably knew more about building concrete houses than anyone else. Though Lambie made sure that the project on Ingersoll Terrace received maximum publicity, only a few reporters showed up on the day they were unveiled. Concrete houses were now "old news." Worse, Lambie and Ingersoll had difficulties finding

FIGURE 26. Thomas Edison standing next to a model of his house.
It was too complicated to build, so he settled for a far simpler design.

buyers for them. Although somewhat plain, the homes were certainly no uglier than others in their price range. Why were the houses not selling?

The reason Lambie and Ingersoll had trouble moving the houses was likely due to Edison: for years the Wizard had trumpeted concrete homes as the salvation of the slum dweller—and who wanted to be known as a rescued slum dweller? Edison's insistence on restricting the profit margins probably rankled contractors as well. In effect, Edison had dampened most people's

FIGURE 27. Edison sitting between two of his phonograph cabinets (the concrete one is on the right). The problems he encountered with his concrete creations prompted him to give up on the idea of houses and furniture made of the material.

enthusiasm for either purchasing or building concrete houses. Ingersoll quickly lost interest in the Union project and withdrew from his partnership with Lambie. The remaining twenty-nine homes were never built.

Frank Lambie would continue building concrete houses, but on a scale much smaller than he had originally envisioned. By 1920, Lambie was selling houses of the simplified Edison design for three thousand dollars each, more than the cost of a comparable wood-frame house. He had apparently given up the dream of low-cost concrete homes through mass production and instead emphasized the advantages of concrete over wood construction. Had Edison done the same and skipped his pitch about saving the slum dweller, most of our houses today might well have been built of Portland cement instead of pine and plaster.

THE ARCHITECT OF OAK PARK

When Thomas Edison announced his intention to build concrete houses in 1906, a thirty-eight-year-old architect in Oak Park, Illinois, had already been designing extraordinary buildings constructed of the material. And by 1917, when Frank Lambie was finally putting up his Edison homes in Union, New Jersey, this same architect was using the properties of reinforced concrete to create buildings, the likes of which had never been seen before. The architect's name was Frank Lloyd Wright.

Until Wright came along, reinforced concrete structures appeared no different from their wood or masonry counterparts, for they were essentially modeled on them. Wright was the first architect since Roman times to recognize that concrete allowed for the creation of completely new forms. Whereas the Romans used unreinforced concrete to create soaring ceiling vaults and domes, Wright employed the great tensile strength of reinforced concrete to build amazing cantilevered structures. He would rewrite the rules of structural design, and his reputation and work would be forever tied to his imaginative use of concrete. As a result, the visual landscape of our world would never look the same.

Wright's contributions deserve a closer examination than those of most other architects of his period. Besides his original and pioneering use of concrete, and the enormous influence his work has had on modern architecture, most of his buildings have also withstood the test of time, at least from the standpoint of aesthetics, far better than the creations of most of his contemporaries, many of which now appear dated, ugly, or both. And, contrary to most geniuses, who usually produce their best work in youth or early middle age, Wright's considerable gifts seem to have blossomed more with each passing decade, reaching full flower in his senior years.

Frank Lincoln Wright was born in Greenfield, Wisconsin, in 1867. His father, William Carey Wright, was an itinerant minister and music teacher with three children who was widowed shortly before meeting Wright's mother, Anna Lloyd-Jones. Anna was a former schoolteacher who belonged to a close-knit Welsh family. Wright's parents quarreled frequently. William's mood swings suggest bipolar disorder, and Anna may have suffered from the same affliction. It was not a happy household. Frank frequently escaped the hostile home environment by withdrawing into a more perfect imaginary world. Bipolar disorder is often a hereditary disease, and Frank's parents may have passed it on to him. It would explain both his brilliance and occasional social disconnectedness.[8] They divorced several years later, while Frank was still in his teens.

Wright strongly identified with his mother's branch of the family and would later change his middle name from Lincoln to Lloyd in honor of them. The Lloyd-Joneses had converted to the Unitarian branch of Protestantism over a century earlier in the old country and had remained true to both their faith and a long tradition of progressive politics (they were fervent foes of slavery and early champions of women's emancipation).

In his biography, Wright asserts that he "had no choice" but to take up the profession that would later make him famous. His loving, though strong-willed, mother had declared that her only son would be an architect while the lad was still an infant. To help direct his young mind toward that goal, she put up prints of the great cathedrals on the walls around his crib and encouraged him to play with elaborate building blocks invented by the German educator Friedrich Fröbel.[9] Wright never questioned his mother's career choice for him, for it proved a perfect fit for his innate talents as an artist, draftsman, and dreamer. He would take great pleasure in designing unique and beautiful buildings until the last days of his very long life.

Wright also possessed an extraordinarily rare gift: an eidetic imagination, the ability to visualize in his mind a complex three-dimensional object in all its details and then view it correctly from all angles—a remarkable attribute he shared with his near-contemporary, the inventor Nikola Tesla. An incredible but well-authenticated story illustrates this exceptional faculty. One day, when an impatient client had called Wright about a long-overdue home design, the architect told him that the plans were ready and that he could come over to look at them. In truth, aside from visualizing the plans in his mind, Wright had yet to do any work on the project. Knowing that it would take his client several hours to reach his office, Wright called in his staff, pulled out three large sheets of drafting paper—one for each story of the large residence—and proceeded to draw the blueprints, explaining to his staff the purpose of each feature as he worked. He drew so fast that his employees had to constantly sharpen pencils as he wore down each to its wooden nub. Wright finished his design within a couple of hours, a feat that most would have deemed impossible were it not witnessed by a half dozen people.[10] The house was the famous Fallingwater, consistently ranked by many architects as the most beautiful home designed in the twentieth century.

In this youth, Wright's lack of academic achievement seemed to cast doubts on his prospects. He never finished high school, he attended college for only a couple of semesters, and he ignored subjects that bored him or had no relevance to architecture. With the possible exception of his adoring mother, few members of Wright's family or small circle of early friends recognized his special qualities. Many suspected that he would, like his father, wind up a shiftless charmer. Shiftless he was not. Armed with nothing more than a few sketches, and all the cheeky courage and boundless determination of youth, Wright went to Chicago and promptly got a job as a draftsman in Joseph Lyman Silsbee's architectural firm. Wright was just twenty years old.

Silsbee was one of Chicago's most successful architects. Though his work generally conformed to the prevailing conservatism of his clients, Silsbee could also design daring buildings when given the freedom to do so, such as Chicago's Lincoln Park Conservatory. To most aspiring architects, the opportunity to work for such a prestigious firm would have seemed like a gift from heaven, but Wright was not like most aspiring architects. A year after being hired, he jumped ship and went to work for an even *more* prestigious firm: Adler and Sullivan, whose chief architect, Louis Sullivan, had been making waves in his profession for almost a decade.

The Great Fire of 1871 had leveled most of Chicago. The rebuilding of America's second-largest city had required the work of not only thousands of carpenters and brick masons but of hundreds of architects as well. Instead of the gradual permeation of new architectural styles to replace the old, a fresh generation of architects would leave their imprint throughout the Windy City. From this new breed would arise the "Chicago School" of architecture, and Louis Sullivan was at the forefront of this new movement.

When Wright came to his firm, Sullivan was at the apex of his career and busy creating a novel and uniquely American architectural form: the skyscraper. The tremendous drop in iron-ore prices that had ruined Edison's mining venture allowed Sullivan to build tall, sturdy buildings utilizing steel frames. It was Sullivan who made the famous remark that "form ever follows function."[11] (That dictum would be taken to extremes by a few later architects who eliminated all ornament from their buildings, insisting that pure functionality was its own beauty.) Some architects would later call Sullivan the "Father of Modernism," but others have argued that it was really Wright's work that represented the first complete break from nineteenth-century stylistic conventions. At the very least, Sullivan was the essential bridge between the old and new forms. Wright could not have chosen a better mentor.

As before, when he had applied for the draftsman's position at Silsbee's firm, Wright brought some of his drawings—now more accomplished—to his interview with Sullivan. Sullivan was struck by the young man's charm, vision, and almost religious reverence for architecture, which Wright saw as the highest expression of all the arts. The young architect's designs were still somewhat derivative, yet Sullivan also detected in them a striking originality. Wright was hired on the spot. Sullivan and Wright had much in common: both men were transcendentalists who had read Emerson and Thoreau and felt a kinship with the German romantics (both revered Goethe's *Sorrows of Young Werther*).[12] Wright always looked back with fondness to his days at Adler and Sullivan and would often refer to Sullivan as his *lieber Meister* (beloved master). He kept in touch with his master long after he became an independent architect, and he would one day write a biography of Sullivan noted for its spirited defense of its subject's work against the criticisms of the "mobocracy."[13]

Wright rose quickly at Sullivan and Adler. Sullivan recognized Wright's inventiveness with house designs and put the young man in charge of all residential work. Wright had a special knack for houses, and the vast majority of his work over the next six decades would be from designing and building homes. His residential work would always provide Wright with a dependable source of income between major projects.

In 1893, Sullivan discovered to his dismay that Wright was accepting private commissions on the side. Wright would later claim that it had been necessary to do some moonlighting in order to support his wife and growing family. (Wright had married Catherine "Kitty" Tobin in 1889 and had two children by this time.) Sullivan regretfully fired Wright.

Wright started his own firm and began working out of his self-designed home in Oak Park, a Chicago suburb. His reputation as a gifted architect grew, as did his client list. The first exhibition of Wright's work was held the following year (1894) at the Chicago Architectural Club, where he received some favorable notices. Soon he had so many commissions that he was forced to hire other architects to help with the workload. It was during this period that he developed his famed "Prairie house," a style that emphasized expanded living space. Living rooms and dining rooms were especially enlarged. To achieve this spaciousness, Wright expanded the horizontal axis of the structure. Flattened cantilevered roofs projecting out from the main building blocked direct sunlight and offered inviting shelter from the elements. The Prairie home was in stark contrast to the then-popular—and unambiguously vertical—Victorian house.

WRIGHT AND CONCRETE

As was typical for his time, Wright's first use of concrete was as a foundation material. Having designed homes for some years, Wright probably knew that most people were not attracted to the material (a fact that Edison and Lambie would learn the hard way). Still, he was fascinated by the potential of reinforced concrete: its tensile strength was ideal for cantilevered designs, and it could be poured into molds to create structural adornments that had the appearance of sculpted stone. Wright probably began creatively experimenting with the material around 1900, but he could not find a way to use reinforced concrete for his early cantilevered houses: a wooden beam more economically addressed the modest load demands of the Prairie home.

The cast-concrete exhibit Wright designed for the Universal Portland

Cement Company and displayed at the 1901 Pan-American Exposition in Buffalo, New York, is the only work by the architect of which no photo, image, or even description survives. Architectural historians have combed through surviving company records (consisting of just three folders held by Indiana University) and thousands of private and commercial photographs from the Pan-American Exposition in hopes of uncovering this "lost" treasure, but all such efforts have so far failed.

We do have a photograph of an exhibit Wright designed for the same company at an industry show held in New York City in 1910, showing two cast-concrete stelae, or building ornaments, and what appears to be a bench in the background. All display the strong Mayan influence that would characterize much of Wright's work in his middle period. Still, so much had changed between 1901 and 1910, both in the architect's personal life and in his artistic endeavors, that many Wright enthusiasts have naturally assumed that the vanished 1901 exhibit would have looked quite different from the 1910 version.

Or would it? The Mayan influence appears shortly after the 1901 Pan-American Exposition in the design of the Robert M. Lamp House (1903) and in Wright's cast-concrete decorative panels in the central court of the Larkin Administration building (1904). Clearly, Wright was exploring the plastic attributes of concrete to create Mesoamerican motifs long before 1910. It is conceivable that Wright simply employed the same molds used for the 1901 exhibit for the 1910 show. Although the Universal Portland Cement Company was a major player in the concrete business (its Indiana Harbor factory would soon surpass Edison's New Jersey operation as "the largest cement plant in the world"), the modest castings at the 1910 show seem more like something a small concrete contractor would cobble together than an exhibit sponsored by one of the world's largest cement manufacturers. It is quite possible that the company asked Wright to design another exhibit for them but allotted only a modest budget for it. Wright agreed and simply used the old 1901 castings; Universal got what it paid for. Unless a photo surfaces from the 1901 show that displays Wright's original exhibit, the matter will remain shrouded in mystery. My guess is that there is no lost treasure, only an old curiosity that had been pulled out of storage and dusted off.

In any case, the years between 1901 and 1910 were Wright's watershed years in regard to his use of concrete. During those years, he would use the material for his first cast ornamentation and, in an odd throwback to classical times, employ ancient Roman concrete wall-building techniques to create some of his most original structures. Most notable of all his achievements during this time was his first monolithic reinforced concrete building.

FIGURE 28. Frank Lloyd Wright was the first architect to explore the ability of reinforced concrete to create entirely new architectural forms.

FIGURE 29. One of Wright's more fanciful creations in concrete: the Mayan Revival Hollyhock House in Los Angeles.

THE UNITY TEMPLE

In late 1905, members of Oak Park's Unitarian congregation approached Wright about designing a new church to replace the one that had been lost to fire a couple of months earlier. Wright's wife had taught Sunday school at the church, and a prominent member of the congregation, Charles E. Roberts, was one of Wright's clients. Roberts liked the architect's work, and he also knew that Wright, while not a regular churchgoer, had grown up in the Unitarian faith. Wright was awarded the commission.

Although monolithic concrete construction was certainly gathering steam by 1905, it was still relativity uncommon. Since Wright's writings preclude any suggestion of outside influences, it is worth examining why the architect chose to build the Unity Temple using this method.

Wright avidly read the industry journals to keep abreast of new developments in building techniques and materials. He knew of the pioneering work being performed with concrete elsewhere, especially in the nearby

FIGURES 30 & 31. Wright's Unity Temple, exterior (*top*) and interior (*bottom*), in Oak Park, Illinois. It is considered by many to be the earliest modern structure.

state of Ohio. As mentioned earlier, the most notable cutting-edge application of concrete up to that time had been the Ingalls Building in Cincinnati, Ohio. Although the skyscraper's design was hardly daring, it did represent the most dramatic leap of faith yet for the material and evidently provided proof to Wright that reinforced concrete had shown itself to be a suitable building material for erecting whole structures.

The building site for the Unity Temple was difficult. The lot was narrow and long, and the quiet country lane that had bordered the first church when it was erected in 1871 had now become a busy street. Sermons were often interrupted by the noise of clanging streetcars and honking automobile horns. The modest budget for the temple was also a challenge: $40,000 (roughly $800,000 in today's inflation-adjusted dollars). This limited outlay included not just the building but the furniture and stained-glass windows that Wright was also expected to design. Many nearby homes had cost more to build. That he was able to accomplish so much, and within such limited means, is another remarkable aspect of the temple and of Wright's resourcefulness.

Wright accepted the commission just a few weeks after returning from Japan, where he had spent some of his time studying the magnificent mausoleums of two Tokugawa shoguns in the city of Nikko. These temple tombs had a *gongen*-style floor plan, a bipartite design that separated the main sanctuary from the worship hall via a kind of loggia called an *ainoma*. Wright liked the arrangement and used a similar floor plan for the Unity Temple, while incorporating a few distinctive touches of his own as well.

The first step was the creation of elaborate forms into which the rebar was placed before the concrete was poured. Careful sanding and bracing of the wood prevented any major blemishes from appearing after the concrete had set. To keep the costs of the molds down, the only concrete ornamentation was a stylized Mayan-like design on the upper half of the exterior pillars. Wright deliberately chose to expose the aggregate of the main walls, which gave their surfaces a pebbly texture. (This was partially obscured after a 1971 renovation that also repaired the aging concrete.) Unlike the Edison homes or the Ingalls Building, the interior concrete casting of the Unity Temple was even more complicated than that of the building's exterior. This is perhaps why the temple took almost three years to construct, instead of the one year originally specified.

Once the concrete had set and the forms were removed, Wright installed the stained-glass windows, painted the interior in attractive pastel colors, and, as a final touch, brought in his custom-designed wooden chairs. The stained-glass windows were exclusively colored with earth tones: natural greens, yellows, and browns. Less successful were the visually appealing but very uncomfortable chairs. (The church would sell off the chairs a few years later—probably at cost—an unwise move, since each now fetches thousands of dollars at auction).

When Wright finished the Unity Temple in 1908, everyone agreed that it was a remarkable achievement, but few would have guessed that the humble church would one day become an icon of twentieth-century architecture.

The temple's exterior had an austere, solemn appearance and provided no hint about the revolutionary nature of the building. One entered the loggia at the street level, but unlike the open *ainoma*, the visitor was required to make a series of right-hand turns to reach the chapel. The circuitous nature of the entrance was reminiscent of prayer mazes and suggested a refuge from the outside world. It also served as a sound barrier to keep street noise to a minimum. Also unlike an *ainoma*, the loggia was divided into two sections: one directed worshipers to the church chapel, while the other led visitors to its community center, Unity House. Since the entrances to each section were on opposite sides of the building, people coming to one function never met those arriving for the other. To further reduce noise in the chapel, Wright eliminated windows on the lower levels. Natural sunlight came through the stained-glass windows in the ceiling and upper-level clerestories. The latter were deeply recessed behind ornamental pillars, further buffeting any unwelcome street clatter. The recessed panels below the ceiling windows were painted gold, which gently diffused the other colors and imparted a relaxing yellow light that gave warmth to the interior. As with his Prairie homes, Wright had taken sharp angular patterns (normally unpleasing to the eye) and arranged them in such an ingenious way as to present a ravishing whole. Although many other architects have tried to imitate this angular technique, none have equaled Wright's mastery. It was sui generis.

Imaginative use of space was also what set the Unity Temple apart from other churches of its day. It was especially evident in the chapel, where worshipers were seated on two elevated reinforced concrete tiers to each side, facing one another across a floor section where the seating was arranged in the standard front-to-pulpit manner. In this way, no member of the congregation was more than forty feet away from the pulpit, and the superb acoustics ensured that the minister's every word was clearly heard.

Since Wright had united a new aesthetic with a new building material, steel reinforced concrete, Unity Temple is considered by many architectural historians to be the world's first modern building. Wright's imaginative design, creative use of interior space, and his thoughtful and quite original solutions in meeting practical needs (sound abatement, lighting, and so on) made Unity Temple one of the first significant architectural achievements of the new century.

Visiting Unity Temple is an experience not easily forgotten. While it lacks the majestic scale of the Pantheon, the temple displays a harmony of form and color that is rarely matched elsewhere. The wonderful orchestration of the horizontal and vertical planes would become a trademark of Wright's work until his autumn years, when a curvilinear style arose unexpectedly from his pen and asserted itself with equal mastery.

WRIGHT'S EMBRACE OF ROMAN WALLS AND HIS TALIESIN RETREAT

After completing Unity Temple, Frank Lloyd Wright withdrew from monolithic concrete construction, at least for the creation of whole buildings, for a number of years. Although the reason for this withdrawal is unknown, it is quite likely that Wright, like many people, did not like the appearance of set concrete. Even though his greatest works are inseparably tied to the material, he often tried or suggested ways to cover or obscure its unpleasantly dull, washed-out gray color. Exposed concrete surfaces in a building's interior could be painted, or their blanched appearance could be masked by clever lighting. Exterior concrete surfaces were trickier. This is perhaps one reason why Wright deliberately chose to expose the aggregate on Unity Temple's exterior walls: it offered the authenticity of real stone versus its bland binding agent. While he continued to expand his use of concrete, for the next two decades Wright mostly restricted the exposed concrete exterior surfaces of his buildings to cast ornamental embellishments.

Much has been written of the Japanese and Mayan influences on Wright's work, but little is known of the inspiration he gleaned from the engineers of ancient Rome. While stylistic influences were predominately Japanese or Mesoamerican, the engineering techniques Wright employed between 1905 and 1925 for building the walls of some of his most important buildings were pure Vitruvian. That his biographers have not discussed this aspect of his work suggests that they have long depended on the architect's own writings to explore his intentions, and Wright was silent on this subject. One must be cautious or even skeptical when reading Wright's two autobiographies and numerous articles. He wrote primarily to promote his work and to project the image of a pure artist blazing a new path, uninfluenced and unencumbered by the aesthetics of the past. Since he opposed the neoclassicism of the Beaux-Arts school, it would hardly make any sense for him to mention that he was successfully employing Roman concrete wall-building techniques to create some of his most striking structures. If we consider that Wright possessed an expansive library, it is almost inconceivable that he did not own one of the several English translations of Vitruvius's *On Architecture* then available.

For some reason, Wright was drawn to Roman bricks, which are flatter and wider than standard bricks. He used them for the first time in his design of the William H. Winslow House (1894) in River Forest, near Oak Park. The Winslow house was Wright's first major commission after leaving Adler and Sullivan, and the broad building with its cantilevered roof is rightly considered the prototype for his celebrated Prairie homes. Wright liked the look of Roman brick masonry on the Winslow house and would henceforth prefer it to standard brick masonry until the 1930s, when production of Roman bricks ceased due to the Depression and a lack of demand. Perhaps the flatter profile of the blocks, combined with the thin mortar lines that Wright insisted on, gave the masonry more of the organic look that he was striving for. Sometimes he would subtly alternate between darker and lighter shades of various Roman bricks within the wall, which suggested the textured appearance of red sandstone.

Wright's Roman concrete-cored masonry wall building technique came later, sometime after the turn of the twentieth century. As discussed in chapter 3, the Romans laid twin courses of mortared brick perhaps a dozen layers in height and separated by a distance of one to several feet. Concrete was then poured between the two courses and tamped down. Another two courses of brick were laid on the older layers, and concrete was again poured between them; the process continued until the desired height of the wall was reached. The bricks were thus securely embedded into the wall's structure. The process was unlike the modern method of first constructing a concrete wall and then adding a veneer of thin brick to cover its surface. Time and weathering can dislodge brick veneers, but it is virtually impossible for the elements alone to dislodge the embedded bricks of a concrete Roman wall. Wright did use brick-veneered walls on one occasion during this time. In 1904, he surrounded the Larkin office building (his first major commercial commission) with a concrete perimeter wall covered with a veneer of brick, the terminus at each end capped by a large brick masonry pier. The Larkin building was torn down in 1950, but one portion of the perimeter wall remains. Virtually all the brick veneer of the concrete wall has vanished (the lone surviving brick masonry pier has been restored). Even if souvenir-seeking scavengers had removed some of the veneer bricks, little effort was needed for the task: lime mortar degrades significantly with age, especially if it is used as a vertical adhesive. Wright probably recognized this shortcoming, and so preferred the Roman masonry-embedded concrete walls.

Wright initially employed Roman walls for his later Prairie houses, the most notable exemplar being the famous Frederick C. Robie House (1906) in the Hyde Park neighborhood of Chicago. The only difference between Wright's Roman walls and those of his ancient predecessors was the vertical insertion of steel rebar in the concrete core. Wright no doubt liked these masonry-embedded concrete walls for the same reason the Romans did: the technique produced a stout wall for less cost than a purely brick one of the same thickness. Satisfied with the results of using Roman walls for houses, he then adapted them for his more ambitious buildings.

Shortly after putting together the Universal Portland Cement Company exhibit in 1910, Wright left his wife and children and took an extended trip to Europe accompanied by his mistress, Martha "Mamah" Borthwick Cheney, the wife of one of his clients. He spent much of his time in Europe preparing a lavishly illustrated volume with a German publisher on his work. To disguise the fact that it was a vanity book, Wright underwrote the publishing costs by paying an enormous fee for the American distribution rights. Titled *Ausgeführte Bauten und Entwürfe von Frank Lloyd Wright* (Studies and Executed Buildings of Frank Lloyd Wright), it greatly furthered Wright's reputation in Europe. The book profoundly influenced a new generation of architects on the Continent, including Le Corbusier, Walter Gropius, and Richard Neutra.

The press eventually found out about the Wright-Cheney affair—it was initially kept secret by embarrassed family members—and a major scandal ensued. Wright's reputation and once-thriving architectural firm suffered as a result. He returned to the United States and affected reconciliation with his wife, while covertly building a large house in the Wisconsin countryside near the town of Spring Green for himself and Mamah. He called the house Taliesin, Welsh for "shining brow" and also the name of a famous Celtic bard. Appropriately, the estate sits on the brow of a hill overlooking a beautiful green valley. The newspapers learned of the construction of Wright's "love nest," and another scandal broke out, again sullying the architect's character and turning off many potential clients. Wright moved into Taliesin with Mamah and simply ignored the moral outcry.[14]

The Taliesin estate was larger than any of the homes Wright had previously designed. He wanted it to be not only a residence but also a state-of-the-art workplace where he and his apprentices could draw inspiration from the pastoral surroundings. Wright designed the estate to include offices, conference rooms, a small theater, and a large drafting studio where a dozen or more architects could work on designs and blueprints under his direction. Local wood and limestone were used in its construction. Two years before building

Taliesin, Wright wrote an article in which he outlined his theories of "organic architecture." Briefly, the principles of Wright's organic style called for a building's design and construction materials to conform to its site. Only colors found in nature were to be used, and the structure itself should possess a "spiritual integrity."[15] Of course, Wright felt free to modify these somewhat vague parameters, but beautiful Taliesin fully conformed to all the elements of organic architecture. It seemed to blend into its surroundings, appearing almost like a rocky outcrop of the hill on which it sat. Taliesin burned down twice but was quickly rebuilt each time. As with all his residences, Wright could never stop from making improvements, renovations, or expansions of one kind or another. Taliesin also served as a test bed for many of his design experiments. As long as Wright lived at Taliesin, the estate would never be truly finished but was always exist in a state of becoming. Becoming *what*, only Wright knew at any given time.

Two major commissions kept Wright occupied and *somewhat* financially secure (the architect was, like his father, a notorious spendthrift) between 1913 and 1923: the Midway Gardens in Chicago and the Imperial Hotel in Tokyo, Japan. Both would share striking stylistic similarities and would incorporate Wright's concrete Roman walls and cast-concrete ornamentation.

The first project, Midway Gardens, was a major entertainment complex in Chicago that would cover a small city block and include restaurants, a beer garden, a large cocktail lounge, and a concert stage for big bands or polka ensembles for Oktoberfest. The Midway Gardens project was awarded to Wright in 1913 and came at a desperate time in the architect's career. Because of the scandals that sprang from his turbulent domestic life, Wright received only two other commissions that year: one for a house, the other for small warehouse. Fortunately, the Gardens project was the largest and most lucrative assignment he had yet received.

A tall, rectangular wall of brick encasing a reinforced concrete core surrounded the Gardens. Strange, cantilevered concrete buildings with a floor plan describing a cross occupied each end of the rectangle and vaguely suggest one of Wright's Prairie homes. A third building, the largest, sat in the center of the complex, with courtyards at each end. Rising above each corner of this central building were four very odd-looking towers capped by crosshatched projections that appeared as if they served some technological purpose, like broadcast aerials, but were really just ornamental flourishes. Upon entering Midway Gardens, the visitor was confronted by a riotous medley of decoration that encompassed Mayan, cubist, and Oriental influences, as well as statues of "sprites" carved by Alfonso Iannelli that could best be described as "proto-Art Deco." The predominating stylistic influence of the Gardens is Mayan, which was especially evident in the rectangular layout of the courtyard and structures and in the cast-concrete decorations that adorned the buildings' interiors, exteriors, and open spaces. Indeed, it was difficult to find a spot that was without a cast-concrete embellishment made to appear like sculpted stone. Since most contemporary observers were not yet familiar with the Mesoamerican influence in Wright's work, they assumed the Gardens to be a cubist creation.[16] It was easily one of the most flamboyant reinforced concrete construction projects the world had seen up to that time.

Although many people—especially Wright—held that architecture was a deeply serious business, there must have been at least a few discerning individuals who took notice of one obvious aspect about the Gardens: one could have fun with concrete. Short of hiring hundreds of stonemasons and sculptors at horrific cost, Wright could not have conceived of the Midway Gardens were it not for the special qualities of concrete. A few years later in Nashville, Tennessee, a full-scale replica of the Parthenon would be built from concrete at a fraction of the manpower, time, and relative cost that had been required for the original.[17] Want a somber church? A pagan temple? A Mayan plaza? Concrete allowed you to construct almost anything you desired—and at a discount, too.

The Midway Gardens project was a definite triumph for Wright. About two hundred thousand people had flocked there within three months of its opening, and the complex quickly paid back its construction costs.

Sadly, Prohibition would make futile the whole purpose behind the Midway Gardens. Fifteen years after the heralded opening in 1914, the magnificent Gardens were torn down, and the rubble was used to create a breakwater in Lake Michigan. Wright took grim satisfaction in learning that the contractor who won the bid to tear down the complex lost money in the venture. Apparently, the stout Roman walls were extraordinarily difficult to demolish.

However, none of this mattered in 1914. While the Gardens project certainly restored Wright's fortunes, its impact on the architect went beyond financial and career considerations: the project, in a way, also saved his life.

On August 15, 1914, while Wright was overseeing some final touch-up work at the Midway Gardens—which had officially opened six weeks earlier—Mamah was playing host to her eleven-year-old son, John, and nine-year-old daughter, Martha. (The Cheneys had amicably divorced, and Mamah had been allowed visitation rights.) After serving lunch to Mamah and the children in the dining room, one of Wright's servants, Julian Carlton, who had been exhibiting strange behavior of late, splashed gasoline around the entrances of the buildings. Armed with an axe, he went into the dining room and killed Mamah and her children, then ignited the gasoline and positioned himself at the outside of the adjoining room. As workers and guests began jumping out the windows—some with clothes on fire—Carlton hacked at them with his axe. In all, Carlton killed seven people and severely wounded two others. Taliesin burned to the ground. Carlton tried to commit suicide by swallowing hydrochloric acid, but he succeeded only in burning the lining of his throat and stomach and was taken into custody. He went on a hunger strike while in jail and died several weeks later.[18]

Although Wright was devastated by the murders, he channeled his grief through work and began rebuilding Taliesin soon after the funerals. In 1916, shortly after completing Taliesin II—though no place where Wright resided could ever truly be called "completed"—another important commission arrived for him: the contract for the Imperial Hotel in Tokyo, Japan.

THE LEGENDARY HOTEL

Frank Lloyd Wright, accompanied by his new mistress, Miriam Noel (Kitty still refused to grant him a divorce), boarded a Japan-bound steamer in December 1916 to begin work on the Imperial Hotel. It was a project that he had been angling for since 1911, when he first learned that the Japanese were considering replacing the outdated and overcrowded Imperial Hotel in Tokyo with a grander edifice equipped with modern conveniences and many more rooms. He had spent several months in Japan in early 1913, showing his designs to Japanese officials and going over the details with them. Although Wright spoke no Japanese, his deep familiarity with the country's art and culture made a favorable impression on his prospective clients. He left the country confident that he had an excellent chance of being awarded the project. His confidence was presumably well founded—three years later the commission for the Imperial Hotel was his.

Wright went to work almost immediately after his arrival in Japan. He met with hotel officials the following day to discuss the logistics of the project, and began making arrangements to set up a Tokyo office through which he could seek other local commissions (he obtained only a few, and all but one were for private residences). Wright estimated that he needed a year to consult with officials and draw up the more detailed blueprints, and a couple more years to oversee the construction work. That would turn out to be an optimistic estimate. In all, Wright would spend over six years in Japan working on the hotel, interspersed by brief trips back to the United States to oversee projects there.

One of the reasons why the Imperial Hotel project was so complicated and took so long to construct was Wright's near obsession about its ability to stand up to an earthquake. When he had submitted his design for the hotel, the 1906 San Francisco earthquake was still a fresh memory. At the same time, an intense controversy had been brewing between two Japanese seismologists. In 1904, Akitsune Imamura, a young seismologist at Tokyo University, believed that a powerful earthquake would likely strike the Tokyo-Yokohama region sometime in the near future. He came to this conclusion after studying historical records going back several centuries and noting a pattern of regularly occurring major earthquakes in the region. Imamura took his findings to his superior at the university, Dr. Fusakichi Omori, at that time the most respected and well-known seismologist in the world. When it came to earthquakes, he was *The Man*. His Bosch-Omori seismograph (built by the German firm Bosch to Omori's exacting standards) was used throughout the world, and his formula for calculating the strength of earthquake aftershocks, called Omori's law, is still used today. Imamura went through his findings with Omori, as well as the probable consequences. The young seismologist estimated that between the earthquake and subsequent fires, the number of casualties could be as high as 150,000, with hundreds of thousands of people injured. Omori was not impressed by Imamura's data, though his counterarguments are unknown. Based on the then-prevailing theories, it is possible that the elder seismologist pointed out that the earthquake groupings may have been coincidental, or perhaps that the groupings themselves were representative of swarms that occurred in cycles that popped up every few thousand years and had already passed for the foreseeable future. To be fair to Omori, seismology was then still in its infancy, and the precise cause of earthquakes had yet to be definitively explained. Disappointed, Imamura mulled over his options and decided that the safety of thousands of Japanese citizens took precedence over other considerations. The young seismologist went over Omori's head and made his findings public. Omori was furious and denounced Imamura's data as flimsy, his findings as unsubstantiated, and his prediction of an imminent major earthquake as alarmist. Imamura was reassigned to a remote back office and given work assignments so elementary that they would have insulted a graduate student. For the next nineteen years, Imamura existed in a professional limbo, chiefly remembered as the fellow who made all those wild earthquake predictions, rather than the once-promising protégé of the revered Dr. Omori.[19]

Though Japanese officials dismissed the notion of an imminent earthquake, the recent disaster in San Francisco reminded them that such an event was certainly within the realm of possibility. Building codes for the many major buildings then going up in Tokyo were strengthened, requiring that large structures be built of either steel frame or reinforced concrete. Unfortunately, these new building codes did not address one obvious issue: the tens of thousands of flimsy wooden homes and apartment houses inhabited by Tokyo's poor and lower-middle classes. These dwellings dominated all the districts, save the city's financial center and the region around the Imperial Palace. Officials felt that to condemn these buildings would inflict too great an economic hardship on the buildings' occupants or landlords, and so nothing was done.

Either Wright had a preternatural sense that a future major tremor might strike Tokyo, or he realized the bad public relations that would result if his building came tumbling down in such a disaster. In any event, he went far beyond the existing building codes to ensure that the Imperial Hotel would be seismically robust. Wright read all he could on the effects of earthquakes on structures. He essentially thought "outside the box." Noting that pipes and wiring conduits were often ripped open by seismic stresses, Wright decided not to follow the standard procedure of embedding them in concrete but instead designed a hollow shaft where they were placed in a loose fashion, giving them the additional "wiggle room" to prevent them from snapping apart during an earthquake. Wright also installed what we would call today "seismic separation joints" every twenty to sixty feet. In other words, the buildings were sectionalized, and each unit was quasi-independent of the others. The units were connected to one another via a lead sleeve that would bend under seismic stress. For instance, if one end of a wing were to sink a foot in an earthquake, it would do so in a benign manner and not cause the rest of the wing to collapse. He also observed that roof tiles—a standard feature of traditional East Asian architecture—were often shaken loose and tossed to the ground by tremors, killing or injuring the hapless people below. He eliminated the tiles and in their place installed thin copper sheets that were nailed to the roof.

Finally, Wright's design called for the hotel to be low to the ground. One of the dangers posed by an earthquake are the lateral forces that can violently shake a structure. To see this phenomenon in action, take two building blocks and put one on top of the other. Then gently move the lower block back and forth. The top block should remain in position. Now try it again with seven blocks stacked one on top of the other, the result of which will likely be the toppling of the stacked blocks. The energy caused by the shaking accumulates as it rises, causing the upper portion of a tall building to sway more than its base. The Imperial Hotel's guest rooms were situated in wings that did not exceed two stories. These wings flanked a slightly taller building that housed the lobby and administration offices. Wright reasoned that the hotel's low stature would keep the lateral forces experienced during an earthquake to a minimum.

FIGURES 32 & 33. Wright's Imperial Hotel in Tokyo, exterior (*above*), and Peacock Lobby (*left*). Wright designed the hotel to hold up well during an earthquake. When the catastrophic 1923 Great Kanto earthquake struck just minutes before the hotel's official opening, the building remained standing while most nearby structures were destroyed.

When he submitted his designs for the Imperial Hotel in 1913, many engineers in the United States and Japan were stressing the importance of *rigidity* in taller buildings to counter the forces of a seismic event, not *elasticity*, as is the case today. Wright addressed both the rigidity and elasticity issues by applying the simplest remedies: keep the buildings low, rigid, and sectionalized.

He also understood that the hotel would be built on soft clay soil, so instead of laying down deep piles for the hotel's foundation, he decided to do something entirely different. He later explained that "[d]eep foundations would oscillate and rock the structure. That mud seemed a merciful provision—a good cushion to relieve the terrible shocks."[20]

In other words, the hotel would ride out the earthquake like a ship in a stormy sea. Wright had recognized the process of soft-soil liquefaction in an earthquake some fifty years before it was described and given a name. He did sink relatively shallow pilings beneath the hotel, but these were meant to stabilize the building and keep it upright during a quake.

The additional work arising from all these safety measures, plus the complexity of the hotel itself, caused numerous delays. It was plain by the end of 1919 that the hotel's original opening date of October 1920 would not be met. The Japanese displayed extraordinary patience, but by 1922, their patience in the face of Wright's repeated postponements was becoming exhausted. That same year, the original Imperial had burned down, and the shortage of adequate hotel rooms in the city had become an urgent issue. The architect reassessed the amount of work still to be completed and then provided a "firm" completion date: the Imperial Hotel could officially open its doors to the public at noon on September 1, 1923.

In August 1923, Wright toured the hotel grounds one last time to inspect his work before returning to the United States. Construction crews were adding a few finishing touches and installing the furniture in the rooms and suites. The same Mesoamerican influences found in the Midway Gardens buildings were here as well, though stronger. Besides the cast-concrete ornamentation, the hotel's Peacock Room, a huge salon for dining and dancing, was directly derived from Mayan architecture. Its contours could best be described as an upended pentagon—what one might imagine a royal audience hall at Chichen Itza would look like. The Peacock Room was Wright's first use of monolithic concrete construction since Unity Temple over a decade earlier. The concrete was painted white and bordered in red. The bases of the pentagonal supports were faced with red brick with additional detailing made from sculpted *oya*, a soft rock similar to soapstone.

From the air, the hotel formed an "H" with a high cross-stroke. The flanks of the H consisted of long two-story wings that held the guest rooms. Between these wings were majestic gardens sporting cast-concrete stelae that stood guard over fishponds stocked with brightly colored koi. The high cross-stroke joining the wings held the lobby, the Peacock Room, and administration offices. The hotel was an extravagant jewel that delighted the eye and offered fresh surprises with every turn of the head.

After inspecting the almost-finished hotel, Wright sailed back to the United States, exhausted but thoroughly pleased with the result and happy that the six-year-long project was finally completed. On the day of its official opening, the Imperial Hotel would become a legend, but not for any of the expected reasons.

DOOMSDAY AT NOON

Shortly before noon on September 1, 1923, as dozens of officials and hundreds of guests, spectators, and reporters gathered for the ribbon-cutting ceremonies at the new Imperial Hotel, the Kanto fault that traverses the Japanese island of Honshu suddenly and violently shifted, causing a massive earthquake that seismologists would later estimate was a magnitude 8.4 on the Richter scale (7.9 on today's moment magnitude scale). Tokyo, Yokohama, and scores of smaller cities and towns suffered widespread damage from the quake and the fires that arose afterward. The rickety wood-frame houses and apartment buildings that officials had not wanted to address in their building codes collapsed by the thousands, killing or injuring their occupants. Innumerable small cooking braziers used to prepare lunchtime meals overturned in the quake, spilling their red-hot coals among the wood rubble and starting perhaps as many as several hundred widely scattered fires in the span of just a few minutes. Many of the people who could extricate themselves from the wreckage tried to flee the flames, but they found their way blocked by other fires and succumbed to burns or smoke inhalation. An estimated 140,000 people died in the disaster.[21] Imamura's warnings had been vindicated, but at a terrible price.

Coincidentally, Dr. Omori was visiting a seismic monitoring station in Australia when the quake struck. As he was examining the station's seismograph, its needle began registering a distant tremor. It took only a short time for the Australian scientists to triangulate the data registered on their machine. They turned with ashen faces to Omori and informed him that a powerful earthquake had struck the area around Tokyo and Yokohama. The seismologist was staggered by the news and immediately returned to Japan. In a noble and touching gesture, Omori apologized to Imamura and recommended him for a key position on the scientific committee that was being formed to study the earthquake and the damage it caused. Omori felt partly responsible for the many deaths and injuries caused by the disaster. He died a few months later, a broken man.

The Imperial Hotel survived, however, and none of its guests or staff were killed or seriously injured. The hotel sank a little, and some flooring was warped in one section, but otherwise the buildings remained intact. Firemen were also able to pump water out of the fishponds to hold off the inferno that arose after the earthquake. A photograph taken shortly after the disaster shows the hotel standing as a kind of island among the blackened rubble. The limited damage was quickly repaired, and the hotel continued to remain in business while much of the city still languished in ruins.

The Imperial Hotel fell to the wrecker's ball in 1968: the large area of real estate in central Tokyo across which it sprawled had simply become too valuable. The facade of the Imperial Hotel's central lobby has been preserved in an open-air architectural park outside Nagoya, Japan.

The Imperial Hotel represented a fine marriage of Wright's skills as architect and engineer. Unfortunately, its value as a positive object lesson in seismic design was never fully appreciated.

THOROUGHLY MODERN MAYAN

Wright made brief trips back to the United States during the construction of the Imperial Hotel to oversee the progress of his projects there. Most of these were palatial residences in Southern California designed in a style that architectural historians would soon call the "Mayan Revival" movement.

Wright strongly objected to any suggestion that his work was influenced by another culture, saying that if Oriental or Mesoamerican architecture bore any resemblance to *his* designs, it simply confirmed their adherence to the same universal aesthetic that all great artists, poets, and architects followed. This was preposterous dissembling. The houses he designed in the Los Angeles area are less "influenced by" and more "imitative of" Mayan architecture.

Nevertheless, these residences are marvelous concrete creations. Some were built using a combination of monolithic concrete and cast-concrete blocks (the Hollyhock House) or constructed almost entirely of the latter (the Ennis House). These fantastical homes are a visual treat, but they represent a kind of dead-end for Wright. While he scorned the Western neoclassicism of the Beaux-Arts school, perceiving it as a creative straitjacket, his inordinate attachment to classical Mayan architecture in the 1920s constrained him artistically as well. The only difference was that one form of neoclassicism was more exotic than the other.

One frustrating problem in discussing Wright's use of concrete in the early and middle periods of his career is the paucity of information that exists on the subject. The architect rarely revealed the nitty-gritty technical details of his work. He was like an artist who enjoys talking about the meaning of his painting but refuses to reveal his brush techniques or the kinds of paints he employs. Still, a careful examination of Wright's structures can lead to some rational, though imperfect, assumptions. For instance, did Wright intentionally expose the heavy aggregate of Unity Temple's concrete walls as an aesthetic statement, or did he simply use too much aggregate? The fine detailing of his cast-concrete ornamentation obviously precluded the use of heavy aggregate, so we know that Wright used only Portland cement and sand (light aggregate) in the mix. The con-

crete of Wright's Mayan Revival houses is quite white and not the usual
faded battleship-gray that he evidently found so distasteful. Titanium oxide
is added to concrete mixes today to give it a white appearance, but this technique was unknown in the 1920s. Wright either used extra lime in the mix
or very white sand, or both. For the Mayan block homes, like the Ennis
House, the relatively small size of the blocks would have ruled out heavy
aggregate, so he likely used Portland cement, white sand, and perhaps a
little extra lime. Using marble dust instead of sand was another method
employed to whiten concrete at this time, but that would have negatively
affected the already marginal load-bearing characteristics of the blocks.
(Concrete block construction of the kind Wright used in the 1920s is now
banned in California for seismic safety reasons.)

Perhaps no other architect performed so much experimentation with
concrete—or likely had as much fun doing so—as Frank Lloyd Wright.

By the beginning of the 1930s, Wright was mainly perceived as an
architect whose best years were behind him. Many believed that Wright,
like his mentors Louis Sullivan and Joseph Silsbee, had advanced his craft
to a certain point, after which he had to yield the torch to a younger generation of architects. Few suspected that Wright still had some surprises up
his sleeve, or that these surprises would rely on pushing the astonishing
properties of reinforced concrete to their limit.

Chapter 8

THE CONCRETIZATION OF THE WORLD

After the Ingalls skyscraper proved that reinforced concrete could be used to construct large buildings, many architectural and engineering firms were quick to jump on the bandwagon. Two interesting structures in the United States that followed on the heels of the Ingalls Building were Terminal Station (a railroad station) in Atlanta, Georgia (1905), and the Marlborough-Blenheim Hotel in Atlantic City, New Jersey (1906). Like many reinforced structures of its day, Terminal Station was built in such a way as to make it look like a stone-masonry building. The Marlborough-Blenheim Hotel was more flamboyant and resembled a Mughal palace. Its architectural style might best be termed "proto-Las Vegas."

By this time, the first reinforced concrete church had been dedicated in Paris, France (1904): Église Saint-Jean-de-Montmartre, designed by Anatole de Baudot. Work had begun on the church ten years earlier but was held up because there were no provisions for reinforced concrete construction in the building codes. Like many of Frank Lloyd Wright's works—and those of the Romans—the church's exterior brick was an integral part of the concrete core.

More and more, roads and highways were being made with concrete. In 1891, American inventor George Bartholomew constructed the world's first concrete street in Bellefontaine, Ohio. It is still with us today. Although more expensive than asphalt, concrete holds up better and requires fewer repairs, and so it was deemed ideal for road construction. The culmination of concrete use was the American Interstate Highway System (1956–1992), the largest use of the material in a civil engineering project up to the time.

Large bridges began to be built of reinforced concrete as well, beginning with the hundred-meter-long (*ca.* 328 ft) Risorgimento Bridge in Rome (1911). The Panama Canal's massive locks were built of reinforced concrete (1914), capping a project that had taken ten years and 5,609 lives to complete,[1] and that was easily the largest single engineering feat the world had yet seen. Most of the larger automobile manufacturers, from Ford to Fiat, began building their assembly plants of reinforced concrete before the 1920s. When World War I broke out in Europe, reinforced concrete bunkers demonstrated their formidable power to hold up against small-arms fire, thrown grenades, and even direct artillery strikes. Dams, which had previously been constructed of masonry or compacted earth, began to be built with concrete. Use of concrete reached its twentieth-century apex in the 1930s with the construction of the behemoth Hoover and Grand Coulee Dams. China's awe-inspiring and costly—both in environmental and monetary terms—Three Gorges Dam project, completed in 2009 in Hubei Province, has surpassed all previous civil engineering projects both in size and in the use of concrete for a single site: an estimated 27,150,000 cubic m (35,510,859 cubic yds) of the material.

The explosion of concrete in the construction industry grew with technical advances in its creation and application. In 1927 alone, the first cement-mixer truck with a rotating horizontal drum (United States), the first prestressed concrete (France), the first "aggremeter"—a large hopper that correctly measured the cement mix and aggregate in volume (United States)—were introduced to the world.

As we have seen, the vast majority of the buildings built with reinforced concrete resembled their wood or masonry counterparts. A young generation of new architects, almost all of them deeply influenced by Frank Lloyd Wright's work, began using the plastic qualities of concrete to create new styles of architecture. Walter Gropius in Germany introduced his Bauhaus ("construction house") buildings, and the Swiss-French architect Le Corbusier (born Charles-Édouard Jeanneret) pioneered the "International Style." German architect Ludwig Mies van der Rohe (born Ludwig Mies), influenced by both Gropius and Le Corbusier, developed a stark form of modern architecture that followed his dictum of "less is more."[2] Austrian architect Richard Neutra came to the United States in 1923 to work under

Frank Lloyd Wright, but after a brief stint at Taliesin—Neutra found working under Wright difficult—he moved to Southern California where he launched a successful architectural career designing homes and buildings that featured his creative use of concrete.

Aside from Frank Lloyd Wright, it seemed as if all the adventurous architects in the United States prior to World War II were European. Most American architects were conservative in their designs, preferring to conform to the tastes of their clients rather than trying to convince them to adopt bold patterns or new forms. By the mid-1930s, the European pioneers began seeing Wright's work as passé, and naturally assumed that his best days were behind him. Almost all creative individuals, whether they are physicists or filmmakers, usually perform their best work before the age of forty; after that, their efforts generally follow a repetitive, if still productive, pattern. In the mid-1930s, Wright was approaching his seventieth birthday. Outside of perhaps a few of his admirers, almost everyone assumed that he would experience the same waning of intellectual and creative skills that usually accompanies a person's biological decline. Surprisingly, Wright would go on to produce his best work in his eighth, ninth, and *tenth* decade of life. It is as if film director Orson Welles had made *Citizen Kane* in his corpulent senescence, or Albert Einstein had published something as important as his general relativity theory during his quasi-retirement at Princeton. It is not so much that Wright remained productive in his later years—many older people continue to keep active in their respective fields—but rather it is the startling originality of his later work that secured Wright's position as one of the greatest architects of all time. And a key component of the architect's audacious autumnal renaissance was his bold and imaginative use of concrete.

FRANK LLOYD WRIGHT RETURNS

After the construction of Tokyo's Imperial Hotel in 1923, Frank Lloyd Wright's architectural practice slowed to a crawl. Wright's first wife, Kitty, had finally granted him a divorce in 1922 after twelve years of separation, which allowed him to marry his second live-in mistress, Miriam Noel. Wright left Miriam after a couple of years and instituted divorce proceedings after he met Russian émigré dancer Olgivanna Lazovich Hinzenberg. He would marry Olgivanna, but not until he had been arrested under the Mann Act for their relationship. The Mann Act was a uniquely American statute against the interstate transport of a woman for "immoral purposes"—usually sex between unmarried but consenting adults. Eventually, Wright and Olgivanna crawled out of the humiliating mess, but the scandal sullied his reputation once more.

Unlike today, "bad publicity" in the 1920s was not "good publicity," and between the scandals and the onset of the Great Depression, Wright could find little work. He had rebuilt his retreat in Wisconsin and took in apprentices. These young aspiring architects were not only required to pay for the privilege of helping Wright design buildings, but they were also expected to perform unskilled labor around Taliesin. The supposed architects-in-training found themselves performing carpentry and mixing concrete for the continual expansions and renovations to which every Wright residence was subjected. Many of the apprentices moved on after a year or two, while those who stayed usually revered Wright with an almost religious awe. Wright liked to be revered. He also liked the money, for he was always spending more than he earned.

One of his apprentices was Edgar Kaufmann Jr., son of Edgar Kaufmann Sr., a successful businessman in Philadelphia and owner of the Kaufmann's Department Store chain, a common fixture throughout many of the Midwest and Mid-Atlantic states at the time. Edgar Jr., who had worked under Wright at Taliesin from 1934 to 1935, knew his father was considering building a small mountain retreat for his family on land he owned in western Pennsylvania, so he convinced him to hire Wright for the job. The result was Fallingwater, briefly mentioned in chapter 5 and consistently ranked as one the most beautiful architectural creations of all time. The most striking features of Fallingwater are the cantilevered reinforced concrete floors and decks that project out over a brook's small waterfall. One little-known aspect of its construction—though known to most Wright scholars—was the architect's insistence that just four lengths of rebar be used for the main deck. Everyone, including Wright's former pupil Edgar Jr. and his own onsite supervisor—another acolyte from Taliesin, Robert Mosher—insisted that this provided far too little reinforcement. Actually, Wright's proposal was a bit insane, as concrete cantilevers require an adequate amount of steel to overcome the material's weak tensile strength. According to Edgar Jr., the number of rebar was quietly increased to eight without Wright's knowledge (he threatened to leave the project if his wishes were not followed to the letter). Even then, the use of just eight reinforcement bars was so minimal that workers were afraid to remove the forms after the concrete had set, fearing a collapse. Luckily, no collapse occurred after the forms were removed, but the decks did slump slightly at the ends. One wonders what would have happened if Wright's four-rebar prescription had been filled.

As with most of Wright's commissions, the time and costs of construction exceeded the original estimates. When Fallingwater was completed in 1937, its final costs totaled $155,000 (approximately $2.5 million in 2011 dollars), over three times Wright's original quote of $50,000. Still, Fallingwater is one of the architect's great masterpieces. *Time* magazine featured both Wright and the house on the cover of its January 17, 1938, issue. For the first time in many years, Wright's work was receiving more public attention than the scandals that attended his complicated lifestyle. The press exposure given to Fallingwater soon brought Wright more clients, especially for upscale homes. His remarkable work and uncompromising nature inspired the reactionary writer Ayn Rand to base her fictional character Howard Roark on Wright in her best-selling novel *The Fountainhead*. Wright liked the book, but when the two finally met at Taliesin, each was disappointed with the other. Rand found a cultlike atmosphere among the apprentices, who gasped whenever the writer disagreed with Wright, while the architect was disturbed by Rand's extreme views, endless pontificating, and—worse—her chain-smoking. At one point, Wright plucked a cigarette out of Rand's mouth and threw it into the fireplace. After that episode, Rand denied that Wright had been her inspiration for Roark, even though her letters tell a different story.[3]

Fallingwater would have been inconceivable without reinforced concrete, as were two more buildings Wright designed that are ranked among his very best: the Johnson Wax Administration Building (completed in 1939) in Racine, Wisconsin, and the Solomon R. Guggenheim Museum (completed in 1959) in New York City. Due to space limitations, it is not possible to discuss all of Wright's later works, but these two best represent his new design aesthetic using concrete.

While building Fallingwater, Wright approached the president of the Johnson Wax Company, Herbert "Hib" Johnson, grandson of the firm's founder, S. C. Johnson, about designing their new corporate headquarters in Racine, Wisconsin, which was a few hours' drive east from Taliesin. Wright knew that another architect had already designed the plans for the company's new building, but he gave it a shot anyway. He invited Hib Johnson over to Taliesin, and they talked for several hours. Wright used his unique combination of charm and impertinence to win over Johnson, who later remarked of the encounter, "If that guy can talk like that he must have something."[4] Despite being insulted "about everything," Johnson liked Wright, and the two men bonded, although it was always a tense relationship. Before introducing Wright to the company's top officials, Johnson told him, "Please, Frank, don't scold me in front of my board of directors!"[5] Wright convinced the directors to pass on the "awful" plans submitted by the other architect and give him the contract instead.

The deal almost fell through when Wright insisted that the buildings not be constructed at Racine, but at a more pastoral setting several miles outside of town. It was the one point about which Johnson and the directors would not budge. Wright's wife, Olgivanna, convinced him to accept the company's demand that the building site remain in Racine. Her argument was both simple and effective: the Wrights needed the money.

As with all Wright projects, there were significant milestone postponements and cost overruns. Hib sent the architect numerous telegrams complaining about the numerous delays and rising expenses, which Wright alternately ignored or calmly answered, citing unexpected construction difficulties that would be more than offset by the finished building's magnificent splendor. Still, such assurances during the Great Depression could not have offered much comfort for Hib and the company's board of directors.

As with Fallingwater, Wright probably expected that most of these complaints would vanish once his clients saw the completed building. And vanish they did.

The Johnson Wax headquarters was remarkable for a number of features. The building had no edges, and all corners were rounded, following the design dictates of the then-popular Streamline Moderne movement, which were first applied to ships, locomotives, travel trailers, and automobiles (both Wright and Johnson owned streamlined Lincoln Zephyrs) to reduce air resistance. Later, the principles of streamlining were used to design stationary objects like buildings and household appliances. The Johnson Wax Building had an especially sleek appearance, enhanced by the tiled exterior bricks covering its concrete walls. Window space was minimal and consisted of thin strips running along the upper portions of each story. Light was supplemented by extensive use of Pyrex® and plastic tubing funneling sunlight from the roof to the building's interior. Walking around the outside of the building offered a different view from almost every angle. Sometimes the smaller upper stories looked like a collection of disks mounted on each other in an offset manner. From another viewpoint, one of the stories was revealed as an ovoid pinched at one end. The pinched ovoid resembled a smooth escapement lever interacting with the round, toothless "gears" of the other stories. Together they looked like some huge, frozen clockwork mechanism.

The most arresting aspect of the Johnson Wax Building was the "Great Workroom," an open administrative hub whose roof was held up by spindly columns made of reinforced concrete that were barely nine inches (ca. 23 cm) at their base but that gradually expanded as they reached the ceiling, where they supported eighteen-foot-wide (ca. 5.5 m) "lily pads" at their apexes, which were also made of reinforced concrete. Unfortunately, the columns violated local building codes. The columns and attached lily pads were considered to be too narrow at their base to support the amount of load they were expected to carry. The building inspectors were doubly alarmed to learn that the columns also contained a hollow that acted as a drain spout for water from the roof—hardly a confidence-inspiring feature from a structural engineering standpoint.

When Wright saw that no amount of arguing or cajoling would assuage the building inspectors' fears, he arranged for a public test of a column to demonstrate its load capacity. A column and its lily pad were set up at the site, lightly braced by wooden beams to prevent sideway slippage. Load after load of large sandbags were then placed on the structure. When the required weight of twelve tons had been placed on the lily pad, Wright urged that more weight be added, and, as a further act of bravado, the architect walked under the structure. The sand bags ran out after thirty tons had been placed on the pad, yet Wright insisted that the loading continue, so loose sand and pig iron were dumped instead. At sixty tons, small cracks appeared, and Wright stopped the loading process. When a crane pulled away one of the bracing timbers, the lily pad and its load snapped off the column and fell to the ground. The impact of 120,000 pounds falling to the ground from a height of almost twenty feet caused a water main buried ten feet below to break. By this time, the building inspectors had already left, but not before approving Wright's dendriform (tree-shaped) columns. The demonstration did more than validate Wright's design: it was perhaps one of the clearest examples of the awesome compressive strength of concrete.[6] Yes, the concrete was reinforced by steel, but the lengths of rebar by themselves would have bent and failed long before the minimum safety load had been reached, let alone the *sixty tons* the column eventually supported before any cracking appeared.

As usual, Wright also insisted on providing his own visually stunning, but notoriously uncomfortable, furniture for the Johnson Wax headquarters. This time, the architect's style was a bit extreme, for he designed the office chairs with just three legs—supposedly to improve the workers' posture. Hib Johnson expressed doubts and asked Wright to sit in one of the chairs. Wright did, and both chair and architect toppled sideways to the floor. Wright quickly agreed to redesign the chairs and add a fourth leg. It must have been a supremely satisfying moment for Johnson.

Despite the delays and cost overruns, Wright had promised Johnson a building to which his workers would look forward to coming each day, and he delivered. A study conducted later showed a 25 percent improvement in employee efficiency at the new headquarters. The Great Workroom where the dendriform columns were installed was almost magical in appearance.

FIGURE 34 & 35. Johnson Wax Headquarters Building and Research Tower (*above*) and the Great Work Room of the Johnson Wax Building (*left*). Frank Lloyd Wright used concrete to create streamlined walls and wild "lily pad" columns. A dramatic field test conducted to demonstrate the weight-bearing capacity of the slender columns confirmed that they could bear many times their specified loads.

FIGURE 36. Wright's Fallingwater House (1935), built for the
department store owner Edgar Kaufmann. It is consistently ranked
as the world's most beautiful nonroyal private residence in existence.

FIGURE 37. Wright's Solomon R. Guggenheim Museum (1959) in New York City. It took over fourteen years to build, thanks largely to Mr. Guggenheim's unwillingness to finalize the construction date, and, later, to bureaucratic red tape and design changes.

The light entering the workroom between the lily pads via the Plexiglas® tubes, supplemented by the strip of clerestory windows just beneath them, imparted both airiness and a nonclaustrophobic sense of submergence, as if looking up from the bottom of a lake to the water's surface above.

The Johnson Wax Building is consistently ranked among the architect's greatest works. Hib Johnson was so taken by it that he commissioned Wright to build the company's research tower a few years later. Like the other building, it sported sleek surfaces and smoothly rounded corners where intersecting angles would normally be found.

WRIGHT'S LAST GREAT WORK

By the 1940s, Wright and his family and apprentices had moved to Arizona, where they set up Taliesin West outside Scottsdale. Wright had come down with a severe case of pneumonia in Wisconsin, and moving to a drier climate seemed the prudent course. As had been the case at Taliesin, the unpaid apprentices performed most of the construction of Taliesin West, though they did receive food and water for the work, although the former was decidedly less exalted fare than what the Wrights dined on.

After the success of Fallingwater and the Johnson Wax Building, Wright expected more commissions. Unfortunately, Wright was often his own worst enemy, and his own words ended up turning on him. Although a proud pacifist, he had trouble understanding the moral depth of critical political issues. As Hitler's Germany was invading its neighboring countries, Wright could only see Britain's efforts to stop the Nazi juggernaut as somehow connected to an attempt to hold onto her empire.[7] He also defended Japan while it was trying to gobble up China. Despite Wright's enormous, and perhaps unequalled, aesthetic sense, he sorely lacked political sense and common sense. Wright's pronouncements alienated many potential clients, especially after the Japanese attack on Pearl Harbor and Hitler's declaration of war on the United States in support of her ally. The war years were lean years for Wright.

Still, there was a patch of green in this literal and figurative desert. In 1943, the board of directors of the Solomon R. Guggenheim Museum, which was then operating out of a rented building in New York City, requested Wright to design a new museum for them on a roughly one-acre patch of land it owned on Fifth Avenue facing Central Park in Manhattan's Upper East Side. Solomon Guggenheim had made a considerable fortune in gold mining and had retired in 1919 to pursue his chief passion, collecting impressionist and postimpressionist art. After 1927, Guggenheim was assisted in this endeavor by Hilla Rebay (born Hildegard Anna Augusta Elizabeth Freiin Rebay von Ehrenwiesen), a German baroness and art connoisseur who had recently moved to the United States. Rebay had studied art in Berlin and was considered one of the top authorities on the impressionist and cubist painters.[8] It was Rebay who convinced Solomon Guggenheim that only Wright possessed the requisite artistic vision for handling such a project.

Wright immediately began work on the museum. Twenty years earlier, he had been approached by wealthy Chicago businessman Gordon Strong about designing a building atop Sugarloaf Mountain in the Blue Ridge Mountains. Strong believed that the spectacular views from the mountain would make it a popular tourist destination for motorists from nearby Baltimore and Washington DC. Wright accepted the commission and drew up plans for a monumental concrete ziggurat ringed by a roadway on which "people sitting comfortably in their own car" could watch "the whole landscape revolving about them, as exposed to view as though they were in an aeroplane."[9] Strong rejected the designs, perhaps because Wright's vision would have cost a fortune to build. (The volume of concrete needed would have been enough to construct a medium-size dam.) Like most ziggurats, the one proposed for Sugarloaf Mountain had a large base that narrowed as the building rose. However, the work Wright had done since the 1920s convinced him that the tensile strength of reinforced concrete allowed the construction of a ziggurat with dimensional *expansion* as it rose from its base; in effect, an upside-down version of a conventional ziggurat. Also unlike the Sugarloaf ziggurat, which was mostly a solid structure with public buildings at its apex, the Guggenheim Museum would be a hollow structure, with an open central court surrounded by sloping galleries. An elevator would take visitors to the top story, where they would disembark and slowly walk down and around the court, viewing works of art as they descended to the ground floor. Wright completed his design for the museum two years later, and, along with Guggenheim and Rebay, unveiled a model of the proposed museum at a press conference in July 1945. That was the easy part. Little did Wright know that hundreds of design changes and fourteen years of bureaucratic wrangling lay ahead of him; or that both he and Guggenheim would not live to see the building—one of the great achievements in twentieth-century architecture—open its doors to the public.

The struggle to build the Guggenheim Museum would require a book of its own to adequately describe. After World War II ended, Guggenheim got cold feet. Suspecting that the real estate market would collapse in New York, he put off building the museum. New York's building commissioners were hardly cooperative. Among the many design alterations they insisted on was a major change to the ziggurat's dimensions, because its upper stories projected out over the Fifth Avenue sidewalk. Pedestrian peace of mind was restored when Wright finally agreed to reduce the ziggurat's expansion. Still, this one alteration required also changing many other details, some major, some minor. The great central skylight, a complicated affair, was made smaller, as was the amount of display space available on the upper stories. Then there were the hundred little details that always need to be addressed when the master blueprint is modified: the position of the elevator, the courses of the plumbing and electrical conduits, and so on. Far more time was spent redesigning the Guggenheim than designing it.

In 1949, while still waiting to see what would happen to the real estate market, Solomon Guggenheim died, which put the project on hold for a while. Hilla Rebay, Wright's chief ally at the Guggenheim, found that many of her duties at the museum were reassigned to others. The bureaucratic wrangling with city officials continued but abated somewhat when Robert Moses, New York's public works czar, who ruled his fiefdom with an iron fist, told the head of the city's Board of Standards and Appeals, "Damn it, get a permit for Frank. I don't care how many laws you have to break. I want the Guggenheim built!"[10]

As the building was nearing completion, a petition signed by many prominent artists was submitted to the museum, declaring that Wright's design would not do justice to their works. Among their understandable complaints was that visitors walking down the sloped galleries would be tempted by gravity to keep moving and not give their works the attention they deserved. Other sore points were that the round walls of galleries presented mounting problems for the paintings, and that the gradual slope would make the works appear slightly askew (apparently, obsessive-compulsive patrons would feel compelled to adjust them—a hopeless task with inclined floors and ceilings). However, their principal worry was that the building's daring design would distract visitors from the art on display. Fortunately, the museum directors, while making soothing noises about artists' concerns, pretty much ignored their complaints. The directors properly regarded Wright's masterpiece as the museum's premier artwork.

Frank Lloyd Wright was luckier than Solomon Guggenheim in that he was able to see the museum in its finished form. The scaffolding finally came down in July 1958, and Wright decided to tour the building. Followed by a few reporters, the architect used his cane to point out various features of the museum. Afterward, Wright consented to be interviewed by journalist Mike Wallace on his evening television show, where he confirmed his belief that he was undoubtedly the greatest architect who ever lived. (Modesty was never Wright's strong suit.)

A few months later, on April 9, 1959, Wright was admitted to St. Joseph's Hospital in Phoenix to have an intestinal blockage removed. It was not considered a very dangerous operation, and Wright seemed to have recovered amazingly well for a man of his age. Then, suddenly and quietly, he died. Thanks to the architect's decades-long insistence that he had been born in 1869 (it was actually 1867), the newspaper obituaries dutifully listed his age as eighty-nine, not ninety-one.

Despite all his considerable personal failings, Wright was one of history's great architects. After his death, only one building would be universally hailed as the equal of Wright's best work. It would be designed at one end of the globe and built on the other. The difficulties constructing it would dwarf the problems that plagued the Guggenheim project, and its cost overruns would make the egregiously low estimates that Wright always gave for his buildings seem like minor rounding errors in comparison. Skyscrapers aside, it would be the last great signature building of the twentieth century and would quickly become emblematic of the country in which it was built.

THE SYDNEY OPERA HOUSE

The story of the Sydney Opera House began in 1947, when Eugene Aynsley Goossens was appointed conductor of the Sydney Philharmonic Orchestra and director of the New South Wales Conservatory of Music. Goossens was considered one of the greatest English conductors, second only to Sir Thomas Beecham, who had mentored him. Prior to coming to Australia, Goossens had established a formidable reputation conducting orchestras in the United States. From 1923 to 1931, Goossens conducted the Rochester Philharmonic Orchestra in New York and taught music at the nearby

Eastman School of Music. In 1931, he succeeded Fritz Reiner as the conductor of the Cincinnati Symphony Orchestra. Among Goossens's many accomplishments at Cincinnati were his recording of the first complete version of Tchaikovsky's Second Symphony and his prodding of Aaron Copeland to write a patriotic fanfare during the Second World War, the result being the composer's famous "Fanfare for the Common Man." When Goossens accepted the Sydney appointment, nine American composers, including Copeland, Ernest Bloch, Roy Harris, and Walter Piston, collaborated on an orchestral piece dedicated to the conductor: *Variations on a Theme* by Eugene Goossens, for which Goossens was allowed to compose the finale. Needless to say, Australians were excited to have such an illustrious conductor coming to their shores.

It is difficult to describe how remote and provincial Australia was in the late 1940s and throughout most of the 1950s. Prop-driven airliners required several fuel stops to reach the island continent, and total travel time by air was usually a strenuous two days. Ocean liners offered the only comfortable means of reaching the country, if we assume that time was not an important factor. On arrival, the traveler faced another problem: aside from the country's natural wonders—many of which involved long rail journeys in passenger cars without air conditioning—there was little to see or do. In short, it was like Kansas with kangaroos. Most Australians will admit that their nation was a pretty boring place during the postwar period, but Eugene Goossens was determined to change all that. He immediately began lobbying for a large, world-class entertainment center that could host both symphony concerts and operas, both of which were then being held in the Sydney Town Hall, a venue hardly acclaimed for its acoustics or comfort.

The New South Wales Labour government gave verbal support for the proposed entertainment center, but it was not until 1954, when NSW premier Joseph Cahill decided to get behind the project, that things began to move. Cahill—who curiously insisted that his last name be pronounced "carl"—recognized that an ambitious cultural center would bring international recognition to the sleepy nation's arts scene. Although Cahill knew that his name would not grace the center, he probably thought that such a building would be a wonderful legacy and certainly better than the usual tribute given an ex-premier: the naming of a street in his honor.

The Opera House Committee was formed to oversee the project. To raise the $7 million the project was expected to cost, it launched a fund-raising campaign to collect private donations. This included selling advance season tickets for the first year's performances and, typical of pre-Enlightenment Australia, the auctioning of kisses, the brave recipients of which were mostly female classical musicians.[11] Only a few hundred thousand dollars were brought in during the drive, so Cahill initiated a lottery to raise the rest of the money.

The site chosen for the complex was the end of Bennelong Point, a spit of land projecting out into Sydney Harbour. At that time, a rather ugly tram depot occupied the site. The tram depot was surrounded by high crenellated walls and a tower that formerly belonged to Fort Macquarie, which had stood guard over Sydney Bay since it was built by convict labor in 1821. The fort's interior had been demolished in 1901 to house the tram depot, but the imposing stone exterior was left in place to scare away any would-be attackers. It was probably the city's most ill-kept secret.[12] By the 1950s, the tram depot was also scheduled for demolition, freeing up a spectacular piece of real estate; indeed, a better location for such an ambitious project could not be imagined. The proposed opera house would stand alone, unobstructed and undiminished by any nearby buildings. Although the center would hold a theater, symphony and opera halls, smaller auditoriums, and a restaurant, the committee members decided to call the complex the Sydney Opera House. They felt that "opera house" sounded grander than "multi-venue entertainment center." Unfortunately, such a name hardly suited the cultural inclinations of the populace at that time. Although Australia produced formidable singers, like the sopranos Nellie Melba and Joan Sutherland, its residents showed little interest in opera. Australians did enjoy orchestral music, but many of them considered the other art form as a kind of regal caterwauling. (Again, the analogy to Kansas comes to mind.) Apparently, both Goossens and Cahill believed that "if you build it, they will come."[13]

Everyone agreed that obtaining a design worthy of the site's location would entail holding an international competition. In September 1955,

Cahill announced with great fanfare that the government of New South Wales would solicit designs from architects around the world for an opera house consisting of two halls, one with seating for between 3,000 and 3,500 people, and a smaller one that would seat 1,200. It would also include a restaurant and two public rooms suitable for large meetings. A panel of prominent architects would judge the designs, which had to be submitted no later than December 3, 1956. The architect whose design was chosen would receive $10,000. Of course, since the winning architect was also expected to oversee the project, his fees would be many times that amount.

Sadly, the prime mover behind the project, Eugene Goossens, would soon be out of the picture. The story behind his departure is not a pretty one.

In England a few years earlier, Goossens had met artist and self-described "witch" Rosaleen Norton. Rosaleen's "art" combined bizarre elements of satanism and eroticism; in short, paintings that only someone like occultist Aleister Crowley could love.[14] The married Goossens soon began a secret affair with Norton. Unfortunately, a reporter from the *Sydney Sun* had infiltrated the witch's coven, and an informant (the reporter?) tipped off the Australian authorities. When Goossens returned to Sydney, the customs officials were waiting for him. As a distinguished and respected celebrity, Goossens was normally waved through customs, but not this time. He was detained, and his luggage was thoroughly searched. Confronted with photos of Norton's curious ceremonies and Goossens's passionate letters to her, Goossens had no alternative but to plead guilty to pornography charges.[15] This was followed by an exposé of the affair in the *Sydney Sun*. Goossens returned to Britain in disgrace, his career destroyed. Goossens had been so admired, and the affair was considered so tawdry, that many people could not bring themselves to write or speak of the scandal. A contemporary chronicler of the Opera House's construction, John Yeomans, wrote simply that Goossens "was met by detectives at Sydney Airport and on 22 March he was fined $200 in a special Sydney court for a Customs offense. His resignation from his Sydney musical posts was announced on 11 April and he left for Europe soon afterwards."[16] Goossens, one of the great lights of twentieth-century music, died a broken man in his native England in 1962.

In total, 223 designs for the Sydney Opera House were submitted. Most came from Australia, Europe, the United States, and Japan, but a few came from such countries as Iran, Ethiopia, and Egypt. Of the four judges, two came from Australia, and one each were from Great Britain and the United States. The most famous member of the panel was the Finnish-born American architect Eero Saarinen. Legend has it that it was Saarinen who rescued the winning design from the trash heap. In truth, he had arrived several days late and mistakenly assumed that the entry had been dismissed. In fact, the other three judges had liked the design in question and had set it aside for the short list, which Saarinen had apparently confused with the rejection list. Still, Saarinen was the design's strongest advocate.[17]

After several weeks of deliberation, the judges reached their decision for the winner (first premium) and two runners-up (second and third premiums). Of course, Mr. Cahill—known as "Old Smoothy" by reporters—called for a press conference to announce the results. Standing next to the chairman of the Opera House Committee, Stanley Haviland, Cahill made a few remarks before Haviland passed him the envelope. Determined to use the event to create as much suspense and publicity as possible, Old Smoothy decided to play the tease. Holding the envelope in his hand, he made a little speech about the importance of the project. He paused, looked at the envelope, then said, "Now before announcing the name of the person who submitted the winning design, I shall run the risk of trying your patience by making one or two general remarks."[18] Cahill's general remarks went on for five minutes. He then took a deep breath, looked at the envelope and said, "And now, ladies and gentlemen . . ." He paused again to enjoy the crowd's anxious tension. "And now," he continued, "before I announce the prize I should like to take your minds back to a meeting two years ago when it was decided this project should be put in hand . . ."[19] This went on for another few minutes before Cahill, perhaps sensing a tomato might be thrown at him or, worse, that he would be strangled en masse by the reporters, eagerly abetted by the newsreel photographers (whose film was running as low as everyone's patience), finally decided to bring his concluding remarks to an end and open the envelope. He pulled out a sheet of paper and announced, "The design awarded the first premium is Scheme

No. 218. The design awarded the second premium is Scheme No. 28. The design awarded the third premium is Scheme No. 62. I'm afraid I haven't got the names of the winners." Haviland stepped forward, plucked a second piece of paper from the envelope and gave it to the premier. Cahill read out: "Scheme No. 218 was submitted by Jørn Utzon—I'm told the correct pronunciation is Yawn Ootzon—of Hellebaek, Denmark, thirty-eight years of age."[20] As Cahill went on to read the names of the runners-up, the reporters, some of whom were knowledgeable men who wrote for the architectural press, wracked their brains trying to recall who Jørn Utzon was. They knew many of the entrants, as well as the winners of the second and third premiums—both respected firms practicing in the United States and Britain—but who was this Danish fellow? And for that matter, where the hell was Hellebaek? Was it a suburb of Copenhagen?

After the press conference, one enterprising Australian reporter consulted an atlas (which must have been fun, since many authoritative atlases at the time did not list the tiny settlement) and then placed an international call—not an easy or inexpensive undertaking in the presatellite 1950s—to a Danish information operator in an attempt to obtain the phone number for a Mr. Utzon in Hellebaek. After getting the number, the reporter reached the Utzon residence, where he detected the sounds of a boisterous party in the background. Utzon's family had learned of his win over the radio while he was walking in some nearby woods. His young daughter ran to give him the news and then coolly informed him that he no longer had an excuse not to buy her a horse. Utzon returned to the house to celebrate the event with a bottle of champagne that he was sharing with his wife and friends. Between the poor connection, the background noise of the party, and Utzon's accented English, the reporter had difficulty understanding the architect. When asked how he would spend the $10,000, Utzon said that he would use the money to come to Australia with his family. Despite the communications problems, the architect's joy was unmistakable. It was a moment to savor, for it would be downhill after that.

The controversy in Australia over the judges' decision began long before Utzon arrived to take up his post. Unlike the other proposals, Utzon had submitted only sketches of the building. Although architectural sketches are more detailed than what a layperson might consider "sketches," Utzon's submission was still less comprehensive than what most other entrants had sent to the competition, a fact that troubled the judges as well. Indeed, Utzon had not even submitted a perspective drawing. An Australian professor of architecture, A. N. Baldwinson, was quickly recruited to draw a color picture of the Opera House in time for the press announcement of the winner. The picture of the completed Opera House had a startling effect on all who saw it. As with all great works of art, it entranced many and repelled a few. Art critic Robert Hughes calls it "the shock of the new."[21] Like Frank Lloyd Wright's finest buildings, the proposed Opera House was not "ahead of its time" but beyond time. The building consisted of a series of vaulted shells, with the largest ones towering over smaller ones, and each of the latter set at a lower angle. The shells suggest different things to different people. Some have compared them to the white canvas of the sailboats in Sydney Habour, others to the budding of a lotus flower, and a few have suggested a series of stop-action frames of some colossal marine bivalve filtering seawater for food, or waves cascading on a beach. Some people have noted Oriental influences, while others suggest parallels to Islamic or Mayan art. Many of the architects and engineers who gazed on the building's enormous shells recognized that it would involve the most complicated large-scale application of reinforced concrete ever attempted for an occupied building. Some weren't sure it could be done. It was probably then that some of them guessed that the $7 million budget, as generous as it was for its time, would not be enough for such an ambitious venture.

As members of the press delved into Utzon's past experience as an architect, they were amazed to discover that the largest project he had designed and supervised was a small housing development in Denmark. If the complexity of engineering endeavors were to be rated on a scale of 1 to 10, a modest housing development would probably be a 2, while the Opera House was undoubtedly a 10. Was this chap up to such a task?

Utzon probably felt some misgivings himself, for he quickly began scouting for a first-rate engineering firm to assist him with the project. He decided on Ove Arup & Partners in London. Born to a Danish family in England, Arup had spent much time in Denmark and spoke fluent English and Danish. All the people he selected to work on other aspects of the building, such as the acoustics and mechanical and electrical installation, were also Danes. To some Australians, this group would come to be viewed as a closed clique resistant to outside influence or even inquiry.

Two chief challenges facing Utzon and Arup were coming up with detailed blueprints to supplement the slim sketches submitted for the competition and figuring out how to construct the great concrete shells. Utzon initially proposed the erection of forms into which the concrete would be poured. This was how almost all large concrete buildings had been constructed since Roman times. However, the complexity of arranging the rebar within the curving form presented problems, as did finding a way to effectively anchor the shells once the forms were removed, for what would hold them up? This led to another messy discovery: if not properly anchored, the failure of one shell might cause the others to collapse like a row of dominos. After all, the shells would literally react like sails under strong wind loads, submitting them to stresses that could cause flexing and dangerous destabilization over time. Besides these two main issues, there were other problems to solve. Still not worked out was the difficulty in arranging enough seating to comply with the original specifications while conforming to the unrealistic large space between rows (three feet) required by the draft proposal. Also challenging was coming up with a way to rapidly and quietly change stage sets between acts, an especially tricky thing to do with such extravagant operas as Verdi's *Aïda* or Wagner's Ring Cycle. Another tough requirement would be providing the requisite number of dressing rooms for so many performers in the limited remaining space. Finally, ensuring perfect acoustics within the halls would be a major headache, since the reverberation cycles had to be precisely arranged. Most of these issues had already been solved in classically designed opera houses, like those in San Francisco, Vienna, and Milan, but not for something as radically different as the proposed Sydney Opera House.

The tram depot was pulled down while Utzon, Arup, and their associates tried to work out the problems and come up with construction blueprints. When two years had passed with no work yet begun on a project that was slated to take four years to complete, the premier started to worry.

Criticism of the Opera House was rising along with its costs. In late 1958, Cahill finally ordered that construction begin as soon as possible, a move that would later have serious consequences. Cahill probably did not know that many of the design details of the very complex building had yet to be finalized, or, more importantly, that no one in Arup's firm or on Utzon's team had yet figured out a way to build the imposing structure. Beginning construction of the building's podium without first knowing how the superstructure of shells would fit upon it was practically giving a notarized guarantee that future obstacles would arise (as they did). Cahill has been unfairly criticized for demanding that work begin prematurely on the Opera House, but he was a consummate politician who knew that once a major undertaking has begun, it becomes much more difficult to call it off. Cahill's move may have saved the project from early termination or, at the very least, prevented the Opera House Committee from falling back on a design submitted by the second- or third-place finalists, neither of which were as thrilling to behold. Australia owes much to Joseph Cahill for his unwavering support of Utzon's design.

Work officially commenced on March 2, 1959. Eight months later, Joseph Cahill died of a heart attack. (Although Cahill did not live to see the Opera House built, he achieved at least one distinction before he died: Australia's first freeway, the Cahill Expressway, was erected and dedicated in his honor the year before his death. Unlike the Opera House, the Cahill Expressway is a rather ugly affair not much beloved by the people of Sydney.)

Several years passed while work on the podium advanced at a snail's pace, which was probably a good thing, since less of it would have to be destroyed or revamped once the engineering problems on the shells had been fixed. Sometime toward the end of 1961, Utzon came up with an idea about how to build the shells. Instead of pouring the concrete into vast molds, portions of the shells could be prefabricated onsite and then mounted on a series of long reinforced concrete ribs. This would securely anchor the shells and at the same time be less expensive than conducting a series of massive pour operations. One obvious problem was that such an approach would also involve a major design change. The parabolic shell of Utzon's original design had to be ditched in favor of one that used two sectionalized portions of a sphere's skin that met together to provide mutual support. It would give the Opera House slightly more severe lines, but the new form was the artistic equal of Utzon's earlier conception. Utzon and the engineers worked on the new design for several months before approaching the Opera House Committee in March 1962 with the unpleasant news that the building could not be built according to the original plans. Utzon and Arup would try to accommodate the new design with the already-built podium columns, but there was no certainty that it would work. The committee accepted the new design, for they knew that there was really no alternative plan that could address all the formidable engineering issues.

Now that the project's major construction obstacle had been solved, work began in earnest. Of course, many issues still remained to be solved. One was the noise around Bennelong Point. The horn blasts of ferries and ships in the harbor sometimes reached 102 decibels, a level of noise akin to a rock group in full cry (although no rock group then could attain such a volume). The blast of such noise could cause the shells to act as a reverse horn, their large openings concentrating the sound as it moved back through an increasingly smaller channel. How could an audience hear the softest *pianississimo* (*ppp*) orchestral passages with such a racket outside? However, this would soon be the least of Utzon's problems.

The conservative Australian Liberal Party, which was then allied with the Country Party (representing mostly rural constituents and now called the National Party), saw the difficulties and rising costs in building the Sydney Opera House as a way to take power in New South Wales. For some years the Liberal Party had done well in the national elections, but New South Wales had stubbornly remained in the Labour camp.[22] They pounced on the Opera House project with all the relish of a dog given a soup bone.

Candidates running on the Liberal and Country ticket made continual attacks on the building, especially on the delays and cost overruns. If elected to office, they promised to finally "put some business common sense into what was happening at Bennelong Point."[23] They also implied that Utzon was profiteering, since his fee was calculated as a fixed percentage of the construction costs; they overlooked the fact that this was a standard practice in determining an architect's fees, and that solving complex problems also entailed more work. Australian architect Walter Bunning, whose very successful firm had submitted a design for the Opera House that had lost out to Utzon's, was happy to assist them. Bunning wrote many articles criticizing Utzon and Arup's work, and there seemed to be no aspect of the structure that he did not find wanting.

The tactic worked. Under the leadership of Robert Askin ("With Askin You'll Get Action"), the Liberal/Country Party coalition came to power in New South Wales in 1965. Once in power, Askin took control of the project from the Opera House Committee and gave it to the Ministry of Public Works, which was now headed by his appointee, Davis Hughes. Hughes called Utzon into his office for a meeting. Hughes, who was not known for his love of the arts, started the discussions by complaining about the project's cost overruns and then said that $30,000 could buy "a lot of culture" in his home district of Armidale. Perhaps Hughes had confused the most impor-

FIGURE 38. The Sydney Opera House in Australia, designed by Danish architect Jørn Utzon. As work was being finished on the building's exterior, Australian officials forced Utzon off the project because of cost overruns. Ironically, far more was spent by the New South Wales government on the building after Utzon resigned. The Opera House ranks alongside the best work of Frank Lloyd Wright and is inarguably one of the world's great architectural creations.

tant cultural center ever built in the planet's southern hemisphere with a North Tablelands community center.[24] When Utzon tried to explain the complexity of the problems that he and his partners were dealing with, Hughes seemed to ignore him and railed on again about the costs. In retrospect, it now seems certain that Askin and Hughes had already made plans to give the Danish architect and his immediate staff the boot.

In politics, you do not fire popular people—and Utzon did enjoy support among many Australians—but instead make it so difficult for them that they have no choice but to resign. Hughes began tightening the screws almost immediately after his meeting with the architect. By this time, the chief engineering hurtle, the construction of the shells, was nearing completion. Sydneysiders enjoyed seeing the Opera House's beautiful form finally take shape, and perhaps Hughes suspected that opposition to the project might soon subside. He refused to pay the $102,000 owed to Utzon's staff. When Utzon complained, Hughes said that he would investigate the matter. After weeks passed with no payment forthcoming, Utzon pressed him again about his staff's salaries, and the minister told him that he was *still* investigating the matter. Knowing that it would be impossible to work without his staff, Utzon resigned. Hughes may have wanted to also dismiss Ove Arup & Partners as well, but calmer heads probably advised the minister that changing the engineering firm in the middle of such a complicated project would have disastrous consequences. That evening, Hughes announced the architect's resignation to the press. Hughes expressed "regret" about the resignation and assured the Australian people that it was "the government's intention to complete the Opera House" and that "the spirit of the original conception" would be fulfilled.[25] The Liberal/Country coalition had delivered on its promise to finally do something about the Opera House. One Australian observed: "Brave words. All that was needed to float the Ark was another Noah."[26]

While the move undoubtedly pleased many grumblers, the reaction of many Australians was outrage. Demonstrations were held protesting Hughes's acceptance of Utzon's resignation and calling upon the minister to reinstate the architect. A petition containing the names of many prominent Australians was submitted to Askin asking that Utzon be reinstated.

Respected architects from around the world also joined in signing a petition of protest. Hughes claimed that Utzon's "resignation was neither sought nor expected by the government."[27] Yes, of course.

Feeling the heat, Hughes agreed to one last meeting with the architect to see if a compromise could be reached. Hughes told Utzon he would be allowed to remain on the project in an advisory capacity, but that he would have no final say in any of the decisions reached by the new team of Australian architects that would replace him. As an ex-businessman, Hughes probably thought it was a fair deal, since the architect would continue to receive a paycheck. But to an artist like Utzon, the offer was an insult. He refused the suggested arrangement and quietly left the country with his family. Utzon never returned to Australia. Although he genuinely liked its people and climate, he could never endure having to look at the altered version of his magnificent creation.

The substitution of an architect does not remove the complexity of the structure he was building. Ironically, far more money would be spent on the Sydney Opera House once it passed into Australian hands to finish (the final bill was $107 million), and it would take eight more years to complete. Queen Elizabeth presided over its official opening in 1973, and it has since become not only emblematic of Sydney and New South Wales but of all Australia. Although its final form may not be exactly what Utzon originally envisioned, it does come quite close, and there is little one can criticize about its beauty, accommodations for audiences and visitors, or its acoustics. The tiled concrete shells reflect the ambient light with startling intensity, whether it be the bright rays of noon or the various reds of sunset. The Sydney Opera House is easily the busiest performing arts center in the world, hosting orchestral concerts, rock performances, and stage dramas attended by well over one million people each year. The formerly opera-phobic Australians now flock to performances of Puccini, Verdi, Wagner, and Lortzing. Well over one hundred million tourists have visited the Opera House since it opened almost four decades ago. In 2002, when the NSW government decided to renovate the structure, Utzon was invited back to oversee the overhauling of its interiors to better conform to his original vision. Utzon decided to remain in Europe, working on the design changes, while his son Jan oversaw the renovation work in Australia. Utzon died of a heart attack six years later, on November 19, 2008, in Copenhagen. He was ninety. If the old saying that buildings are an architect's best headstone, Jørn Utzon could not have asked for a more beautiful monument than the Sydney Opera House. *Requiescat in pace, architectus.*

Chapter 9

THE BAD NEWS

An illusion is not the same an error. . . . We call a belief an illusion when wish-fulfillment is a prominent factor in its motivation.

—Sigmund Freud, *The Future of an Illusion*

Aside from conflicts arising from politics or religion, or those insidious hybrids of the two, few things have been so costly in both material and human resources over the last century as the general misapprehensions about reinforced concrete that persisted until the second half of the twentieth century. Sometimes the harm inflicted is subtle and accrues gradually, such as the slow and inevitable deterioration of steel-reinforced concrete over a span of decades. In such cases, the harm is financial: a reinforced concrete bridge, wharf, or sewer pipe must be continually maintained and eventually demolished and replaced, since repairs after a certain point are either impossible or prohibitively expensive. If a decision is made to save such a structure (e.g., a reinforced concrete building that has important historical or artistic value), the repairs can be expected to cost many times the original construction costs in inflation-adjusted currency. Other times, the harm is less subtle and more devastating, such as the collapse of a reinforced concrete structure in an earthquake and the resultant loss of dozens of lives.

In the late nineteenth and early twentieth centuries, the Portland cement industry made many claims about reinforced concrete. These assertions included that it was a "permanent"[1] building material, that it was "fireproof,"[2] and that it was "earthquake-proof,"[3] none of which are true.

This last claim has had the most devastating consequences, for thousands of lives have been lost in reinforced concrete buildings previously believed to be able to withstand strong temblors.

All three claims were probably sincerely believed to be true at some point in reinforced concrete's early history. The Pantheon certainly demonstrated the longevity of *Roman* concrete, and modern architects and engineers eager to use the material frequently pointed to the ancient domed building as confirmation of its endurance. Concrete is unquestionably fire-resistant, especially in comparison to wood; and although its resilience to earthquakes had yet to be proved before the 1906 California earthquake and fire, reinforced concrete certainly *seemed* like it would hold up well to tremors. Since the permanence of concrete was the last of these three myths to be discredited, let us look first at the other two claims.

"FIREPROOF" AND "EARTHQUAKE-PROOF" CONCRETE

The fire resistance of concrete is one of its many advantages. You are far less likely to die or be seriously injured in a fire that occurs in a reinforced concrete building than in a wooden structure. Moreover, this fire resistance will provide you with more time to evacuate a concrete building on fire because, while the contents within such a structure (furniture, drapery, and so on) are flammable, the body of the edifice is not. The fire resistance of concrete is easily demonstrated: take a concrete cinder block and try to ignite it with a flame. You will quickly discover—if you weren't already aware of it—that concrete is incombustible. However, *incombustible* is not the same as *fireproof*. Take the same concrete cinder block and toss it into a kiln—a concrete factory's clinker kiln would be an appropriate example—and see what happens to it after a half hour's exposure to intense heat. It will crumble away. This is called "exfoliation" and demonstrates the inability of concrete to withstand sustained exposure to high temperatures. Brick, born in fire, is pretty much immune to all but insanely high temperatures. This is why traditional bread and pizza ovens are made of brick; if they were made of concrete, they would quickly fall apart.

Because the early advocates of concrete construction were competing against brick construction, they decided to claim that concrete was "fire-proof." Their motive was simple: concrete must appear competitive with brick, which is, for all practical purposes, fireproof. Indeed, the building codes of many American cities required brick construction in their business districts to prevent the rapid spread of conflagrations, such as those seen in the Great Chicago Fire and in early San Francisco, the latter city having been devastated by fire at least four times *before* 1906. Just because brick is fireproof does not mean that a brick building cannot be severely damaged in a fire, since its contents are also usually flammable. However, a burning brick building helped nineteenth-century firefighters control the spread of the flames, especially if the adjoining structures were also of brick construction. Hence, the rationale behind the building codes. Reinforced concrete buildings also offered similar protection against the spread of a fire, but they were not ranked alongside brick structures as "fireproof" in municipal building codes until the beginning of the twentieth century. Needless to say, this irked advocates of concrete construction. A brick building did enjoy one advantage over its reinforced concrete counterparts: even if completely gutted by a fire, the remaining brick shell could often be cleaned and renovated at less expense and in less time than it took to build a new structure. Hundreds, if not thousands, of historical brick buildings remain with us today as proof of their resilience to fire.

A reinforced concrete building can also recover from a blaze, as long as the fire is not too intense or widespread. One early example, frequently pointed to by the concrete lobbyists, was Ernest Ransome's Pacific Coast Borax Refinery, built in Bayonne, New Jersey, in 1897, which survived a fire that took place five years after its construction. However, a closer look at the Borax blaze clearly shows why the building held up so well. The fire was at its most intense in the storage area below, and much of the heat there was conducted up through the freight elevator shaft, evidently sparing most of the building from exposure to extreme high temperatures.[4] A melted iron winch—some accounts report it to be brass—was proudly displayed as proof of the blaze's intensity, but this trophy appears to have come from the freight elevator shaft, where the heat was concentrated. Another unusual feature of the Borax facility was its double-wall construction that had an insulating air gap between the walls. This likely restricted most of the damage to the interior portions of the walls. (Ransome rarely used his ingenious double-wall construction, probably because of cost considerations.) While the structure was refurbished and used afterward, the Borax firm built an adjoining one of similar size of steel frame construction several years later. Either the second building was erected to accommodate business expansion, or it served as evidence of some nervousness on the company's part.

It is hard to say exactly when concrete's advocates actually discovered that their beloved material was not really fireproof. If I were to hazard a guess, this probably occurred around the same time that a few of their number began conducting "tests" that allegedly confirmed the fireproof claims made about concrete. Critical information appears to always be missing from these early test reports, such as how long and how close the concrete was exposed to the heat source. These remarkable voids in the data suggest that those conducting the experiments knew well what to steer clear of. The tests were so worthless that even some concrete advocates felt compelled to point out their flaws. In his 1912 book *Fire Prevention and Fire Protection as Applied to Building Construction*, Joseph Freitag wrote that the experiments "placed too much emphasis on load-bearing characteristics *before* [my italics] the tests" and that "the question of doubt lies in their qualities *after* [my italics] the fire test." It is not clear whether Freitag's observations represented his frustration or his keen sense of irony.[5]

Nevertheless, industrial propaganda usually wins out over an absence of scientific data. In the 1911 *Encyclopaedia Britannica*, the article submitted for "concrete" states that reinforced Portland concrete cement was at that point "regarded as one of the best fire-resisting materials known. *Although experiments on this matter are badly needed* [my italics], there is little doubt that good steel concrete is very nearly indestructible by fire." Apparently, the lack of "badly needed experiments" did not deter that august publication from accepting the claims made by the concrete industry.

These claims about concrete's fireproof nature were being made in Europe as well. One example is Emil Mörsch's *Der Eisenbetonbau—Seine Theorie und Anwendung* (*Reinforced Concrete Construction—Its Theory and Application*), published in Germany in 1906. Mörsch claimed that reinforced concrete provided "absolute fire resistance" (*absolute Feuersicher-*

heit).[6] When fire resistance becomes "absolute," we are now talking about "fireproof." It is a dangerous word game but one that was practiced by most concrete lobbyists around the world during this period.

Claims made about the earthquake-proof qualities of reinforced concrete buildings—like that made of the Sweeney Observatory in Golden Gate Park—were less frequent, simply because earthquakes were something few people thought about, but fires were an almost daily occurrence in major cities.

The dubious assertions about the "fireproof" and "earthquake-proof" nature of concrete should have been laid to rest after the 1906 San Francisco earthquake and fire, which provided a "real-world" test that demonstrated concrete was neither. Instead, the concrete lobbyists twisted the data in that catastrophe to "prove" that reinforced concrete had stood up well against the twin evils of tremor and fire. Because of this deception, many people around the world would die in the course of the following century in buildings that they thought were immune to collapse from the violent movements of the earth.

THE 1906 DISASTER TEST OF REINFORCED CONCRETE

In the early 1970s, Gladys Hansen, the city librarian of San Francisco, began looking into the deaths related to 1906 earthquake and fire. She had received a number of letters from people whose relatives disappeared after the disaster but were not listed among the 375 names of known victims in the city's official death toll of 478. Obviously, thought Hansen, these people must be among the 103 unidentified victims. She began going through various hospital records, newspaper articles, and unpublished letters and eyewitness accounts in an attempt to match the names provided by the families with those in these source materials. As the months and years went by—Hansen is a very dogged researcher—she slowly came to realize that not only was the number of fatalities much higher than the official death toll, but the accounts from the unpublished material—many from unimpeachable sources—were completely at odds with the authorized version of events given by city and military officials after the disaster. By the time Hansen and former San Francisco fire chief Emmett Condon published their remarkable book *Denial of Disaster* (1989),[7] Hansen had uncovered 1,800 deaths. In Hansen's research since then, she has discovered approximately 1,600[8] more deaths related to the disaster, and she believes that the total number of fatalities—which can never be known with any exactitude—is probably higher. The unpublished letters, memoirs, monographs, and one military report (stamped "Top Secret" and released through the Freedom of Information Act) describe a natural catastrophe of more horror and destruction than anyone had previously realized or suspected. Almost as disconcerting was the discovery that municipal and military officials—with the exception of the US Navy—had handled the situation with such utter incompetence that they actually contributed to the disaster's death toll and damage. Further research performed by other individuals have filled in missing details and are in general accordance with Hansen's account of what happened in 1906. Besides *Denial of Disaster*, there are Philip L. Fradkin's *The Great Earthquake and Firestorms of 1906*, and Dennis Smith's exceptional work *San Francisco Is Burning*, both published in 2005. Smith uncovered a riveting unpublished monograph by a naval officer who, along with other officers, sailors, and marines under the direction of Lt. Frederick Freeman USN, helped the San Francisco fire department successfully fight off the flames in several areas of the city.

It is not our purpose here to explore once again the many horrors connected with this pivotal event in California history but rather to look at one aspect of it: how the disaster was used by concrete advocates to spin a story that had no basis in reality.

THE POST-1906 ENGINEERING STUDIES

When a portion of the San Andreas Fault, encompassing most of Northern California's coastal region, shifted on the morning of April 18, 1906, it caused one of the largest earthquakes recorded in the state's history. Because modern seismographs did not exist at the time, we can only guess at its strength, although most authorities agree that it probably would have registered between 7.8 and 8.25 on the moment magnitude scale (8.2 on the

Richter scale). The believed epicenter of the quake was a couple of miles off San Francisco's western shore. Within a few minutes of the quake, fires caused by ruptured gas conduits and downed electrical power lines sprang up throughout the city, especially in the so-called South of Market region, where wooden lodging houses and single-family homes were damaged or destroyed by liquefaction of the soft soils on which they had been built. Tragically, the earthquake also destroyed many of the water mains in this same region of the city. The city and the military responded to the lack of water by trying to dynamite "breaks" ahead of the flames in an attempt to halt their expansion, but the use of explosives only caused more fires. The fires merged into several massive firestorms that seemed impossible to stop. Fortunately, after three days of hell, the westerly winds turned easterly, blowing the flames over the already incinerated portions of the city, thus allowing the firefighters to finally snuff out the conflagration.

Shortly after the 1906 disaster, several commissions were formed to study the earthquake and the structural damage it caused. One was organized by the United States Geological Survey (USGS); another was an independent committee chaired by Stanford University president David Starr Jordan; while a third, the most ambitious, was formed and led by Andrew Lawson, a professor of geology at the University of California at Berkeley. Today the three are usually referred to by historians as the USGS, Jordan, and Lawson reports of the 1906 quake. The first two commissions addressed both geologic and seismic engineering issues but mostly concentrated on the latter. Though the Lawson report looked at the temblor's effects on the structures and landscape of Northern California, it mainly focused on the geologic aspects of the earthquake. For this reason, we will just concentrate on the reports of the first two, the USGS and Jordan commissions.

THE USGS REPORT

There were three civil engineers on the USGS committee formed to look at the damage caused by the 1906 disaster: Capt. John Stephen Sewell of the Army Corps of Engineers; Frank Soulé, professor of engineering at the University of California at Berkeley; and Richard Lewis Humphrey, a materials engineer working at the USGS. (The pioneering geologist Grove Karl Gilbert addressed geologic issues in the report, a subject we will not explore here.)

In reading the USGS report, it is clear that all three men were favorably disposed to concrete. (Humphrey would later become president of the National Association of Cement Users, now the American Concrete Institute). While all three grudgingly agreed that the failure of brick buildings was primarily due to poor mortar and/or untied walls (masonry walls not bolted to their frames), they could not resist trumpeting the virtues of reinforced concrete at the expense of masonry construction, claiming the latter was especially vulnerable to earthquakes.

One deeply embarrassing fact was that the reinforced-concrete Sweeney Observatory—a major San Francisco landmark built by Ernest Ransome in the late 1880s—quickly collapsed in the earthquake. Soulé never mentions this in his article. Humphrey, to his credit, does, but he also observes that the shale aggregate made the concrete "very inferior" and attributes the observatory's collapse to ground settling caused by the quake. Humphrey does try to put an upbeat spin on the observatory's destruction. Contrasting its collapse with the failure of two masonry buildings in the park, he writes, "No brick or stone structure could have withstood the shock so well."[9] In other words, the shaking was so severe, and the sandy ground so poor, that little else in the park could have resisted such forces. Sewell mentions the observatory only in passing, admitting that he made no examination of its ruins "except from a distance."[10] Curiously, nowhere in these reports is the landmark Sweeney Observatory mentioned by name. Humphrey refers to it as a "cyclorama" (a name I cannot find in any of the pre-1906 San Francisco postcards or tourist brochures that make mention of the building), and Sewell simply calls it a "circular observatory." Virtually all the prominent structures in the city are referred to by their proper names, especially the Bekins Van and Storage Company building, which, according to Humphrey, was "the only example of the pure type of reinforced concrete [sic] in the city" and was, of course, a survivor of the disaster.[11] It is strange that Humphrey, who mentions the "cyclorama" just four pages before his remarks on the Bekins warehouse, has now apparently forgotten it, even though the landmark edifice in Golden Gate

Park was, unlike the *brick* and concrete Bekins warehouse, truly a "pure" reinforced concrete structure.

Humphrey was not alone in his twisting of the facts: remarks by all three engineers in the USGS report are incontestably biased in favor of reinforced concrete construction and against masonry building. Sewell notes that the "great utility of reinforced concrete in earthquake shocks can not be denied"[12] and that a "solid monolithic concrete structure of any sort is secure against serious damage in any earthquake country," unless "it should happen to lie across the line of the slip [seismic fault]; in that case the damage might be fatal, or it might not, depending altogether on the amount of slip and the intensity of the forces that accompanied it."[13] As with many engineers favorably disposed to concrete, Sewell could not resist a jab at the "opposition of the bricklayers' union and similar organizations" that had "prevented the use of reinforced concrete in San Francisco for all parts of buildings. This action of the labor unions will cost the city a good deal, and, should it be continued, will cost a great deal more in the future."[14] By "all parts of buildings," Sewell is referring to the brick curtain walls that were then being installed in large steel-frame office buildings instead of reinforced concrete ones. In his summary, Sewell writes that "reinforced concrete proved itself superior to brickwork beyond any doubt" in the disaster.[15] Soulé confirms this view in his article and, like Humphrey, points to the Bekins Van and Storage Company's warehouse as "the only building of considerable size in the city made of reinforced concrete," and that it had resisted "the action of the earthquake and fire. In this building the concrete acted as a perfect fireproofing protection for the steel."[16] That the Bekins facility was the only reinforced concrete building in San Francisco (again, the Sweeney Observatory has been completely forgotten), Soulé blames on "the opposition of certain labor unions to the use of this material in place of brick and stone."[17]

THE JORDAN REPORT

The chairman of this commission, David Starr Jordan, was a respected academic whose specialty was zoology—specifically ichthyology, the study of fishes. He left the geological issues regarding the 1906 disaster to the three seismologists on his committee to handle. Jordan also gave a free hand to the sole engineer on his panel, Charles Derleth, to deal with the structural damage report. Derleth must have relished his position as the sole authority for this field of study, as there was no one to disagree with his views or contradict him. For reasons not clear—perhaps he read widely on the subject—Derleth seems to have considered himself an expert on seismology as well. In the Jordan report, he devotes dozens of pages to purely geologic matters. This must have infuriated the real geologists on the panel: Dr. Fusakichi Omori (whom we discussed in chapter 7), Grove Karl Gilbert (who also served on the USGS committee), and John Casper Branner, professor of geology at Stanford University. (In reading Derleth's published excursions on geology, one finds only an unreflective mind parroting the then standard texts.) Even if Jordan did not know much about civil engineering matters, he should have at least restrained Derleth from deviating from his realm of expertise and allowed the more qualified authorities, like Dr. Omori, to deal with the geologic questions.

While agreeing with the USGS findings that most of the damage to brick buildings was due to poor mortar and/or untied walls, Derleth cannot resist remarking that "brick buildings are not capable of withstanding heavy earthquake vibrations."[18] In short, brick masonry construction per se is not seismically robust. He also states that the "most general destruction by earthquake in San Francisco was observed in ordinary brick buildings."[19] Later, he remarks that the "prime requisite for a structure to withstand earthquake shock is elasticity; that is, the ability to return without serious damage to its original shape and position after being distorted." Derleth writes that a "wooden frame and the steel-frame building answered this requirement. To an almost equal extent the reinforced concrete building does so also. But structures of brick and stone built of blocks . . . do not answer the requirements of yielding and elasticity to any desirable degree."[20] This elasticity is certainly important in very tall buildings, but all those in San Francisco at the time of the quake were built of steel frame, and not brick. There were no buildings in San Francisco like the Ingalls skyscraper to test the "elasticity" of reinforced concrete during a strong earthquake. Derleth does admit that "some brick structures made a good showing," but he immediately poisons the observation by remarking that they "are the exceptions that prove the rule. For every brick building that withstood the shock, it is easy to give a number of examples of complete failure."[21] A few pages later, Derleth writes that "there were no reinforced concrete buildings in San Francisco because before the fire there had always been successful opposition to their introduction." Derleth was wise enough not to mention the Bekins warehouse, which was a reinforced concrete and *brick* building. However, Derleth apparently shared Soulé's selective amnesia, for he never mentions the Sweeney Observatory, a large reinforced concrete structure that collapsed in the quake.

WHAT *REALLY* HAPPENED TO THE REINFORCED CONCRETE BUILDINGS IN THE 1906 DISASTER

When the earthquake struck San Francisco on April 18, 1906, there were five, possibly six, buildings that were wholly or largely constructed of reinforced concrete. It seems that "opposition" to reinforced concrete construction was only marginally effective in the city. The five known structures were the large grain annex belonging to Globe Mills on San Francisco's Old North Waterfront;[22] a small office building on Sutter Street (the references to this last structure do not provide its name[23]); the previously discussed Sweeney Observatory; the Alvord Lake Bridge (actually a tunnel) in Golden Gate Park; and the "pure reinforced concrete" Bekins Van and Storage Company warehouse on Mission Street. The sixth possible building will be discussed later.

As we have seen, the Sweeney Observatory in Golden Gate Park collapsed in the earthquake. The little Alvord Lake tunnel, a pure reinforced concrete structure in the same park, survived nicely. It is odd that this fact is not mentioned in the USGS report. Perhaps it was left out because it would contradict Humphrey's insinuation that, while the observatory collapsed in the park, little else could survive as well. In any event, tunnels rarely suffer more than superficial damage in earthquakes, so its survival was unremarkable. The Globe Mills Annex did survive the earthquake but was leveled by the post-tremor fires, decisively proving that concrete was not "fireproof." It is not clear whether the building on Sutter Street was leveled by the temblor or by the subsequent firestorm. The Globe Mills Annex and Sutter Street building are not mentioned in the USGS or Jordan reports. It is as if the buildings had never existed. Their fate evidently did not fit the picture the concrete lobbyists wanted to paint.

This leaves us with the Bekins Van and Storage building, a structure that contained—despite assertions to the contrary—probably the least concrete of all the buildings mentioned. To Sewell's credit, he does point out that the warehouse was not really a true reinforced concrete building because it had brick walls, but he cannot resist mentioning that the "walls were badly damaged by the earthquake, but that the concrete was absolutely uninjured."[24] Both assertions are not true. The brick walls were badly cracked in places, but they remained standing. The fires burned much of the contents on the first floor, and the heat was strong enough to exfoliate much of the concrete ceiling (second-story flooring) above, exposing the rebar beneath.[25] (This damage was described by Humphrey as "slight blistering,"[26] which is like claiming that Jack the Ripper only "slightly scratched" his victims.) Fortunately, there was nothing stored on the second floor that might have caused the weakened ceiling below to collapse. Importantly, the Bekins warehouse was still under construction at the time of the disaster, and only two of its planned six stories had been built. Consequently, it was exceptionally strong for a structure of its size. Were it not for its brick walls, it is quite possible that the Bekins warehouse would have been reduced to a pile of concrete fragments and twisted rebar, like the Sweeney Observatory, the Sutter Street building, and the Globe Mills Annex.

There may have been a sixth reinforced concrete building in San Francisco at the time of the 1906 earthquake and fire: Ernest Ransome's Arctic Oil Works building, the world's first large commercial structure built of reinforced concrete. I have not been able to determine its fate, despite searching through directories and insurance records. It certainly doesn't appear in any records I've been able to uncover that were published after the disaster. However, I find it interesting that Ransome mentions neither the Arctic Oil Works building nor the Sweeney Observatory in his reminiscences published six years later, though both were certainly landmark buildings in the history of reinforced concrete. Once again, it is as if the buildings had never existed.

If we leave out the Arctic Oil Works building, three of the five buildings wholly or largely made of reinforced concrete in San Francisco were completely destroyed by either the earthquake or the fire, while the concrete in a fourth, the Bekins Van and Storage warehouse, was severely damaged. In short, 80 percent of these buildings in the city were damaged or destroyed in the disaster, hardly a ringing endorsement of reinforced concrete construction.

Since only one surviving example of a "reinforced concrete" building, the Bekins Van and Storage warehouse, could be found or, rather, deemed acceptable by the commissions' investigators in San Francisco—the city closest to the quake's epicenter—salutary survivors had to be sought farther afield. Stanford University was an excellent choice, for it had both reinforced concrete and sandstone masonry structures, and the performance of each could be contrasted. The sandstone masonry was severely damaged in places, especially the Stanford Memorial Chapel, which was devastated when one of its towers fell down and crashed through its roof. The two reinforced concrete structures, the Girls' Dormitory (1891) and the Leland Stanford Junior Museum (1894)—both excellent examples of George Percy and Ernest Ransome's work—survived with only superficial damage.

Much was also made of a reinforced concrete bell tower at Mills College in Oakland, a city that suffered far less damage during the quake. Called El Campanil, the tower was designed and built by architect Julia Morgan two years earlier. (A plaque placed at the tower's base when it was completed in 1904 incorrectly states that it is the "first reinforced concrete structure west of the Mississippi."[27] It was preceded by Ransome's Arctic Oil Works building, the Sweeney Observatory, and the Alvord Lake tunnel, all built over fifteen years earlier.)

The news of the San Francisco disaster shocked the nation. The photographs taken by reporters allowed into the city after the fires were extinguished showed vast fields of rubble that was once downtown San Francisco. Since much of the rubble was brick (the wood would have vanished in the fires), many concrete enthusiasts saw an opportunity to use the

California catastrophe to promote the "earthquake-proof" and "fireproof" qualities of the building material. The propaganda campaign on concrete's behalf began before the engineering commissions were formed to study the damage caused by the disaster. An editorial in the June 1906 issue of *Cement and Engineering News* observed that "the American cement industry has grown up through a mass of prejudice, the last vestige of which was overthrown and buried by the splendid showing made by concrete in the San Francisco earthquake and fire. Even that class of architects who cling to the ideas and material of construction of their ancestors with Chinese tenacity, have either fallen into line or have been stricken dumb by the splendid mass of constantly accumulating testimony favoring concrete as an ideal material of construction."[28] Concrete's advocates would point to the rubble photographs of postdisaster 1906 San Francisco for years to come, even though such photographs show the city not only after it had endured the earthquake but also after three days of firestorms and continual dynamiting by soldiers and firefighters. The contradictory evidence presented by other observers, including a few of the more honest advocates of concrete construction, was ignored or suppressed.

The vast majority of the "rubble" photographs taken of San Francisco after the disaster were snapped by tourists visiting the devastated city after the fires had been put out. We have already mentioned these. Of the small remainder, most were taken while San Francisco was still burning. A few photographs were taken by individuals just after the quake but before the fires had gathered strength and destroyed large sections of the city. These photographs are quite interesting, for they show the actual damage caused by the earthquake before the flames obscured the evidence. In the South of Market region of the city, we see severely damaged wood-frame homes and lodging houses built on alluvial soil. Sometimes the damage is total; in other words, one cannot find an uninjured structure. In some of the photographs showing downtown San Francisco, isolated damage can be seen in a few spots; in others, none at all. While much was made of the destruction of San Francisco's graft-built City Hall in countless postfire photographs, the prefire photographs show that most of the buildings around the structure survived with little structural damage or none at all, such as the large wood-frame Mechanics Pavilion just across the street, or the masonry Hall of Records just next to City Hall. Soon after these photos were shot, people turned their lenses on the smoke clouds of the first fires. Depending on where the photos were taken, the foreground images show a variety of different locations, such as the Mission and Western Addition districts of the city. Here we also see very little or no damage to the wood-frame and masonry structures, which were obviously aided by firmer soil in those areas. Surprisingly—or perhaps not—these important photographs were not used in the USGS and Jordan reports, the exceptions being two shots taken from the South of Market region where the destruction was general and mostly confined to wooden structures.

Architect and insurance executive John R. Freeman came across these postearthquake/prefire photographs while he was compiling his highly regarded book *Earthquake Damage and Earthquake Insurance*, published in 1932.[29] He tells us that a "prominent" engineer to whom he showed these photographs "protested that there was a danger that their exhibition might lead property owners to believe that the old standards were good enough, and that the publication of these photographs would delay the enactment of safer building laws."[30]

Freeman's research showed that well-built brick structures held up well in the 1906 quake. This has been confirmed in other studies conducted since Freeman's book came out. Dr. Robert Nason, who studied the damage caused by the 1906 disaster for the USGS in the 1970s, reports that "in general, brick buildings did very well in the earthquake, which I found rather surprising."[31] Besides Freeman's and Dr. Nason's work, Stephen Tobriner in his centennial damage report of the 1906 disaster for the Earthquake Engineering Research Institute writes, "It is challenging to confront the surprisingly good performance of many brick buildings in San Francisco in 1906 considering the prevailing belief in the engineering profession today that such buildings are almost dangerous."[32]

It is relatively easy to define a seismically robust brick building. Basically, it is one for which good mortar was used for the masonry, with walls bolted to its frame, and without unanchored cornices or freestanding projections— such as false fronts (popular in nineteenth-century buildings) or towers. Many engineers, both American and Japanese, who came to California to

study the structural damage caused by the 1906 earthquake and fire, found
the biggest offender to be poor mortar. Dr. T. Nakamura, who accompanied
Dr. Omori to San Francisco, reported that "dishonest mortar—corrupt con-
glomeration of sea sand and lime—was responsible for nearly all the earthquake damage in San Francisco."[33] Architect and author F. W. Fitzpatrick
made similar observations after inspecting the disaster's aftermath, noting in
the demolished walls of one brick building that "sand and water constituted
the mortar."[34] Fitzpatrick was also one of the few people who noticed the
"oases in the desert," and what others called the "islands among the ruins."[35]
These were small pockets within San Francisco's burned region, usually just
several city blocks large, in which the firefighters were able to successfully
beat back the flames. (Most of these pockets are due to the valiant efforts
made by Lt. Freeman—no relation to Charles Freeman—and the men under
his command.) These rescued pockets demonstrate even more effectively
than the photographs how well buildings held up to the earthquake in the
portion of the city devastated by fire. Virtually all the buildings, most of
them brick, were still standing after the quake, and many were in what might
be called "pristine" condition. Both the evidence from these pockets and the
prefire photographs flatly contradict Derleth's assertions that the surviving
brick structures were "exceptions that prove the rule" and that for "every
brick building that withstood the shock, it is easy to give a number of examples of complete failure."[36]

One would be tempted to say that the fallen brick buildings were the
exceptions that proved the rule, and that would be true, at least in San Francisco. San Francisco was struck by a powerful quake in 1868. Poorly built
brick buildings were severely damaged in that tremor, and this taught the
conscientious builders in the city a valuable lesson about what to avoid in
the future. However, in a city where building codes were often ignored and
only sporadically enforced, the lessons learned from the earlier quake did
not prevent the construction of some substandard masonry buildings in
San Francisco. The situation was far worse in other Northern California
towns, like Santa Rosa, which, unlike San Francisco, where poor brick construction was generally the rule rather than the exception, experienced substantial damage to brick buildings. Fifteen miles south of Santa Rosa is the

older town of Petaluma, incorporated in 1858, which actually dates back to Mexican rule. It is also closer to the San Andreas Fault. Petaluma had experienced previous quakes, and perhaps this explains why the brick structures there were built with better care, and, as a consequence, why the city suffered far less damage in 1906 than its neighbor to the north.

It could be that even poorly built masonry structures are preferable to poorly built reinforced concrete buildings during an earthquake. "More important than the question of damage is that of lethality," explains Dr. Nason. "In the majority of cases, the failure of a poorly built brick building represents a danger to those outside the structure, such as pedestrians, since part of a wall will collapse and shower bricks onto the sidewalk. However, the building usually remains standing and those inside are generally unharmed. In the failure of a poorly built reinforced concrete building, you often have a catastrophic collapse that kills or injures many of the people inside it."[37]

If it had been known that three of the five reinforced concrete structures in San Francisco had been destroyed in the 1906 disaster and a fourth damaged, it is unquestionable that reinforced concrete construction would have suffered a temporary setback. However, this setback would have resulted in better reinforced concrete structures and building codes, and would have saved many lives in future earthquakes. Instead, the advocates of reinforced concrete construction would point to the rubble photographs and quote the USGS and Jordan reports for almost a century. As a result, much mischief would be done.

THE "PERMANENCE" OF REINFORCED CONCRETE

Iron and steel reinforcement of buildings did not come into general use until the nineteenth century. Marc Brunel conducted a number of important experiments on iron-reinforced masonry using barrel bands and Roman cement in the 1830s, proving that the tensile strength of such structures was greatly improved. A few decades later, masonry buildings constructed on iron frames were being built, and their numbers increased by the century's end. With a drop in steel prices, steel quickly replaced iron

FIGURE 39. Crowd gathers in downtown San Francisco on the morning of April 18, 1906, to watch the approach of the fire. Photo was taken after the earthquake but before the inferno obliterated the eastern half of the city.

FIGURE 40. A man walking across the rubble caused by the 1906 disaster. Rubble photos were used by the concrete industry to show the "failure" of masonry buildings in the catastrophe. In fact, it was the *reinforced concrete* buildings in San Francisco that proved to be the most vulnerable in the earthquake and the subsequent fires.

frames and, for concrete, iron rebar. As we have seen, reinforced concrete seemed to offer an advantage over steel frame: it is a hybrid material that combines the compressive strength of concrete with the remarkable tensile strength of steel. Because of this, less steel was needed, and so less expense was involved in constructing a structure. Reinforced concrete bridge construction became especially popular for small to medium-size bridges, since the steel rebar within the concrete was "sealed" against the elements and, so it was reasoned, protected against oxidation. Advocates of the material would point to rust-free rebar that had been embedded in concrete for ten years or more, while steel exposed to the elements over the same course of time would be covered in the grit of oxidation. It was known that high alkalinity discourages iron oxidation, and since freshly poured concrete is highly alkaline, it seemed as if the two were made for each other. The seeming impermeability of reinforced concrete made it appear to offer the perfect solution for maritime building, specifically the creation of wharves and piers, which had up to that time been mostly constructed of wood, a material subject to shipworms and slow deterioration.

With seemingly perfect rust protection, how long could a reinforced concrete structure be expected to last? For centuries at least, it was believed, and some architects and engineers in the first quarter of the twentieth century thought that a one-thousand-year life span was a reasonable estimate.[38]

This optimism, based on the sealing properties of cement and its alkalinity, seemed well founded. Unfortunately, this confidence did not take into account iron's extraordinary passion for metamorphosis. No other commonly used metal is more determined to return to its natural state than is iron. In iron's case, its usual natural state on the surface of the earth is iron oxide, what we call rust. Iron is extremely "reactive." In other words, water, air (which carries a percentage of water), and chemicals like sulfur or chlorides (salts), greatly accelerate the oxidation process. Iron is particularly susceptible to seawater, since it consists of both water and sodium chloride. The only immutable forms of iron known are iron asteroids floating in outer space, since they are isolated from these terrestrial agents. Iron oxidation is also the reason why so many swords and spearheads from the Bronze Age have survived, but almost none remain from the Iron Age, even though far more were made in the latter period. Simply put, iron just wants to vanish into a reddish-brown dust. The same is true of steel, which is between 88 to 98 percent iron. As for "stainless" steel (a hard steel made with at least 10 percent chromium), anyone who has found a lost tool or tablespoon that had been left outside for a couple years can attest to its vulnerability to rust as well. This is why it is so important to continually maintain our steel infrastructure, like steel cantilevered or suspension bridges, by painting them, and then repainting them, and then sandblasting the old layers of paint away every decade or so, and beginning the painting process all over again. If this isn't done, the bridge runs the risk of losing its structural integrity and, like London Bridge in the nursery rhyme, falling down.

In reinforced concrete, the oxidation process is especially insidious. The rust-inhibiting alkalinity of concrete gradually vanishes after the completion of the curing process. After a while, there is insufficient alkalinity to prevent rust. At this point, the only thing preventing the steel from oxidizing is the seal provided by the concrete itself. But, to paraphrase a popular saying, "Nature always finds a way in." In concrete's case, the stresses it is subjected to can cause cracking; such stresses include the seasonal freeze-and-thaw cycles in colder climes, the constant vibrations and loads placed on a reinforced concrete bridge, or wind pressures on a tall reinforced concrete building that cause it—as intended—to slightly bend and oscillate. Cracking can often occur in concrete during its setting and curing period as well. Finally, concrete can crack for no readily apparent reason; it's simply the nature of the beast. (One popular saying among builders is "If it ain't cracked, it ain't concrete.) However, some of these cracks will allow air and moisture to penetrate the concrete and reach the rebar. Once this happens, a natural process begins that, unless it is temporarily arrested (it can never be stopped), will have serious consequences. As the rebar rusts, several things happen. Not only is the amount of "good" steel reduced, but the diameter of the rebar expands to as much as fourfold its original diameter, causing more cracks and, in due course, pushing out chunks of concrete. Remember, reinforced concrete is a hybrid material whose structural integrity relies on the vigor of its twin components, steel and concrete, to provide its tensile and compressive strengths. When one or both are compromised, a reinforced concrete structure runs the risk of no longer being able to serve the purpose for which it was designed. In some cases, it simply collapses, but in most instances, it is usually condemned and torn down long before it literally falls apart. Unlike a bare steel structure, like San Francisco's Golden Gate Bridge, where trouble spots can be seen (regular maintenance can even keep most of these from developing further), the rust in reinforced concrete is usually hidden from view. Its first appearance usually comes in the form of a large crack with a brown stain running down beneath it, indicating that the harm is already well under way.

Not surprisingly, the first examples of rebar corrosion were noticed in the earliest reinforced concrete wharves and docks, especially those built in seawater ports, since they were exposed to both sodium chloride and water. Builders had long recognized that iron or steel was vulnerable to seawater, and since seawater had been used in the cement mix of some of these structures, it was reasoned that this must have caused the problem. To prevent corrosion from happening in the future, engineers agreed that concrete cement should be mixed with freshwater if it was to be used in a saltwater environment. The freshwater mixing of the cement did seem to work at the time, but it was soon evident that this was only a stopgap measure and not a permanent solution, since reinforced concrete piers that had used freshwater in their mix also began to corrode. The reaction of engineers to this disturbing development was to launch investigations into the composition and/or application of the concrete. Obviously, the concrete being used was not dense enough, or it was not properly mixed, or there was not enough concrete covering the rebar, and so on. Various solutions were proposed, including a suggestion by the former 1906 investigator on the USGS committee, Richard Humphrey, that more iron oxide (!) be used in the mix.[39] As noted, the life span of reinforced concrete piers was extended by some of these measures, but, as with all attempts to halt the corrosion of steel reinforcement bars, such precautions could only put off the inevitable. Today, reinforced concrete structures built in marine environments still do not last very long, and the best built today are given a service life of roughly fifty years, though some corrode in far less time.

While engineers struggled to find a solution to the seawater corrosion of reinforced concrete, the material's advocates sought to allay any public fears on the matter. That indefatigable booster of concrete, Richard Humphrey, wrote in 1917—long after the problem had become evident around the world—that he had "never found an authenticated instance where concrete had disintegrated from the chemical action unless it had been permeated by seawater *before* [my italics] it had set."[40] In other words, the seawater could not penetrate concrete if its cement had been mixed with freshwater before setting. Humphrey also added that, "in tropical waters where there is no frost action . . . concrete of reasonable density is unaffected."[41] (It would later be proved that seawater corrosion of reinforced concrete is most severe in tropical regions, despite the lack of freeze-and-thaw cycles.[42]) Evidently, the most important thing to Humphrey was to assure everyone that the problem lay not in the material but in its application.

By now, it should be obvious which station this train is headed for. It wasn't long before corrosion also popped up in freshwater docks made of reinforced concrete, and, though the corrosion progressed at a slower rate than that seen in their seawater cousins, it continued apace as well.

Another problem that manifested itself soon after reinforced concrete roads became commonplace was the rapid rebar oxidation seen in colder climes where the cracking caused by freeze-and-thaw cycles combined with winter deicing salts to accelerate the corrosion process. As the years passed, and rebar corrosion began appearing in reinforced concrete structures far from any marine environments or regions where deicing salts were used, people in the industry shrugged their shoulders and blamed "bad concrete." There was something to this verdict. In the early days of the twentieth century, salt was sometimes mixed with concrete—something Vitruvius warned against two thousand years earlier—to supposedly prevent cracking. Naturally, the salt accelerated the corrosion of the rebar. One practitioner of this crack-proofing process was Ernest Ransome, who began salting his concrete sometime around the beginning of the twentieth century.[43] It is no coincidence that most of his buildings are now gone. (Two exceptions are the little Alvord Lake tunnel in Golden Gate Park and the Leland Stanford Junior Museum, both built before his chloride-dosing days.) Because of salting and the inept application of reinforced concrete by inexperienced contractors in those early days, it was not unreasonable in the 1940s and 1950s for an architect or engineer to assume that a corroding reinforced concrete structure had been built using bad cement or sloppy methods.

Still, the corrosion kept on popping up. One culprit, suggested by some authorities almost a century ago—but soon forgotten—was the condition of the rebar at the time it was used in the concrete. Until a few years ago, most rebar delivered to a worksite was already covered in rust. Obviously, this gave oxidation a head start once the concrete's alkalinity began fading after the curing process. The use of rusted rebar is still common: I recently witnessed a reinforced concrete building being constructed with corrosion-coated rebar. Until the latter part of the twentieth century, it was a common belief among builders that rust was actually beneficial in that it helped the concrete "adhere" to the steel.[44]

By the 1960s and 1970s, older reinforced concrete structures began

FIGURE 41. Disintegrating reinforced concrete floor slab. As the steel reinforcement bar oxidizes, its diameter dramatically expands, destroying the surrounding concrete as it, in turn, is destroyed by rust.

FIGURE 42. Disintegrating concrete wall. The middle stripe is merely a rust stain, the rebar itself having already vanished through oxidation. Once a certain corrosion point is reached, repair is difficult or impossible, and the structural integrity of the building is severely compromised.

fragmenting all over the world. The most vulnerable of these—the majority of the pioneering saltwater piers and wharves—had already disintegrated or been demolished by this time, many before the mid-century mark. Occupied buildings have generally fared better, since the interior portions are temperature-controlled and, if air-conditioned, have dryer air as well. Nevertheless, most of these early buildings are no longer with us. If a reinforced concrete structure has some historical value, it can be "saved" if the corrosion has not progressed beyond a certain point. However, measures taken to save such buildings are very expensive. Such measures involve taking a jackhammer to the concrete to expose the corroded rebar before brushing the rebar with a steel brush to remove the rust. Then, a layer of new concrete, the color of which must precisely match the original when set, is applied over it. If the concrete has decorative flourishes, a mold must also be made to match the original pattern. These rescue measures are obviously less complicated and expensive when they are used to save, for example, a column supporting a freeway overpass, which usually has no decorative motifs, and where no one is much troubled if the patch does not quite match the original concrete. (Of course, with the passage of time, the steel rebar under the patch will eventually begin to rust again.)

Corrosion-fighting measures in a historic building like a reinforced concrete church are a different matter. Fine wood may grace the interior portions of the concrete walls, or the concrete walls themselves could also be adorned with interior flourishes. Frank Lloyd Wright's beautiful Unity Temple (1908) is one example. In 1971, repairs were made to its concrete walls to fix the damage caused by rebar corrosion. Of course, it was a stopgap measure, and now the corrosion within the concrete has progressed to a point where repairs to the church will cost an estimated $11 million (some estimates place the cost as high as $20 million). In inflation-adjusted dollars, this would be more than twelve or, using the higher estimate, twenty-three times the original cost to build the church. Of course, this priceless architectural gem should be saved, and the effort will probably allow Unity Temple to endure for another century or more. However, as with almost all steel-reinforced concrete structures, we will one day reach a "Washington's Axe" situation. For those unfamiliar with the term, it refers to a joke: A tourist is visiting a museum that proudly displays the axe used by a young George Washington to chop down the proverbial cherry tree. The tourist asks the museum's curator if the axe is really the same one George Washington used. "It certainly is," replies the curator. "Of course, over the years we had to replace its handle three times and the axe head twice, but it's the same one, all right."

Besides Unity Temple, many other historic reinforced concrete buildings have had expensive restorations, such as Ludwig Mies van der Rohe's Tugendhat House (1930) in the Czech Republic, or Le Corbusier's Villa Savoye (1935) in France. (The deterioration of the latter has been ascribed to its occupation by foreign forces during World War II—both German and American—but while the house was certainly trashed during this period, such treatment probably had little impact on the natural forces of oxidation.)

Among the few early reinforced concrete structures still with us are the William E. Ward House (1876), in Port Chester, New York, and the Alvord Lake "Bridge" (1889) in San Francisco's Golden Gate Park. Both are hardly inspiring examples of the longevity of reinforced concrete. The Ward House is heavily cracked, with a particularly large fissure running from the top to the bottom of the structure. The larger cracks have been awkwardly filled in with concrete. The Alvord Lake Bridge (actually a tunnel) has deep cracks as well (you can put your hand into several of them). The Alvord Lake structure is likely still standing because of the earthen berms supporting it on each side.

The Ingalls Building (1903) in Cincinnati, Ohio, still stands, and to some, this is a mystery. Its temperature-controlled interior would have stalled, but not stopped, the rebar corrosion that has doomed so many other old reinforced concrete structures. Why has the Cincinnati sky-scraper survived so long? Perhaps because the Ingalls Building, unlike most of its reinforced concrete contemporaries, has its exterior clad in masonry. Did this cladding protect the concrete's rebar against environmental harm? Certainly, the question should be studied. Another factor that contributes to the building's longevity is its location: seismically quiescent Ohio. Older examples of reinforced concrete construction, as demonstrated in the 1906 earthquake and many others since then, do not take kindly to shaking. However, while a few reinforced concrete structures have passed their first-century mark, it is highly unlikely that they will endure to see their second. The exceptions being, of course, those buildings that undergo very extensive—and vastly expensive—restorations.

Another splendid survivor is Frank Lloyd Wright's beautiful Fallingwater House in Pennsylvania. It recently underwent an expensive restoration. Although the house is thirty years newer than Unity Temple, restorers were still surprised to find so little corrosion of the rebar in the massive central deck that looms out over the river. It may be remembered that only eight lengths of steel rebar were used in the deck's construction. The reader will recall that Wright had specified that only four lengths of rebar be used, but the more circumspect builders secretly used eight instead. One possible reason why so little corrosion was found is that each rebar is surrounded by a substantial amount of concrete, making it less likely that a crack would penetrate sufficiently to reach the steel and so allow the ingress of air and water. Still, the eight rebar proved hardly enough to support the deck, and it continued to droop over the years. Workers refurbishing the structure had to drill through the concrete and install post-tensioned cables to restore the deck to its original design. Nevertheless, it is still only a matter of time before corrosion begins its devilish work.

Reinforced concrete dams have done better. Because of the massive volume of concrete used in their construction, many of them, such as Hoover Dam, are still undergoing the curing process, thus forestalling corrosion. (It will be interesting for our descendents to discover whether the tremendous weight of these dams will continue to put off the rebar's corrosion expansion.)

There is another possible reason for the longevity of the Ingalls Building, the Edison houses, and Fallingwater. Assuming that the construction work was competent and no salt was used in the mix, could it be that the concrete cement used for these buildings was far better than the modern "high-strength" cements introduced before World War II? To have suggested this possibility a few years ago would have induced laughter and indulgent grins among engineers. Today, no one is laughing or grinning. Indeed, the question has already been answered, and it points to another sad and costly chapter in the story of reinforced concrete.

BETTER, STRONGER, FASTER

Concrete cement is a product, and, as with all products, the manufacturer seeks to make it better in order to remain competitive with industry rivals. In the early 1930s, a new cement was introduced that had more impact on the concrete industry than the development of the automatic transmission did for automobile manufacturers, or, even—forgive the cliché—than sliced bread had on the baking industry. It was sold under a variety of different brand names, but it gradually assumed the generic moniker of "high-strength" concrete cement. The stuff was wonderful: it cured in less time than regular cement, offered greater compressive strength, and more water could be used with it, making it more malleable and easier to pour. As we have seen, concrete sets rapidly, but curing takes longer. It is during this later process that the concrete assumes most of its compressive strength. The older cements developed compressive strengths of only 3,000 lbs/square in (*ca.* 20.7 MPa) after seven days, while the high-strength versions could offer from 4,500 lbs to 5,400 lbs/square in (*ca.* 31–37.2 MPa) in the same amount of time. This meant that reinforced concrete construction could proceed at a much faster rate. A bridge, building, or highway could be built in less time and at lower cost. It is no wonder that the construction industry quickly adopted the new cement. After all, what was not to like about it?

Actually, plenty. We are still on the same train, and it is headed toward the same hellish station, but now we will arrive ahead of schedule.

In 1944, the US Public Roads System (now the Federal Highway Administration) undertook a meticulous survey of two hundred reinforced concrete bridges across several states. The results of the study were alarming. The older bridges were in far better shape than the younger ones. Of the bridges built before 1930, 67 percent were found to be in good condition. In contrast, only 27 percent of those built after that year were deemed to be in good condition.[45] (Remember, the survey took place only fourteen years after 1930!) These results were confirmed in two separate studies conducted in the 1950s.[46] One would think that the data accrued in these studies would result in major changes in the industry's cement formulations. Think again. High-strength cements became more popular than ever. It was not until 1987, when a report by the US National Materials Advisory Board detailing the accelerated decay of reinforced concrete structures was released—by which time the crumbling infrastructure had been noted by all but *some* of the blind—that the construction industry and federal and state governments thought that maybe the problem should be looked into.[47] Not surprisingly, it turned out that high-strength cement was the culprit. Reinforced concrete structures built using the high-strength mix tended to crack at a faster rate than the older cements, allowing water, air, and chemicals to reach the rebar, and so allowing the corrosion process to begin years before it normally would. Concrete specifications and building codes were finally changed[48] in the late 1980s and early 1990s, when the proven durability of concrete mixes to be used in highway construction became a requirement. Things changed for the better—almost fifty years after the problem was first noted.

While we have made improvements to cement formulations and concrete-building methods, the world we have built over the last century is still decaying at an alarming rate. Our infrastructure is in especially terrible shape. One in four of our bridges are now either structurally deficient or structurally obsolete.[49] The service life of most reinforced concrete highway bridges is fifty years, and their average age is forty-two years. The "ride quality" of our roads is substantially deteriorating as well. At the same time, the tonnage of freight transported on our roads and highways is steadily increasing. Between 1980 and 2005, both automobile and truck VMT (vehicle miles traveled) roughly doubled, although highway lane miles grew by only 3.5 percent. It is no wonder that potholes are far more common now than ever before, increasing road accidents and reducing the life of the tires, shocks, tie-rods, and axles of our automobiles.[50]

Besides our crumbling highway system, the reinforced concrete used for our water conduits, sewer pipes, water-treatment plants, and pumping stations is also disintegrating. The chemicals and bacteria in sewage make it almost as corrosive as seawater, reducing the life span of the reinforced concrete used in these systems to fifty years or less, depending on the exposure factors and the kinds of sewage involved.[51]

The American Society of Civil Engineers (ASCE) rates America's infrastructure a "D" grade. To simply bump it up to a "B" would cost us an estimated $2.2 trillion, a figure that is likely rising each day as we continue to put off facing the problem.[52]

It is now recognized by all civil engineers—at least those whose veins and arteries do not now run with formaldehyde—that steel-reinforced concrete is hardly the "everlasting" building material it was once touted to be. Indeed, its life span, certainly shorter than masonry, is probably less than that of wood.

While we're laying on the bad news with the figurative mortar trowel, here's another interesting fact: around the same time we began suspecting that the longevity of reinforced concrete was overrated, we also started to take notice of our worsening air quality. Outside of automobiles and coal-fueled power plants, the manufacture of concrete cement is the largest contributor of CO_2 emissions into the atmosphere.[53] Even more troubling is that all this steel-reinforced concrete that we use for building our roads, buildings, bridges, sewer pipes, and sidewalks is ultimately expendable, so we will have to keep rebuilding them every couple of generations, adding more pollution and expense for our descendents to bear.

The Romans built structures for the ages. Some of their bridges are still being used today, and instead of people and oxcarts, they now bear the loads of cars, trucks, and buses. If the Romans had used steel-reinforced concrete—which they did not have—to build their beautiful bridge in Alcantara, Spain, the bridge would have to have been rebuilt at least sixteen times by now. Can we, with our considerably more advanced technology, build a structure that, barring intentional destruction, lasts a couple of thousand years or more? And could such a structure be built of reinforced concrete? Finally, if we could build a reinforced concrete structure with a two-millennia life span, would it be prohibitively expensive? The answers to these questions in their respective order are yes, yes, and no.

Chapter 10

THE GOOD NEWS

NEW CONCRETE CEMENTS

Sometimes, though less frequently than we would like, one solution appears that neatly solves two problems at once. As we have seen, the tremendous volume of clay and limestone being kilned each year pours millions of tons of CO_2 into the atmosphere, both from the burning of fossil fuels to cook the material and from the material itself. This is especially true of limestone, which generates a phenomenal amount of CO_2 when it is transformed in the kiln from calcium carbonate to calcium oxide. Likewise, the wide-scale adoption of high-strength/low-durability concrete cement in the mid-twentieth century has proven disastrous to our infrastructure. Something obviously had to be done about these troubling situations. Fortunately, a simple solution came to the fore that helped address not only both issues but a third as well.

The steel industry and coal-burning power plants have been generating a tremendous amount of solid waste products for years. The steel industry produces millions of tons of slag, and the coal-burning power plants generate an equal or greater amount of fly ash. Slag is that portion of iron ore that is left after the metal is smelted. To aid in the process of separating iron ore from mineral impurities, lime and magnesite are added, and these become components of the slag as well. Fly ash is the lighter portion of coal ash that was previously allowed to fly out of the smokestacks of power plant furnaces. With the enactment of environmental laws in the 1970s and 1980s, coal-powered power plants were forced to capture this ash with electrostatic precipitators or particle filters. In the case of both slag and fly ash, the material was either piled up in nearby heaps or assigned to landfills.

As with the millstone refuse in Andernach, Germany, some three hundred years ago, people discovered that the chemical composition of slag and fly ash made the combination ideal for producing cement. Even better, while the Andernach chips were suitable only as a pozzolanic element, slag and fly ash have both pozzolanic and cementitious components.[1] In other words, they can replace not only much of the kilned clay in Portland cement but much of the kilned limestone as well. That's not all: because of the high percentage of silicates in some fly ash, it can also be used as a filler to replace some of the sand used to make the concrete. Adding gravy to this good news is that when slag or fly ash is mixed with Portland cement, the result is a high-performance product that has both high compressive strength *and* long durability, thus, no early cracking and premature rebar corrosion to worry about.

The only downside in this otherwise upbeat story is that most fly ash and slag suitable for cement production is still not being utilized for this purpose. Conventional Portland cement still predominates, as well as the pollution and wasted resources that come with its production. The cement industry lobbies hard to block any government legislation or Environmental Protection Agency (EPA) regulations that it feels would limit its freedom to do as it pleases.[2] This is often the case: industry will always lobby for or against anything that it sees as furthering or countering its perceived interests. However, narrow and short-term commercial policies often conflict with the wider public good, as in the case of concrete cements. The US government has been successful in specifying fly ash concrete cement in a number of construction projects, but such measures have had little effect on the private sector, where standard Portland cement or, worse, its old high-strength/low-durability counterpart, can still be used.[3] Taking into account the proven costly and/or dangerous flaws of the latter substance, an outright ban should be seriously considered. As the old saying goes about the squeaky wheel getting greased, the public must make it plain to its elected representatives that a shift to greener cements is in everyone's best interests, including those of the industry that would produce the material. (Surely Portland cement manufacturers would not want to see a return to steel-frame or—heaven forbid!—*masonry* construction.)

NEW REINFORCEMENT BARS

Although the use of the green cements and stainless-steel rebar can double the life span of reinforced-concrete buildings, one basic problem remains: the inevitable corrosion of the steel that eventually compromises their structural integrity.

One method by which the corrosion of steel rebar can be—theoretically—indefinitely postponed is cathodic protection. The corrosion of iron or an iron-based alloy is an electrochemical process. Electrochemical processes pervade the natural world. (As you read this page, countless electrochemical actions are taking place in your brain.) In the case of corroding steel rebar, a small current is generated, with the corrosion patch serving as the positive (+) pole and the closest noncorroding area of the rebar serving as the negative pole (-). In other words, the action is similar to that seen in a battery, with the different portions of the rebar acting as both the anode and the cathode. If unchecked, the rusting anode corrodes and expands, shrinking the cathode until none of the latter remains (or the building fails). All that is left is pure rust. The concrete around the rebar is—especially when moist—the electrolyte, the medium that allows the flow of this current.

The electrochemical properties of rust have been known for a long time,[4] as well as the ways by which it can be managed, both passively and actively. An example of the passive method is connecting another, more vulnerable ("less noble") metal to the one being protected.[5] If you own a standard home water heater, there is a rod inside called a sacrificial anode, usually made of aluminum or magnesium, to which the current is passively directed. This rod rusts instead of the steel of your water tank. Because of the diverse properties of the electrochemical process in various corrosion environments, different sacrificial metals are employed. One that might work well in freshwater, such as aluminum, may not do as well in saltwater, so another is used, such as zinc.

Because steel rebar is especially vulnerable to chlorides, zinc is often used as the sacrificial anode. One excellent passive protection method is coating the rebar in zinc. This zinc coating protects the rebar from corrosion. Should corrosion begin somewhere on the rebar, the zinc continues to draw it away from the steel,[6] even after much of the zinc has been "sacrificed" in the process. Of course, once the zinc coating has vanished, corrosion then begins to attack the steel, so this is not a permanent solution.

The "active" form of protection seems to offer a more permanent solution, but it is also more complicated and costly. A direct current (DC) of electricity is sent into the concrete, making the rebar a cathode. An electrical lead is connected to the rebar that draws the current away to a DC rectifier box powered by standard alternating current (AC).[7] This method has been employed for steel pipelines and ship hulls for many years, but it is more problematic when the steel is buried in concrete. It is best employed at the time of construction, although some reinforced concrete structures can be retrofitted with the devices. The connection points and rectifier boxes must be continually monitored and maintained, just as a reinforced concrete bridge must be, but this process does add costs and an extra box to maintenance checklists. Active cathodic protection of rebar also adds about 15 percent to the cost of an average freeway bridge.[8] For these reasons, some engineers are not especially drawn to the active form of cathodic protection.

NONFERROUS REBAR

A popular response of late to reinforced concrete's corrosion problems has been the use of stainless steel rebar. While stainless steel rebar does last longer than the standard mild steel version, perhaps adding a decade or two to the concrete's life span, it will eventually corrode as well. Again, it is the iron within the stainless steel that ultimately dooms it. Since at least the 1970s, scientists have been intensively researching methods by which the iron element can be completely eliminated in the rebar. By the late 1980s, the products of this research were beginning to come to market. One is GFRP (glass-fiber reinforced polymer) rebar. In tension strength, it is stronger than steel at one-fourth its weight. It is immune to many chemicals to which steel is vulnerable, such as chlorides. Because it does not conduct electricity, GFRP rebar is obviously resistant to electrochemical corrosion as well, and so will not rust. Its nonconductivity is especially useful in some applications. For example, MRI (magnetic resonance imaging) scanners in hospitals are highly sensitive to ferrous metals, including the steel rebar in the walls. Tollbooths using radio-controlled toll-collection devices, airports with radio or compass calibration pads, and high-power voltage transformer vaults can also react with the rebar buried in the concrete.[9] GFRP rebar helps counteract these problems. For the same reason, you may obtain better cell phone reception in a GFRP-reinforced structure than one using steel reinforcement. GFRP has been used for roadbeds and bridge decks. The initial data indicate that it will greatly extend the life of such structures, while at the same time significantly reducing maintenance costs.[10] This latter advantage is important, for while the steel-reinforced concrete structures built today will last longer, they will still need regular maintenance to check for corrosion—and costly repairs once it is found. The tests so far conducted on roadbeds and bridge decks using GFRP-reinforced concrete show that it should last a very long time, certainly longer than its steel-reinforced equivalents.

The physical characteristics of GFRP rebar are different from those of steel. While its tension strength is almost two times that of steel at one-fourth its weight, it is less elastic.[11] Another drawback of GFRP rebar is that it cannot be bent at the worksite to accommodate the elaborate latticework required for columns and other architectural forms. GFRP rebar can be ordered prebent for a construction project, but the small variances that can occur at the worksite may not conform to the ideal found in a blueprint. For this reason, GFRP rebar has been used where straight lengths of rebar are needed, and where the structure's component calls for compressive strength, such as the aforementioned roadbeds and bridge decks.

Likewise, the newer carbon fiber rebar now coming into the market seems to display similar virtues and drawbacks, and more testing is still needed. One application of this technology is the use of carbon fiber grids in precast concrete blocks or panels for sectionalized construction. Because of the strength and lightweight nature of carbon fiber, thinner panels of concrete can be cast, further decreasing the weight. The weight factor may be a major design consideration if, for instance, a building is planned in an area of soft soil. This virtue of being lightweight is shared by GFRP rebar, and handling either rebar is far easier for the workers than the traditional steel versions.

One material that holds much promise is aluminum bronze. Cold-drawn aluminum bronze alloys are of equivalent strength to the mild steel used in most rebar. It does not corrode away and is 35 percent cheaper than stainless steel, one of the most popular varieties of rebar now being used to fight corrosion.[12] Aluminum bronze alloys have been used in the maritime industry for decades. They hold up well in seawater, which steel does not do (unless, of course, the latter is charged by an electric current to provide cathodic protection). For example, the bronze equipment and massive propellers of the RMS *Titanic* will likely be the ship's only metallic survivor after a thousand years have passed. Copper-based alloys like bronze develop a microfine film of corrosion that protects it from further corrosion—often seen as a green patina on the metal.[13] Examples of this patina film can often be seen on bronze statues, some over two thousand years old, which have endured to this day because of the seemingly unlimited life span of this alloy. Classic bronze consists of copper combined with tin. Aluminum bronze alloys mostly consist of copper, combined with 5 to 11 percent aluminum and smaller amounts of nickel, manganese, and iron as well, though the corrosion properties of the latter are suppressed by the larger mass of the alloyed metals.

The tests performed on this alloy have proved very promising, but there doesn't seem to be much interest in the material, even though it offers the potential of providing a "forever" rebar that can also be bent on the worksite.

One argument against aluminum bronze rebar is that its price would rise as demand increases, since copper is less common in nature than iron (aluminum is at least as abundant as iron). Assuming this would be the case, and that the price for the alloy rises and is one day as expensive as stainless steel, let us do some cost comparisons, since calculations have already been performed comparing stainless-steel and standard (mild-steel) rebars. The construction costs shown are an arbitrarily chosen average; some bridges would be far less expensive, and others, far more expensive.

	Bridge Construction Costs	Life Span	Total Costs (over 500 years)
Standard rebar	$56,000,000	75 years*	$336,000,000 +
Stainless steel rebar	66,640,000	110 years*	266,560,000 +
Aluminum bronze rebar	66,640,000	500+ years†	66,640,000 (initial cost only)

*Assumes additional maintenance costs during each bridge's life span. This has not been calculated, since environmental factors, such as whether the bridge was built in a dry desert or a humid marine climate, would greatly affect such costs.

†May last two millennia, substantially increasing accrued savings. Maintenance costs are also unknown and not calculated, but they likely would be minimal in comparison to mild-steel or stainless-steel-reinforced concrete bridges.

These are conservative estimates based on the increased life span of the new fly ash concrete cements. The 19-percent-higher construction cost for the bridge when using stainless-steel rebar is amply returned by the structure's increased life span. However, this gain is relatively small in comparison to that offered by aluminum bronze rebar. Actually, we do not know how long the third bridge would last—it might be two millennia or more. In any case, it would be a very, very long time. The concrete might crack, and perhaps small chunks would fall off during the centuries, but these would likely be cosmetic deformities that could easily be patched at minimal expense. Assuming that the replacement costs for these bridges would be the same as the construction costs (we will ignore adjustments for an unknown inflation rate), one sees the enormous savings accrued over the following centuries. Not calculated are the enormous amounts of pollution generated to manufacture the cement and the tremendous waste of resources entailed in rebuilding that same bridge over and over again.

DO WE REALLY NEED REINFORCEMENT FOR ALL CONCRETE STRUCTURES?

Perhaps the most controversial solution to the problems presented by reinforced concrete is to simply eliminate the reinforcement completely. I am on dangerous ground here. While I am not necessarily advocating such measures, there are enough examples—both ancient and recent—of this kind of construction to allow me to play the role of devil's advocate.

As noted in chapter 3, the Pantheon offers a perfect paradigm for the durability of *unreinforced* concrete, and there are other instances nearer at hand and time. George Bartholomew's *unreinforced* concrete street in Bellefontaine, Ohio, has lasted over a century, during which time it required less maintenance than other nearby streets. It is now a pedestrian zone, but this change was made primarily to preserve the original concrete surface that would have been otherwise obscured by a fresh layer of modern concrete poured on top that would keep the roadway up to spec. Since concrete has enormous compressive strength, why is reinforcement needed for a street or highway, particularly if either is well enough bedded so that fractures do not lead to lateral displacement of their parts? In our world of steel-reinforced concrete, cracks are feared—and rightly so. A crack can allow the ingress of air, water, and salts, and this can lead to the corrosion of the steel rebar, endangering the structural integrity of the roadway. However, cracks in unreinforced concrete are usually benign. This is not to say that unreinforced concrete streets and highways will not need to be patched or resurfaced every so often, but the costs of these measures are far less than the expense of roadway replacement.

The open-minded engineer would say at this point, "You *might* have a point there, but the use of unreinforced concrete for other applications would not be suitable. A bridge, for instance, would require the tensile strength of some kind of reinforcement, whether that be steel or some other material, such as aluminum bronze alloys or polymer-carbon fiber composites." Yes, reinforced concrete would be preferred for most construction work, but perhaps concrete bridge building does not *exclusively* require such reinforcement.

In southern England there is a remarkable structure, the importance of which has not been widely recognized. The Hockley Viaduct is an elevated rail platform that was part of a line connecting Didcot, Newbury, Winchester, and Southampton. Completed in 1891, it provided a second independent line to the Southampton Docks in order to break the monopoly then held by the London & South Western Railway. Like some Roman bridges, it is combination of masonry and concrete, mostly the latter. In fact, it looks very much like a Roman aqueduct and has thirty-three arches. During both world wars, the viaduct was extensively used to transport military personnel and equipment to Southampton, the main embarkation point for France. The viaduct provided an especially vital link during World War II, when it was completely closed to passenger traffic to allow the transport of the mountains of war material sent to Southampton for the Normandy Invasion. In the year prior to D-Day (June 6, 1944), sixteen thousand train cars traveled across the viaduct, many of them carrying heavy tanks and artillery pieces.[14] The transport of this equipment no doubt exceeded the load capacities envisioned by the viaduct's nineteenth-century builders. The Hockley Viaduct was closed under the "Beeching Axe," the informal name given to the British government's reorganization of the country's railways under the direction of Dr. Richard Beeching in the 1960s. The reorganization closed many lines deemed "unproductive," including the one to which the Hockley Viaduct belonged. What makes this brick-clad concrete structure so interesting is that it has *no reinforcement*. The beautiful viaduct stands as a testimonial to the strength of ancient building methods applied in the Industrial Age. It is easily the most important concrete structure to have survived from the nineteenth century, not only for its beauty, but also for the lessons learned from its construction methods and its remarkable durability. Unfortunately, this splendid viaduct has been subject to much abuse since the 1960s. In some places, the adjoining walls have been demolished to pilfer the bricks. Since English Heritage refused to grant the Hockley Viaduct landmark status to ensure its preservation, a small group of local volunteers must continually paint over the scrawls of vandals, repair the damaged spots, replace stolen copings, and pull weeds to prevent them from taking root in the mortar seams and causing further injury.

We are so attached to steel and steel-reinforced concrete construction that the idea of building an *unreinforced* roadway or bridge is inconceivable to most engineers today, yet such structures undeniably possess much longer life spans and lower maintenance costs than our corroding modern structures. The concrete and masonry Aelian Bridge (now called the Ponte Sant'Angelo) in Rome, built in 134 CE by the emperor Hadrian, is doing

FIGURES 43 & 44. Views of the Hockley Viaduct in England. The structure was built of *unreinforced* concrete in the 1880s and, given minimum protection and maintenance, may last thousands of years.

just fine after nineteen hundred years, and, if given minimum protection, the Hockley Viaduct in England should also last as long. Though such construction methods may rarely be employed today, they should at least remain on the table as an option.

THE WORLD WE HAVE BUILT

I once heard an engineer who, while talking about his involvement in constructing a bridge that will have a one-hundred-year life span, concluded his remarks by saying, "By the time that thing fails, I'll be long dead." Considering that most existing bridges have a service life of fifty years, he was proud to have built something that would endure twice as long. I experienced a similar sense of the despondency when reading about the construction of the Pentagon Memorial dedicated to the 184 victims killed at the Pentagon and on American Airlines Flight 77 on 9/11. Its builders confidently predicted that it would last over a century.[15] Think about that: a memorial that will last only a little longer than the life span of a healthy person. What is the point of a memorial that will mostly be viewed by contemporaries who already have firm recollections of the tragic event it memorializes? Compare this to the many bronze and granite memorials in our nation's capital built in the nineteenth and early twentieth centuries that will, like their Roman and Greek predecessors, probably endure millennia. Such are the values of this world we have created, one in which we have come to accept the short life expectancy of not only our infrastructure but of our memorials as well.

We have built a disposable world, and we pride ourselves on being able to extend its existence a bit more, rather than seeking ways to make it permanent. We can always tear down a "permanent" structure if it stands in the way of an important public development, or, given a revival of a now largely vanished sense of aesthetics, to replace it with something more beautiful.

One altruistic belief beloved by Americans is that the world we leave to our children should be better than the one we found. Apparently, we have confused "better technology" with a "better world," for we have done a sorry job with everything else. Our principal legacy for our descendents is a soaring national debt and a corroding infrastructure. And the two are not entirely unconnected.

Let's go back and look at that comparison of the various bridges built with different rebar. Take the savings accrued by building a permanent bridge (aluminum bronze rebar does not necessarily have to be involved) and compare it to the "extended life" of a stainless-steel-reinforced concrete bridge, and multiply that by six hundred thousand (the number of rail and highway bridges in the United States). The savings in bridge construction over this five-century period in the United States alone would be just under $120 trillion, over three times the total current public debt of the *entire planet* (approximately $39 trillion as this book goes to press).

The concept of nonpermanent construction is a recent one. Before the advent of reinforced concrete, major buildings and bridges were built to last a very long time. One can walk around many European cities—particularly those that had not been subject to Allied or German bombing during World War II—and find oneself surrounded by buildings constructed centuries ago that will likely last centuries more. Look at the extraordinary beautiful ancient structures in Prague in the Czech Republic, particularly the splendid Charles Bridge (*Karlův most*) designed and built by Peter Parler in 1357. The old builders of Prague would have been struck dumb with amazement if they had been asked to construct buildings that could only last a century.

We do not need to go back to masonry construction to achieve the same durability for our buildings that our ancestors simply took for granted. We have the tools to do it now with reinforced—or *unreinforced*—concrete. A good first step would be to put into place a transition period from steel to nonferrous rebar for most construction work. In the meantime, we can begin utilizing GFRP rebar for roadbeds and bridge decks, and aluminum bronze rebar for other purposes. Since copper is a finite resource, something similar to the "X Prize" (the "X Prize" was awarded for the first successful privately financed space vehicle) should be put forward to encourage the development of a strong, enduring artificial rebar that can be bent at the worksite.

We can do this. In fact, we cannot afford *not* to do this. Depending on which course we take, our descendents will either thank us or curse us.

The Collapse of the Champlain Towers South... and How to Avoid Another Such Incident from Reoccurring

Around 1:20 am on June 24, 2021, a few of the sleeping residents of the Champlain Towers South in Surfside, Florida, were awakened by loud noises. Some describe them as "crack, crack, crack sounds." One woman thought that some sort of nighttime construction work was underway and was understandably upset.[1] Another woman, Cassie Stratton, was standing on the balcony of her apartment and talking to her husband, Dean, on the telephone. Her description was more visual than aural. As they were speaking to one another, Ms. Stratton suddenly exclaimed, "Honey, the pool is caving in. The pool is sinking into the ground!"

Dean, who was no doubt puzzled by the odd image his wife was describing—*The water in the pool was sinking? The pool itself was sinking?*—said, "What are you talking about?"

She blurted out, "The ground is shaking! Everything is shaking!"

According to Dean, she then made a "blood-curdling scream" before the line went dead.[2]

The time was approximately 1:25 a.m. when Cassie Stratton cried out. At that same moment, almost one hundred other residents of the Champlain Towers were also about die. First one section—the one with the Stratton's condo—crashed to the ground, followed a few seconds later by another. Nearby people who rushed to the scene of the collapse heard some people screaming in the rubble. By the time first responders had arrived, the cacophony of cries had diminished to one woman calling for help. After a while, even that solitary voice went silent.[3]

The lucky ones were those residing in the one north-facing wing of the Towers that remained standing. Disturbed by the horrendous crash, the residents living there put on robes and walked out into hallways that disappeared into the night and had been blanketed by the fine debris from the enormous dust cloud that arose from the collapse. These people would be rescued by first responders over the next few hours. The sole human survivors who were inside the rubble of the collapsed sections, a fifteen-year-old boy and his mother, were pulled out. The woman soon died, but her son survived with no serious injuries.[4]

If you have reached the end of *Concrete Planet* and are now reading this epilogue, you can guess what likely happened to the Champlain Towers South: rebar corrosion compromised a crucial load-bearing section of the structure—likely the walls and ceiling of the garage—and thus precipitated the collapse. The various photo and video evidence posted online that show the Towers' parking garage virtually disintegrating from rebar corrosion in the weeks and days and hours prior to the disaster seem to confirm this assumption.[5] Once the garage came apart, the entire building that sat on the parking level collapsed as well. A competing theory—that a sinkhole was responsible for the condo's destruction—was ruled out after the rubble was removed and no sinkhole discovered.[6]

Aside from the poignant personal stories about the victims of the collapse, much press attention was also focused on the seamy side of the tragedy. The period of the late 1970s and early 1980s in South Florida were remarkable for the widespread corruption that permeated both the construction and the county agencies responsible for enforcing the building codes. The most notorious example being the plain white envelopes containing money that were clipped to the construction permit applications submitted to county building inspectors for their approval.[7] While none of the parties who conceived, designed, or built the Champlain Towers South (completed in 1981) were indicted for such egregious offenses, most had what might be politely termed as *less-than-blameless pasts*.[8] These are all fascinating stories, but that is not topic of this epilog. I have reserved my feelings about such matters to a brief aside, *Musings about our Future*, at the end of *Errata and Explanations*.

The focus on corruption, while important, obscures the fundamental natural forces at play that produce such tragedies. I will mostly restrict myself to examining the likely causes of the collapse and what can be done to prevent future catastrophic failures. For example, a well-designed and constructed reinforced concrete building can also suffer the same fate as the Champlain Towers South. An ill-designed and constructed reinforced concrete building might easily outlast the well-built one. It all boils down to three factors: the building's *geographic location*, its *immediate environment*, and *how well the building is maintained*.

The first factor, geographic location, is an important consideration. An air-conditioned reinforced concrete building located in an arid desert might last a century or more. The low humidity inside and out helps prevents steel corrosion. However, the Southeastern United States and Hawaii are often hot, humid, and wet. This is not conducive to the preservation of steel rebar. Cracks in the concrete—all concrete eventually cracks—allow moisture in to do its nefarious work.[9] (Elsewhere in the United States, specifically colder climes, freeze-and-thaw cycles are a damaging factor on reinforced concrete structures.) The Southeast is also regularly visited by hurricanes. These mega-storms often produce vast amounts of rain and/or major tidal surges that frequently cause widespread flooding. At the time they strike, the public's general attention is on their immediate effects: people killed or displaced, the vast expenditure needed to rebuild the areas struck, etc. Almost nothing is mentioned about the possible deadly long-term consequences of the flooding.[10]

A towering reinforced concrete office building, hotel, or condominium may be "restored" after having been partially submerged in water for a few days or weeks. The water is pumped out, broken windows and damaged doors replaced, fresh paint applied, new carpeting laid down, etc. (As one post-disaster cleaning company puts it: "We make it look like nothing even happened.") Nevertheless, there may be a ticking time bomb in the first story and garage and, possibly, second floor as well: wet rebar beginning the corrosion process. It may progress very slowly before reaching a critical stage. After that, a load-bearing member in the lower stories will no longer be able to support the floors above and the structure will collapse—assuming it is not first condemned, evacuated, and then demolished. In the *Summary Report on Building Performance / Hurricane Katrina, 2005*, published by FEMA in 2006. the problem is referred to in a rather oblique manner. Buried in the many pages contained in the report, I found this single sentence: *The long-duration flooding* [caused by Hurricane Katrina] *led to a moisture entrapment within the walls and floors of flooded buildings, which could impact the structural integrity of building materials over time* (4.1.3)[11]

This low-key, no alarm bells, matter-of-fact description is similar in tone to engineer Frank Morabito's comments in his two inspection reports from 2018 that he submitted: one to the Champlain Towers South Tenants Association[12] (although few members actually saw it); the other to the City of Surfside Building Department presumably to gain official approval for the work. (He never expressed any alarm or suggest the evacuation of the residents. To be fair, Morabito's inspection was conducted three years before the Towers came crashing down. During this period, the situation— already very bad—continued to deteriorate.)

The second and third factors—*immediate environment* and *maintenance*—are also important variables. The Champlain Towers South was situated very close to the ocean and sat on ground with a high-water table that mostly consisted of seawater. Another factor in its immediate environment was the leaking swimming pool. While water is bad for rebar, the dissolved chlorides of seawater and the chlorinated water of a swimming pool are far more damaging and accelerate the corrosion process.[13] When water with dissolved chlorides permeates sections of the reinforced concrete, the rebar could disintegrate in a matter of years instead of decades. The sealant properties of the concrete in the pool and garage walls probably served as adequate protection for its rebar during the first decade or so, but obviously this protection declined as the concrete cracked and water intruded. The last images we have of the Towers' garage are from a video posted by the *Miami Herald*.[14] It was taken by someone across the street from garage entrance just before the building buckled and fell. It shows a torrent of water pouring into the garage from its ceiling, below which, rubble can be seen on the floor of parking level. Nothing on earth could have saved the Champlain Towers South at that point. It had reached the end of its life. Meanwhile, most of its occupants were asleep, and almost all were unaware that anything was seriously amiss. Yes, there were problems, but they were scheduled to be fixed soon. And didn't that Surfside official tell them that the Towers were in "very good shape"?[15] Cassie Stratton, who watched the building collapse around her, was the only one who knew *exactly* how bad the situation had become.

We have looked at *geographic location* and *immediate environment*, now let us look at perhaps the most crucial factor of all, the one that can collectively save us hundreds of billions of dollars and thousands of lives: *maintenance.*

Many of us have put off maintenance issues, like a car repair, at some point in our lives. Perhaps we had a low-paying job, or were starving students, or faced big medical bills. In any event, we decided to wait until we had enough money for getting new brakes, tires, or whatever else was needed. We could afford to wait and try not to use the car too much in the meantime. Once we had the funds, we got the car fixed. The cost of replacing a part didn't change much in the interim.

In the case of buildings, their timely maintenance is very inexpensive compared to putting off needed repairs. However, if the owner(s) or manager of a reinforced concrete condo or apartment building postpones the work until it's almost too late, the costs can go up by three or four orders of magnitude. How inexpensive is a corrosion repair if it's caught early? It can run a couple thousand dollars or less to repair one small corrosion patch.[16] Wait five years before repairing it and it will have grown and now might cost thirty thousand dollars to fix. Wait another five years, and you could be looking at three-hundred thousand dollars to fix the problem. If you wait twenty years, you might be looking at multiple millions—assuming by that point the whole building hasn't already come crashing down. Even if the Champlain Towers South condominium owners had decided to perform repairs shortly after receiving Morabito's inspection report, it would have still cost them close to ten million dollars[17], or just under four orders of magnitude more than a corrosion repair performed in 1995 (at 1995 prices:

$1,000). Instead, they waited another three years *after* the inspection report until the recertification requirement became mandatory in 2021. By that time, it was too late.

Signs of budding rebar corrosion are easy to spot. It is usually a crack in a wall, often with a brown stain around or underneath it. It is vitally important to fix the problem at that point, when the corrosion is just beginning to grow. Just make sure the work is correctly done—most corrosion repairs are not. (I have personally watched rebar corrosion repairs being performed and viewed others online. Not a few people performing these repairs were, to put it politely, ill-instructed in their craft. For example, some only remove the corrosion *from the side of the rebar facing them*, which leaves the corrosion on the other side free to keep growing. Ceresit, one of the firms that manufactures products for rebar corrosion repair, has an excellent video on YouTube demonstrating how to do it properly. Indeed, it is the most competent demonstration that I have yet seen of the procedure. (Even vintage rebar can be seen in the section being repaired.) It can be viewed at https://www.youtube.com/watch?v=1nla2aj-YZI. (I have no connection of any kind to Ceresit, nor do I even know anyone who works there.)

It is also important to point out that corrosion does not often run rampant throughout the entire lattice work of rebar inside the concrete of a building. In a majority of the photos taken of the fallen remains of Champlain Towers South, most of the exposed rebar sticking out from the building's fragments appears uncorroded and some seem almost pristine. All it takes is for the corrosion to destroy the rebar *in one critical section*. This cannot be over-emphasized. In structures very close to the ocean with a high-water table, the chances are good that the corrosion damage will occur on the first floor or underground garage, which is any reinforced concrete building's most vulnerable part. Sadly, this vulnerability is shared by many buildings along South Florida's coast.

The Surfside disaster was not the first of its kind in the Miami area. The Drug Enforcement Agency (DEA) building in Miami also suffered a partial collapse in 1974.[18] A leaking roof had caused rainwater to pool and the rebar to corrode until it finally failed. The section of the structure that failed was the upper rear, where the employee parking garage was. The top story collapsed upon the ones beneath, causing more damage and casualties. Seven people were killed and fourteen injured. Had the failure happened on the lower levels—as in Surfside—most the upper stories would have followed, causing far more deaths and injuries.

One result of the DEA building's collapse was that building codes in Miami-Dade and Broward Counties instituted a rule whereby a building must be inspected and re-certified 40 years after its completion.

What became known as the 40-year-recertification was passed at a time when most of the local buildings were relatively new. Besides, forty years was a distant point in the future for many of the people who lived in 1974. In short, it was essentially a minor revision to the building code that would neither upset nor inconvenience property owners or developers—though a few did suggest making it *50* years instead.[19]

Champlain Towers South was 37 years old when its condo association received the very unfavorable inspection report and were 40 years old at the time of their collapse. At this point, I believe most of us can agree that building inspections should *not* skip the first four crucial decades. And, as we have also seen, it is sometimes too late or too expensive to fix any advanced deterioration by a building's 40th year.

In the wake of this very tragic—and very *public*—catastrophe, Miami-Dade County officials began frantically perusing recent building inspection reports to discover if there might be other structures in danger of collapse. Within days, one large condominium in North Miami, the Crestview Towers, was evacuated.[20] This action was based on a six-month-old inspection report that city officials and Crestview residents saw for the first time on the day the evacuation order was given. The occupants were permitted just two hours to pack up their essentials and leave. (The Crestview Towers were/are in bad shape and, prior to their evacuation, the owners had preferred to pay fines rather than perform the more expensive and long-deferred repairs.) In the following weeks, residents of another condo and two apartment buildings were also ordered to leave their homes.

For reasons not entirely clear, another large condominium in very evident structural distress, the Port Royale, was not evacuated. Frightening photos in the *Miami Herald* of one of Port Royale's exterior crumbling away from rebar corrosion, while others showed walls of its parking garage also spalling away for the same reason. (As we've seen, corrosion on the lower levels—especially the garage—should immediately raise red flags.) As if to assuage the worried residents of the forty-year-old Port Royale, spindly steel rods—they appear to be individual rods of rebar inserted into bases made of god-knows-what material—were installed to support the garage's ceiling. When one of the Port Royale's condo owners expressed safety concerns, Amarilys Oliu, the building manager, reportedly pointed an aggressive finger at him and said, "Be careful of what you say or you'll be sorry."[21]

As I write this, the Port Royale's website is still up, and still enticing folks to move in. Here's what it reads:

> ***Sophisticated living meets refreshing locale.*** *Located just steps from beautiful Miami Beaches, The Port Royale offers a robust resort-style lifestyle and direct access to white sand beaches. What we love most about our building is our friendly and intimate community, as well as being centrally located among the best restaurants, retail shops and upscale entertainment venues that South Beach has to offer. What a perfect place to call home!*[22]

(The Port Royale resident to whom Ms. Oliu was determinedly counseling, Marash Markaj, decided not to follow her guidance about zipping his lips. He talked to reporters at the *Miami Herald*, told them what he found in the garage and, no, he doesn't feel sorry. He has a frightened daughter who kept asking him whether they were safe or not and he felt that condo's management wasn't giving him any real answers.)[23]

Other buildings are being looked at as well. Obviously, no public official wants *another* mass-casualty structural collapse to occur on his-or-her watch.

So, how many large, older residential buildings are there in South Florida? Thanks to the work of some intrepid journalists—Brittany Wallman, Lisa J. Huriash, Spencer Norris, Susannah Bryan, and Mario Ariza—at the *South Florida Sun Sentinel* in a cover story published in the Sunday, July 11, 2021, issue, we have a rough answer, albeit, for just two counties.

Poring over thousands of records in the County Appraiser's office, they came up with some distressing figures. Of the 1,499 apartment buildings and condominiums with occupancies of 100 or more in Palm Beach and Broward Counties, 47.4% of them were 40 years old or *older* and 24.8% were 30–39 years old. Including both groups, that's close to eleven hundred buildings in just those two counties that probably need to be looked at. The *Sun Sentinel* also has a zoom-in map showing their locations (*https://www.sun-sentinel.com/local/fl-ne-condo-building-ages-broward-palm-20210712-66twcriuzfa5dfrr76vxch2vry-story.html*), and most of the red dots—representing buildings 40 or more years old—are clustered in the best locations right near the ocean. I imagine many of them also have vintage swimming pools as well. As someone with marketing savvy once put it: *What a perfect place to call home!*

As briefly touched upon, South Florida has a reputation for being lax in enforcing the building codes and being patient with those owners whose condos and apartment complexes are not in compliance and *just needed a little more time*—read: *years*—to make the necessary repairs. A June 28, 2021, *Miami Herald* editorial bewailed the "sordid history of shoddy building practices" and the "slipshod construction and look-the-other-way enforcement" that have plagued South Florida for many years. This sad state of affairs, which continues to this day, will no doubt be the subject of future books. (No, I won't be writing one. South Florida is blessed with many talented journalists who can better handle that task.)

While the memory of the Champlain Towers South collapse with all its horrific casualties is still fresh in our minds, concerned Floridians should act. However, it is safe to say that there are not a few property owners or managers are hoping people will lose interest in the problem with the passage of time. They're betting that Floridians will be too distracted by the latest celebrity scandal, the latest rankings of the Miami Dolphins or Florida Panthers, or the latest coup attempt in Washington to pay any attention to what they are quietly doing behind the scenes: working with their political cronies to pass nice-sounding "reforms" in the building codes that, as in 1974, will actually accomplish very little. They will probably call it something like "The Safe Building Practices and Maintenance Act" and push it through before the more knowledgeable people have a chance to look at it and discover that it should have instead been named, "The Pretty Much the Same Building and Maintenance Practices as Before Act." (Commercial property owners and developers are among the biggest political campaign contributors in the Sunshine State.)

So, what should a *real* Safe Building Practices and Maintenance Act have in it?

(1) According to Florida law, condo associations are required to set asides reserves of no less than $10,000 for deferred maintenance purposes. However, the condo associations can waive this requirement if a majority of their voting members decide otherwise. Many do so. Pass laws to eliminate this loophole. A law or regulation isn't a law or regulation if one can decide not to observe it.

(2) Require that all reinforced concrete buildings be inspected at least once every ten years for any corrosion damage—and not skip the first four decades! (The first inspection can be a simple visual one. Little corrosion takes place in the first ten years and any damage found will likely be storm related.) Any corrosion damage spotted by residents or management must be immediately reported to the county and repaired within sixty days, even if the building recently passed an inspection. Have the law mandate that the person doing the corrosion repair photograph each step of the repair with a camera or smart phone. The steps are: 1. Removing the concrete from all sides of the corrosion spot. 2. Using manual or mechanically driven steel brushes to remove *all* the corrosion. 3. Coating the rebar with a sealant for future protection. 4. Replacing the concrete and smoothing the surface to match the surrounding elements prior to re-painting. (Without these photographic safeguards a lazy slob might could just pound away at the cracked wall, fill it in with concrete cement, and tell the building owner that he fixed the corrosion. This may sound cynical to people outside South Florida, but for South Floridians, they've seen far worse work from a few local contractors.)

(3) Require all owners or co-op owners of a building with three-or-more stories to report all single-unit or multi-unit flooding incidents within thirty days. This must include seemingly minor events, such as a resident accidently allowing his-or-her bathtub water to overflow. The consequences may be minimal or deadly, but it's best to be sure. Five years later, or in the next scheduled inspection—whichever comes first—this unit and the connecting walls and/or ceilings of the adjoining units should be subjected to light ultrasound inspection to detect any resulting corrosion, unless the corrosion has progressed to point where it becomes visible (cracked walls, rust stains, etc.), when it must be immediately repaired.

(4) Required all condo owners and residents be instructed in the dangers of rebar corrosion and especially the economic consequences of not fixing them immediately. (If people knew how much the costs of corrosion repairs would balloon if not performed early, they would be far more pro-active in fixing these problems at an early stage.)

(5) Require the use of hot-dip galvanized rebar for all new reinforced concrete building construction of buildings that have three or more stories. This is steel rebar that has a sacrificial plating of zinc. If corrosion appears, it will go after the zinc, consuming it first before attacking the steel. This will significantly increase the lifespan of both the steel and the building. It will also save building owners a considerable amount of money if their property can last eighty years instead of forty or fifty. However, we must also recognize that this measure may simply put off the building's inevitable fate, for after all the zinc has corroded away, the steel will become the next meal.

SUGGESTIONS

(1) Push to have Florida become a world leader in the adoption of non-ferrous rebar. Bronze-aluminum doesn't have corrosion

problems, and it has the same strength as steel rebar. GFRP (glass fiber-reinforced polymer) rebar, is strong and loved by construction workers (it is much lighter than steel, so they don't have to worry about throwing out their backs while carrying it around). While not a universal replacement for all rebar applications as is bronze-aluminum, GFRP can still be used to replace steel in a host of building applications.

(2) Consider going back to steel-frame construction for high-rise buildings until rebar substitutes can be used. Yes, it uses steel, but unlike reinforced concrete buildings, the corrosion of one part of the steel girder does start a cascade of other problems. Rebar corrosion destroys both the steel *and* the concrete. I know of no steel-frame building that has collapsed from corrosion issues—and I have looked hard to find evidence of one. Not that they are invulnerable to corrosion, but it will probably take a *very* long time to manifest. The downside is that steel-frame construction cannot be used in the same creative ways as reinforced concrete. (For this reason, most architects are not crazy about the older method.)

(3) During the above-suggested temporary return to steel-frame construction, consider allowing reinforced concrete structures to be built *if* they completely clad in another material. I found a common element in in old reinforced concrete structures that have improbably stood the test of time, such as W. P. Anderson's *Ingalls Building* (1903) in Cincinnati, Ohio, the first reinforced concrete skyscraper; Anatole de Baudot's *Église Saint-Jean de Montmartre* (1897–1804) in Paris, and Jørn Utzon's *Sydney Opera House* (1959–1973). Each is still standing, and each was covered in either stone, brick, or tile. This outer cladding seems to have protected the rebar from rain and humidity. It seems to work. Let's try it.

Let's construct buildings that last for centuries, like our ancestors did. There are bridges still standing from the early Industrial Age, the Renaissance, and even the Roman Empire that now carry cars and trucks, instead of people and oxcarts, simply because they did not use a metal that eventually disintegrates into brown dust.

SPECULATIONS ABOUT OUR FUTURE

As sea levels rise in Florida, and faster than previous models have suggested[24], Floridians have had to mentally adjust to the changes this will inevitably bring to their state. It's a touchy subject. Atlantic University recently studied the probable effects of a sea level rise of just two feet on Florida's most heavily traveled roads and highways. (Less trafficked roads were not studied.)[25] The study is fascinating, both for what it predicts, and does *not* predict. It found that 5.5 percent of the most heavily trafficked highways (252 centerline miles) will be under water, and a storm surge from a Category 5 hurricane would temporarily flood 20 percent or more. What's not mentioned in the study—for it was not within its purview—was the effect this would have on the less-trafficked roads and streets. A joint studied conducted by the *South Florida Sun Sentinel* and *Miami Herald* to look at the study's figures and extrapolate its probable effect on those less trafficked roads, found that 445 miles of roads in the state would be flooded by 2040, and 1,600 miles of roads by 2060.[26] What was not extrapolated from either study was the probable number of people displaced by the sea level rise, although it would likely be in the hundreds of thousands, if not millions. Of course, with rising sea levels in Florida one can expect ever larger and more catastrophic tidal surges from hurricanes. This will have a terrible effect on already-aging reinforced concrete structures. No wonder it is a touchy subject.

In the not-too-distant future, sea water will cover the southern-most part of the state and some of its present coastline. About 80 percent of the state will still remain, but the area most affected will be its crown jewel, the Florida that everyone wants to see and experience. The Florida of *Miami Vice*, of *Scarface* (the more recent one with Al Pacino), of *Key Largo*, etc. The part of Florida that I visited back in the 1990s and enjoyed immensely.

In other words, the bottom fraction of the state that includes the Miami-Dade, Broward, Palm Beach Counties and the Florida Keys. Can Floridians do something about it?

I lived in the Netherlands for several years. It's a clean, cozy country populated by pleasant people. I was especially impressed that much of that nation had been reclaimed from the sea. It's fascinating to see farms and towns where halibut and octopi once thrived. Three quarters of the Netherlands is either beneath sea level, or less than one meter above it. The Dutch had been successful keeping out the sea for centuries. At least, until 1953, when a colossal storm pummeled Northwestern Europe. It was a freakish combination of extraordinary high tides and high winds. It had one recorded wind gust of 136 mph. In terms of *sustained* wind speeds, it was comparable to Category 1 hurricane. (Although other estimates have given the 1953 storm a higher category wind number, these seemed to have confused the maximum *gust speeds* with maximum *sustained wind speeds*. By the time it reached south Britain and the Netherlands, its sustained wind speed was clocked at 70 knots, or 81 mph or 130 km/h.) The 1953 storm caused massive damage and hundreds of deaths in Great Britain and Belgium. However, the worst hit country was the Netherlands. Between 1800 to 2000 people were killed, ten percent of the country was flooded, and thousands of buildings and homes were washed away.[27]

Since 1953, the Dutch people have done everything possible to prepare for the next storm of that scale. Today, they are confident that their system of dikes, dams, and man-made land will hold up against another once-in-every-five-hundred-years storm. If you were to ask a Dutch hydro-engineer if their current defense system of dams, dikes, made-land, etc. would protect them if a similar storm struck, he-or-she would probably say something like, "We believe so, but we won't know until a storm of that size strikes us again."

Now, if you were to ask that same engineer if their system could cope with a similar or stronger storm making landfall in Holland once or twice a year as they sometimes do in Florida, he-or-she would probably say "no" or fall back on "I'm not sure." One Dutch friend told me that much of the country's western farmland would be reduced to a swamp. In others words, the Dutch system is fine for what it is designed to do: guard against the rare freak storm, not multiples of such storms. Such a system would not work in Florida, which is four times the size of the Netherlands, has perhaps six to eight times the coastline, and experiences *multiple* powerful storms.

Even if the geography were more favorable to Florida, there is also the question of cost, which also brings up the subject of taxes. The people of the Netherlands pay high federal and provincial taxes, and the provincial governments of those districts most threatened by flooding bear much of the costs for the Dutch sea defense system.[28] It is unlikely that most states in the Southeast would be willing to pay for such a system, or even *share* such a massive expense with Federal Government. When Houston, Texas, was devastated by Hurricane Harvey in 2017, there was still almost $1 billion dollars of unspent recovery money left over from Hurricane Ike (2007) because Texas city and county governments had been reluctant to raise the matching funds required by federal agencies.[29] In contrast, the people of the Netherlands are willing to pay the costs of keeping the North Sea in check because, unlike the United States, there is no room in their small country for shifting around large numbers of internally displaced people.

Florida isn't alone. The coastal areas of Texas, Alabama, North and South Carolina, Louisiana, Maryland, Virginia, and parts of Pacific western states are just as vulnerable. An especially knotty problem would be how to preserve or manage the Potomac River and Chesapeake Bay, which is shared by Maryland, Virginia, and Delaware.[30] It is estimated that lower Washington, D.C., will likely see record flooding by 2040, and more of the city may be underwater by the following century. Its magnificent monuments and buildings will probably have to be removed and reassembled on higher ground within 150 years, if not before.[31] New York's subway tunnels will probably have flooded by that time, perhaps replaced by the elevated streetcars that were once so common in the nineteenth century.[32] The San Francisco Bay Area will also look different. Pricey real estate in the towns of lower Marin County, like Larkspur, Tiburon, and parts of Sausalito will no longer be expensive, because they will be under the Bay. As for San Francisco itself, its northeastern and eastern districts will be inundated—including major tourist attractions, like Fishermen's Wharf and the Ferry Building. Like the state of Florida, about 20 percent of the city will be gone.[33]

Sadly, the principal response to rising sea levels in the United States will be the mass relocation of people to higher ground. Altitude, not waterfront ocean views, will soon be the principal determinator of what people will be seeking in the coastal regions.

I've grown philosophical about climate change. We've done this to ourselves, so now we reap the consequences. Unfortunately, while humans can adjust to such changes, thousands of species will disappear as result. Many are already gone.

I remember an incident back in September of 2020 that unnerved me. The smoke of several massive fires blew down from the north and turned day into night in San Francisco. I couldn't understand why my alarm clock had gone off when it was still dark outside. I put on my pants and shirt and walked outside. The sky was red, but everything below was as dark as night. If you looked at a house across the street, it was merely a black silhouette with almost no details that could be described. The streetlights, which work on timers, were naturally turned off. (This is why all the photographs published of this phenomenon were greatly overexposed, otherwise you would only be able make out the red sky and nothing below.) It was as if I had awoken on another planet. Of course, by then I could smell the smoke and knew the cause. Still, I had never before seen nor read about anything like this singular spectacle. Then it hit me: the world is changing and it will not be same as we remembered it. I knew this already, but on that day, it hit me like a visceral blow. The state is getting hotter, dryer, and is being gradually carbonize. The magnificent conifers of California, such as the Sequoia and Redwood trees, will eventually drift north or disappear. Water-preserving plants, like cacti, will replace them. Agricultural output will greatly diminish. Californians will just have to adapt to these changes.

Floridians will adapt as well. Like Californians, they have no choice. You cannot use prayer or positive thinking to will it away.

As a youngster, one of my favorite rides at Disneyland in Anaheim in the early 1960s was the Submarine Voyage. You would board a facsimile of a submarine that would take you to fascinating underwater locales, like the *Graveyard of Lost Ships* or *Atlantis*. Decades later, I wanted to take my wife on this goofy ride, but it had closed by then. I was heartbroken.[34]

In a century or so, there will probably be underwater scuba diving tours of the submerged remains of Mar a Lago and Miami. High-rise buildings will tower over the water that has now engulfed their first floors. The crown jewel of Florida will remain, but it will be different. It will be like a giant Atlantis, and people will visit it in droves, using glass-bottom boats or with air tanks strapped to their backs. And it would be better than the Disneyland submarine ride because it will be a *real* lost world.

Sadly, this tourist marine Mecca will not last long. The seawater will slowly dissolve the rebar of the first stories, just as it did with the Champlain Towers South, and one-by-one, the towers will collapse into the sea. Underwater tourism there will be banned as too hazardous, but by that point there will be few visitors: viewing rubble is less interesting.

The next stage will be more fascinating. The huge piles of rubble from the collapsed buildings will passively collect flotsam and sand. Gradually, small islands will form and plants will take root. By that time, the remains of countless broken windows will have been transmuted into millions of smooth pebbles of sea glass which will dot the small beaches of these islets, which may slowly combine to form larger islands. Nature will have once more transformed South Florida into a natural jewel.

And if humans are still around, the real estate developers will come to create their own version of paradise.

ERRATA AND EXPLANATIONS

THE MYSTERY OF THE ROTARY KILN IN COPLAY, PENNSYLVANIA

After *Concrete Planet* came out, one reader pointed out that the first rotary kiln in the United States for concrete cement production was not at the Coplay Cement Company in Coplay, Pennsylvania—as stated in my book—but by another firm in the same town. This is true, and I regret the error. I was misled by a historic marker that reads: *FIRST CEMENT – David O. Saylor was the first to make portland* [sic] *cement at Coplay in the United States in 1871. First use of the rotary kiln to manufacture cement on a commercial scale also was here Nov. 8, 1889.* [I think a colon or word is missing in the last sentence.]

So, who did install the first rotary kiln for concrete cement production in the United States? Examining the application request on the National Register of Historic Places Nomination Form for the marker that was eventually put up does indeed mention that a nearby "competitor" had "developed a rotary kiln that used powdered coal for fuel" in 1889—a decade before Coplay Cement adopted the process in 1899. However, Coplay's competitor is not mentioned by name. In short, the historic marker is wrong about the date and implicitly, though not directly, steers the viewer to a cement company that shares the town's name.

So, which competitor was it? Actually, the indignant reader who brought up this issue did mention another company's name, but ten years later I cannot find her email. After more sleuthing, I discovered that the company was the Keystone Portland Cement Company (also in Coplay). However, Keystone did not install their rotary kiln in 1889, but rather 1892.

Sometimes history's landmarks—even modest ones—-are not so much deliberately altered for sinister motives, but are instead simply smudged by a series of errors. I humbly suggest that the citizens of Coplay replace their historical, but mistaken, marker. I have changed the text to incorporate this new information.

WERE FRANK LLOYD WRIGHT CONTRIBUTIONS TO THE ACCEPTANCE OF REINFORCED CONCRETE IN THE BUILDING INDUSTRY OVERRATED?

One criticism that a few readers brought up was that I had devoted too many pages to Frank Lloyd Wright. I disagree. It is sad fact that many of history's most brilliant people, such as Richard Wagner, Thomas Edison, and Frank Lloyd Wright, were very flawed and often very unpleasant individuals. However, Frank Lloyd Wright was the first architect who saw that reinforced concrete could be creatively molded to construct buildings that previously could only be imagined. The influence of his designs was profound and worldwide. Wright was *the* major pioneer in the creative application of reinforced concrete. Period.

He was also a jerk.

MORTISE AND TENON CONSTRUCTION

I instead typed in "mortis and tendon" by accident. My ancient copy of Word, which I was still using in 2010 for sentimental reasons, didn't flag it. Pure laziness, an over reliance on my word-processing software, and editors not versed in carpentry terms, were my downfall. (I think my version was the original Word for Windows circa 1989). I now have the latest version of Word running on Windows 10 and it is—to use an arcane expression— really swell.

DOES THE AUTHOR HAVE A GRUDGE AGAINST REINFORCED CONCRETE?

Absolutely not. Actually, I think that concrete is one of the greatest inventions of all time. The problem is that we did not completely understand the serious problems arising from the way we used steel to reinforce it. Because the problems took decades to manifest—which were initially blamed on bad workmanship—a century passed before it dawned on us what they were. Even then, we lived in denial about the seriousness of those issues which, in short, is that existing steel-reinforced concrete structures need to be carefully maintained and that steel rebar must one day be replaced by non-steel alternatives.

STEEL VERSUS IRON. WHICH IS WORSE?

In my research after the publication of *Concrete Planet*, I discovered that I had been too hard on iron. Here's a slanderous quote from my book (page 319, paragraph 2): "No commonly used metal is more determined to return to its natural state than is iron. In iron's case, its usual natural state on the surface of the earth is iron oxide, what we call rust."

Oddly, no readers called me to task for that statement. Apparently, it is a common misconception. I had confused the corrosion properties of iron with steel. Steel is overwhelmingly iron, but to it is added a smaller amount of carbon, typically 0.002% to 2.14%. The irons used in the early nineteenth century were either *wrought iron* or *cast iron*. In comparison to the longevity of these two irons, steel has an almost ephemeral lifespan. Steel quickly begins corroding when exposed to the elements. It's like a racehorse that can't wait to get out of the gate. Steel really does want to return to its natural state: iron oxide powder. On the other hand, while wrought iron or cast iron also rusts, the rust often acts as an outer protective layer, like the benign green oxidation seen on bronze. Hundreds of beautiful cast iron frame or wholly cast-iron buildings survive to this day. The Eifel Tower is just one example of many. Maybe we should look at this long-neglected metal with fresh eyes. I've kept my error in the text of the paperback edition as a reminder to me of the evil of making assumptions.

DID CONCRETE PLANET EXAGGERATE THE DANGERS OF STEEL-REINFORCED CONCRETE?

Not a few readers thought that I had exaggerated the future dangers of steel-reinforced concrete buildings. To these people I can only say, *I wish you had all been right, and that I had been completely wrong. Sincerely.*

I was particularly distressed by headline I ran across the other day: *Structural engineer says no cause for Houston high-rise dwellers to panic following Florida building collapse.*[35] A few days after the collapse of the condo in Surfside, a couple people in Houston were interviewed by reporter Taisha Walker from local television station, KPRC. A professor at the school of architecture at the University of Houston, Joseph Colaco, did his best to ease the minds of Houstonians. "Generally, buildings don't collapse all of a sudden," he told the reporter. "There are warning signs. There will be cracks in the floors, cracks in the columns, there will be spalling of the concrete."[36] (There were "warning signs" for years before the Champlain Towers South finally fell down. It was just that no one thought to do anything about those signs until it was too late.)

Then Prof. Colaco explained how things are "done differently in Houston." He spoke about the 40-year recertification stipulation in South Florida, and the regular ten-year inspections after that. "There's really no reason to inspect a building after it's been completed," he assured her. He did qualify that astounding statement by once more referring to "tell-tale signs." The reporter apparently did not ask him if such signs should be reported or whether their quick repair should be mandated by law.

Ms. Walker also interviewed a Mister McKinney (the TV name for the local celebrity historian, Mr. R. W. McKinney) McKinney told her that that there many residential high-rises in Houston built about the same time as the one in Surfside. McKinney also informed her that they even used the similar concrete and materials. "There are thousands and thousands of buildings all over the United States and they stood up very well,"[37] he told her.

There is something both disarmingly refreshing and deeply disturbing about a confidence unsullied by bitter experience or informed contradiction. I hope Prof. Joseph Colaco and Mister McKinney visit the Miami area one day soon. They should ask about the old DEA office building, see the site where the Champlain Towers South once stood, and visit the evacuated apartment buildings and condos where so many people once lived. They can walk around the Port Royale, and viewed its damaged walls, and visit its spalling garage. On second thought, they should skip the garage: it might be vulnerable to gravity issues.

I would then ask them to return to Houston and take a closer look at those aging reinforced concrete residential high rises. Look for corrosion spots, especially those at street level or below. When you find a few, ask yourselves whether what had transpired in South Florida won't take place elsewhere? Like Texas? For example, in a city that honors a great American and a great Texan—*no, it's not Austin I'm thinking about.* Here's another hint: it's a city that has been flooded dozens of times since it was founded, and *four times* since reinforced concrete high-rises started going up there in large numbers in the late 1960s and early 1970s.[38] Yes, it's Houston I'm referring to. Be thankful Houston has not yet experienced such a tragedy, and please call for the enactment of measures to ensure it does not.

I retain some hope that South Floridians will institute reforms, but am less optimistic about Houston, where they do things *differently*.

TROUBLED SLEEP

Ever since the collapse of the Champlain Towers South, I often find myself thinking about the next such incident, its tally of preventable deaths and injuries, and sometimes can't sleep nights. I think about all those reinforced concrete residential high rises planted right next to the ocean in South Florida. I think about the similar buildings—some flooded two or three times—in Houston.

I imagine a building there where the corrosion is growing in the garage or first story, eventually spalling the concrete, but isn't repaired because maintenance has been deferred so long that the owner, let's call him Fred, considers the repair costs too high. Let's say that this building is a 22-story mixed business/residential high-rise called the Radcliffe Tower. And I imagine Fred mulling over his choices and then doing something very bad. He performs only cosmetic repairs, filling the spalled concrete cavities with quick setting cement. He then paints all the afflicted walls and, as a dollop, adds beautiful granite veneers to the lobby, along with brass fixtures coated with clear sealant to keep their near-golden glow for a long, long time. He then tells the building's residents and businesses that wants them to be 100% satisfied with newly renovated Radcliffe Tower. He also promises to promptly address any problems or complaints they might have. The online review stars increase from three to five. Fred then sells the building to some gullible real estate tycoon for 270 million dollars.

I imagine the corrosion gradually spreading underneath the fresh paint and quick-setting concrete cement until eventually—say, five years later—it destroys a crucial load-bearing portion of the building and brings the entire Radcliffe Tower down, killing hundreds of men, women, and children, many of whom will are not identified until weeks later by DNA tests.

The collapse happens long after the previous owner has moved on to other things. He is now a capital partner in a successful hedge fund. Thanks to his most recent marriage to a woman who lives in Lucerne, Fred has decided to become a Swiss national. In addition to their beautiful chalet above Lake Lucerne, they also have a full-floor Haussmann apartment on the Champs Elysée in Paris, where they are entertaining dinner guests a few hours before the disaster takes place. They joke and laugh over their plates of cassoulet with duck confit and toast one another with glasses filled with 1970 Mouton Rothschild or 1995 Dom Pérignon.

The following morning, when Fred learns the learns the sad news, he assumes a somber mood and tells his friends and family back in the States— mostly by email—that he was devastated upon hearing what had happened to the Radcliffe Tower. He had spent so much money and time on extensive repairs so that the Radcliff would *never* suffer a fate like this. Maybe the contractor didn't do his job properly. That must be it.

His muted mood slowly diminishes over the next few weeks and vanishes entirely after press coverage about the collapse drops to near zero. A few months later, authorities investigating the disaster say that they would like to talk to him about certain details they've uncovered. For instance, the company he hired to do the repairs on his former building. It was headquartered in Panama and closed down just eight months after it was incorporated. Even stranger, its only recorded job was the work performed on the Radcliffe Tower.

Fred's Swiss lawyer advises him to only answer written questions. The attorney also insists that he carefully examine and perhaps edit Fred's responses before the document is sent back to the investigating agency. Fred agrees. Eventually, the matter seems to have been forgotten, but Fred is advised by his lawyer to not return to United States, for there may be a sealed arrest warrant waiting for him. Fred agrees.

Fred has not cut off all ties with his former homeland. His hedge fund still makes considerable contributions—all 100% tax write-offs—to celebrated non-profit charities headquartered in the United States, like the *World Wildlife Fund* or *Make a Smile Foundation,* which he makes sure to tell all his friends and family when the subject appears relevant to what they're discussing at the moment. (It is bad form to boast about such things.) Fred also keeps one foot, so to speak, back in the US: a nine-figure index fund with a prominent Wall Street investment firm. Naturally, it is legally held by a shell account in the Cayman Islands which, in turn is owned by another shell account in Lichtenstein, which is less than a two-hour-drive from Lucerne. (This keeps his money safe from civil suits initiated by relatives of those who died in the collapse.) And he still hosts visiting friends and family from the States every summer. And they are all well aware of the one topic that must never be brought up in his presence: that terrible tragedy back home. And when they come to visit, his regional accent—which has been mostly MIA from his spoken English since he moved abroad—will return with gusto, and even sounds a couple decibels louder than normal.

And all the visitors will count themselves damn lucky to have Fred as a friend or relation.

Three years after the collapse of Radcliffe Tower, Fred obtains a pardon from President of the United States, an old golfing buddy. Of course, some people in the media claim that the pardon it is likely due to Fred's contribution of 1 million dollars to the president's election campaign, and another million to the spectacular inaugural festivities.

Fred doesn't care. He never has.

Meanwhile, all over the world, hundreds of hidden time bombs continue to tick away. Corrosion is a natural process, oblivious and uncaring of what measures we decide—or not decide—to take. It just slowly consumes steel like a vampire does blood. It does what it does. Relentlessly.

NOTES

CHAPTER 1: ORIGINS

1. Department of Materials Science and Engineering, University of Illinois–Urbana-Champaign, "History of Concrete—A Timeline," http://matse1.matse.illinois.edu/home.html (accessed June 1, 2011).

2. Jennifer Viergas, "Early Weapon Evidence Reveals Bloody Past," Discovery News, March 31, 2008, http://dsc.discovery.com/news/2008/03/31/earliest-weapon-human.html (accessed May 12, 2011).

3. Steven Mithren, *After the Ice: A Global Human History, 20,000–5,000 BC* (Cambridge, MA: Harvard University Press, 2006), pp. 89–90.

4. Charles C. Mann, "The Birth of Religion," *National Geographic*, June 2011, pp. 41, 45.

5. Edward G. Nawy, *Concrete Construction Engineering Handbook* (n.p.: CRC Press, 1997), pp. 21–29.

6. A. Hauptmann and Ü. Yalcin, "Lime Plaster, Cement and the First Puzzolanic Reaction," *Paléorient* 26, no. 2 (2000): 61–62.

7. William Ury, "A Journey to Harran," Interreligious Insight, http://www.interreligiousinsight.org/October2005/WilliamUry10-05.html (accessed May 2011).

8. Gen. 11:31; 12:4–5.

9. Gen. 11:28, 11:31.

10. Louis Ginzburg, "In the Fiery Furnace," *Legends of the Jews*, April 27, 2011, http://www.theologicalhistory.com/?p=1001 (accessed May 1, 2011).

11. "Jewish, Christian, Muslim History in Sanliurfa, Turkey," Vagabond Journey, http://www.vagabondjourney.com/209-0236-jewish-christian-muslim-history-in-turkey.shtml (accessed June 1, 2011).

12. Dastan Rashid, "Xenophon and the Kurds," Department of Archaeology and Ancient History, Uppsala University, Finland (2005), http://www .arkeologi.uu.se/ark/education/CD/Cuppsats/Rashid.pdf (accessed September 2010; link now unavailable).

13. Livy (Titus Livius), *History of Rome since the Foundation* (*Periochae 106:53*), transl. Jona Lendering, corrected by Andrew Smith, http://www.livius .org/li-ln/livy/periochae/periochae106.html (accessed May 12, 2011).

14. Chris Scarre, *Chronicle of the Roman Emperors* (London: Thames & Hudson, 1995) p. 145.

15. Ibid., p. 173.

16. "Battle of Harran," Medieval Times History, http://www.medieval times.info/medieval-battles/battle-of-harran.html (accessed May 10, 2011).

17. Klaus Schmidt, "Göbekli Tepe, Southeastern Turkey. A Preliminary Report on the 1995–1999 Excavations," *Paléorient* 26, no.1 (2000): 45–46.

18. H. Hauptmann, "The Urfa Region," *The Neolithic in Turkey: The Cradle of Civilization*, ed. M. Özoğan and N. Basgalen (Istanbul, Turkey: Arkeoloji ve Sanat Yamlari, 1999), pp. 65–86.

19. K. Kris Hirst, "The Development of Archaeological Method: History of Archaeology, Part 5," http://archaeology.about.com/cs/educationalresour/a/ history5.htm (accessed April 17, 2011).

20. Patty Jo Watson, *Robert John Braidwood, 1907–2003* (University of Chicago Press Office, January 15, 2003), http://www-news.uchicago.edu/ releases/03/030115.braidwood.shtml (accessed June 2, 2011).

21. *Wikipedia*, "Stuart Piggot" http://en.wikipedia.org/wiki/Stuart_Piggott (accessed June 2, 2011).

22. Patty Jo Watson, "Robert John Braidwood, July 29, 1907–January 15, 2003," in *Proceedings of the American Philosophical Society* 149, no. 2, http://www .amphilsoc.org/sites/default/files/490208.pdf (accessed August 17, 2011).

23. J. Mellaart, *Catal Huyuk—A Neolithic Town in Anatolia* (New York: McGraw-Hill, 2000).

24. Bill Broadway, "Earliest Woven Cloth Dated to 7000 BC," *New York Times*, August 26, 1993.

25. Elif Su, "Another Time, Another Life—Çatalhöyük," *Skylife*, August 2006, http://www.turkishairlines.com/en-INT/skylife/2006/august/articles/ catalhoyuk.aspx (accessed December 11, 2010).

26. Jonathan Last, "Çatalhöyük—1999 Archive Report," Çatalhöyük—

Excavations of a Neolithic Anatolian Höyük, http://www.catalhoyuk.com/ archive_reports/1999/ar99_11.html (accessed April 12, 2011).

27. Garth Bawden, "Theories of Middle East Sedentism," University of New Mexico, http://www.unm.edu/~gbawden/328-theory/328-theory.htm (accessed June 2, 2011).

28. "Who Are the Kurds?" *Washington Post*, February 1999, http://www .washingtonpost.com/wp-srv/inatl/daily/feb99/kurdprofile.htm (accessed June 2, 2011).

29. Ibid.

30. "Kurdistan—Turkey: Insurrection," Global Security, http://www.global security.org/military/world/war/kurdistan-turkey-insurrection.htm (accessed June 2, 2011).

31. Robert J. Braidwood and Linda S. Braidwood, "The Joint Prehistoric Project: 1999–2000 Annual Report," Oriental Institute of the University of Chicago, http:// oi.uchicago.edu/research/pubs/ar/99-00/prehistoric.html (accessed June 2, 2011)

32. Stephen Kinzer, "Turkey Catches Rebel Leader," *New York Times*, February 17, 1999.

33. Manfred Heun et al., "Site of Einkorn Wheat Domestication Identified by DNA Fingerprinting," *Science* 278, November 14, 1997, pp. 1312–14.

34. Guillaume Perrier, "Le premier ministre turc annonce un plan de développement de la région kurde," *Le Monde*, May 30, 2008.

35. Mithren, *After the Ice*, p. 89.

36. Ibid., pp. 88–89.

37. "Nevali Çori," The Middle East Explorer, http://www.middleeast explorer.com/Turkey/Nevali-Cori (accessed June 2, 2011).

38. Ibid.

39. Schmidt, "Göbekli Tepe, Southeastern Turkey," p. 46.

40. Andrew Curry, "Göbekli Tepe: World's First Temple?" *Smithsonian Magazine*, November 2008, http://www.smithsonianmag.com/history-archaeology/ gobekli-tepe.html (accessed November 12, 2010).

41. Klaus Schmidt, "Göbekli Tepe, Southeastern Turkey. A Preliminary Report on the 1995–1999 Excavations," *Paléorient* 26, no. 1 (2000): 47.

42. Charles C. Mann, "The Birth of Religion," *National Geographic*, June 2011, p. 39.

43. "Flores Man," Nature News, http://www.nature.com/news/specials/ flores/index.html (accessed June 2, 2011).

44. "Zuerst kam der Tempel, dann die Stadt," Aus dem Hollerbusch, May 30, 2011, http://hollerbusch.wordpress.com/2011/05/30/zuerst-kam-der-tempel -dann-die-stadt/ (accessed June 2, 2011).

45. "Robert, Linda Braidwood, Pioneers in Prehistoric Archaeology Die," *University of Chicago Chronicle*, January 23, 2003.

46. Y. Garfinkel, "Burnt Lime Products and Social Implications in the Pre-Pottery Neolithic B Villages of the Near East," *Paléorient* 13, no. 1 (1987): 74.

47. John Webb and Marian Domanski, "Fire and Stone," *Science* 125 (August 14, 2009): 820.

48. Julia Jackson, James P. Mehl, and Klaus K. E. Neuendorf, *Glossary of Geology*, 5th ed. (Alexandria, VA: American Geological Institute, 2005), p. 371.

49. Author's practical, if unscientific, experiments to create lime. In order to test the theory of the "casual discovery" of lime (see J. D. Frierman, "Lime Burning as the Precursor of Fired Ceramics," *Israel Exploration Journal* 21 [1971]: 212–16), I decided to conduct some informal tests. I obtained some hard limestone from the Santa Cruz Mountains, approximately 65 km (*ca.* 45 miles) south of San Francisco. I broke up the larger pieces of limestone with a hammer and put the fragments into an iron pot with a lid, which I then covered with charcoal in the basin of my bar-beque. I lit the charcoal and added more during the course of the day. I allowed the barbeque to cool down overnight before looking inside the pot. To my astonish-ment, the rocks were unchanged. I decided to break the limestone into smaller pieces, making sure that none were more than two inches thick (some were smaller). I then repeated the experiment, keeping the coals going all day (at least twelve hours), and again allowing the whole to cool overnight. The next morning I found no change to the stones. By this time, the heat had almost destroyed the connection points where the aluminum legs joined the barbeque basin, as well as the lid to its cover. Still determined to make lime, I decided to give it another shot. I came to the conclusion that the iron pot was not conducting the heat properly— though it should have—and so decided to bury the limestone pieces in the char-coal. Not wanting to see my injured barbeque collapse while filled with hot coals, I detached the lid from its cover and the aluminum legs from its basin, and set the whole thing in an old Radio Flyer® children's wagon, propping it up with ceramic flowerpots. I now had to use pliers to take the lid off and put it back in place. After burying the limestone chunks into a pile of charcoal, and keeping each stone widely spaced from the others to make sure that it would be thoroughly baked, I lit the charcoal. I kept the coals going throughout the day, adding more every couple

of hours. After twelve hours, I retired for the evening, confident that I would finally find the rocks transformed into lime the following morning. Needless to say, this was not the case. Further research confirmed that calcinating limestone is no easy affair. I am reasonably certain that lime could not have been discovered by simply building a campfire in a limestone declivity. Besides, the surrounding stone would have acted as insulation.

I commend Dr. Frierman for taking the time to test the validity of a widely held assumption. However, he used chalk for his experiments, and not the hard limestone found at the late Paleolithic and early Neolithic sites. Nor could I discover chalk deposits anywhere near the excavations. Although there are chalk deposits in western Turkey (near Pamukkale and on the Black Sea coast), these are hundreds of kilometers away from the upper Tigris and Euphrates. While flints and obsidian rocks were traded over long distances in the Neolithic period, there is no record of chalk having been traded, and the large amount needed to produce useful amounts of lime would have required a transport method that had not yet been invented: pack animals.

50. Paul J. Krumnacher, "Lime and Cement Technology: Transition from Traditional to Standardized Treatment Methods" (master's thesis, Virginia Polytechnic Institute, February 5, 2001), p. 7.

51. Karen Rhea Nemet-Nejat, *Daily Life in Ancient Mesopotamia* (Peabody, MA: Henrickson, 2002), p. 190.

52. "Information for the Media," National Lightning Safety Institute, http://www.lightningsafety.com/nlsi_info/media.htm (accessed June 2, 2011).

53. "Where Lightning Strikes Most," Weather Questing, http://www.weatherquesting.com/where-lightning-hits.htm (accessed June 2, 2011).

54. "Floods, Rains Continue to Cause Havoc in Pakistan," *Earth Times*, August 11, 2010, http://www.earthtimes.org/articles/news/338896,continue-cause-havoc-pakistan.html (accessed June 2, 2011).

55. Reuters, "Lightning Strike Kills 68 Dairy Cows in Australia," *Planet Ark*, November 5, 2005, http://www.planetark.com/dailynewsstory.cfm/newsid/33303/story.htm (accessed June 2, 2011).

56. C. Suetonius Tranquillus, *The Lives of the Twelve Caesars*, trans. J. C. Rolfe, book 4 (Cambridge, MA: Loeb Classical Library, 1914), p. 28.

57. Rushton M. Dorman, *The Origin of Primitive Superstitions and Their Development into Worship* (New York: self-published, 1881), pp. 263–66.

58. Lin Chen, *Dadiwan Relics Break Archaeological Records*, China.org, http://www.china.org.cn/english/culture/48220.htm (accessed June 3, 2011).

59. Ivana Radovanovic, "'Deep Time' Metaphor: Mnenomic and Apotropaic Practices at Lepenski Vir," *Journal of Social Archaeology* 3 (2003): 46–74. See also the excellent *Wikipedia* entry for Lepenski Vir: http://en.wikipedia.org/wiki/Lepenski_Vir (accessed June 3, 2011).

CHAPTER 2: TOWERING ZIGGURATS, CONCRETE PYRAMIDS, AND MINOAN MAZES

1. Florence Dunn Friedman, "The Underground Relief Panels of King Djoser at the Step Pyramid Complex," *Journal of the American Research Center in Egypt* 32 (1995): 3–9.

2. W. B. Emery, "Preliminary Report on the Excavations at North Saqqara, 1964–65," *Journal of Egyptian Archaeology* 51 (December 1965): 8. Note: Imhotep would receive many titles of veneration. Few pharaohs enjoyed such fame and respect after death.

3. Friedman, "The Underground Relief Panels of King Djoser at the Step Pyramid Complex," pp. 3–20.

4. Bernard Erlin and William G. Hime, "Evaluating Mortar Deterioration," *APT Bulletin* 19, no. 4 (1987): 8.

5. "Exploring the Pyramids," *National Geographic*, http://www.national geographic.com/pyramids/pyramids.html (accessed June 3, 2011).

6. Joseph Davidovits, "X-Ray of the Pyramid Stones," *Science in Egyptology, Proceedings of the Science in Egyptology Symposia* (Manchester, UK, 1984): 511–20.

7. Joseph Davidovits and Margie Morris, *The Pyramids: An Enigma Solved* (New York: Hippocrene Books, 1988).

8. "Portland, Blended, and Other Hydraulic Cements," http://www.ctu .edu.vn/colleges/tech/bomon/ktxd/baigiang/CONCRETE/Chap.2/Chap2.pdf, p. 21 on pdf file (accessed August 11, 2010). Note: The pale granite of San Francisco City Hall is also often mistaken by visitors for cast concrete, which it closely resembles. Ask any SF tour-bus driver.

9. John Noble Wilford, "Scientist Says Concrete Was Used in Pyramids," *New York Times*, November 10, 2006, http://www.nytimes.com/2006/11/30/science/30cnd-pyramid.html (accessed June 3, 2011).

10. Ibid.

11. M. W. Barsoum, A. Ganguly, and G. Hug, "Microstructural Evidence of

Reconstituted Limestone Blocks in the Great Pyramids of Egypt," *Journal of the American Ceramic Society* 89, no. 12 (December 2006): 3788–96.

12. Michel Barsoum, "The Great Pyramids of Giza; Evidence for Cast Blocks," http://www.scribd.com/Chris8157/documents (accessed June 3, 2011).

13. Colin Nickerson, "Role of Concrete in Ancient Pyramids Debated," *Boston Globe*, May 8, 2008, http://www.signonsandiego.com/uniontrib/20080508/news_1c08pyramid.html (accessed June 3, 2011). Story ran in the *San Diego Union* on May 8, 2008.

14. Jana Varcova, David Koloušek, and Jana Schweigstillová, "Pyramids and Geopolymers?" *Keramický zpravoda* 27, no. 2 (February 27, 2011): 5–13. Note: Besides noting the daunting challenge of transporting so much natron (hundreds of tons) to the pyramid worksites, the authors also review the chemical evidence.

15. Dipayan Jana, "The Great Pyramid Debate. Evidence from Detailed Petrographic Examinations of Casing Stones from the Great Pyramid of Khufu, A Natural Limestone from Tura, and a Man-Made (Geopolymeric) Limestone," *Proceedings of the Twenty-Ninth Conference on Cement Microscopy* (Quebec City, Canada, May 20–24, 2007): 207–266. Note: Jana's exhaustive analysis should have ended the debate, unless one takes a position of extreme skepticism and doubts the ultimate provenance of the examined stones (the casement stone came from the British Museum, the Tura limestone from Egypt, and the polymeric concrete from Mr. Davidovits). However, the provenance question was resolved the following year by Liritzis et al. (see below), who examined samples taken directly from monuments (I. Liritzis et al., "Mineralogical, Petrological and Radioactivity Aspects of Some Building Material from Egyptian Old Kingdom Monuments," *Journal of Cultural Heritage* 9, no. 1 [January 2008]). Note: This study subjected Egyptian monument stones to spectrographic and X-ray analysis. In addition, the scientists found numerous fossils of small sea creatures in the limestone that match those at the nearby limestone quarries in both species and distribution patterns. (Fossils are common in limestone, but not after it has been processed into concrete—the exception being those impressions, unquestionably of *Homo sapiens*, found in front of Grauman's Chinese Theatre in Hollywood, California.)

16. Mark Lehner, "Some Observations on the Layout of the Khufu and Khafre Pyramids," *Journal of the American Research Center in Egypt* 20 (1983): 7, 12–13.

17. John Noble Wilford, "Scientist Says Concrete Was Used in Pyramids," *New York Times*, November 10, 2006, http://www.nytimes.com/2006/11/30/science/30cnd-pyramid.html (accessed June 3, 2011).

18. "The Minoan Civilization of Crete," BBC Online, http://www.bbc .co.uk/dna/h2g2/A765146-54 (accessed June 4, 2011).

19. "Minoan Crete," Timeless Myths, http://www.timelessmyths.com/ classical/crete.html (accessed June 4, 2011).

20. Vickie James Yiannias, "What's New Is Old Again," *Greek News Online*, September 29, 2008, http://www.greeknewsonline.com/?p=9171 (accessed June 4, 2011).

21. Richard A. Lovett, "'Atlantis' Eruption Twice as Big as Previously Believed, Study Suggests," National Geographic News Online, August 23, 2006, http://news.nationalgeographic.com/news/2006/08/060823-thera-volcano .html (accessed June 4, 2011).

22. Ibid.

23. *Wikipedia*, "Linear A," http://en.wikipedia.org/wiki/Linear_A (accessed June 4, 2011).

24. Colin F. MacDonald and Jan M. Driessen, "The Drainage System of the Domestic Quarter in the Palace at Knossos," *Annual of the British School at Athens* 83 (1988): 235–58. See also Vincenzo La Rosa, "A Hypothesis on Earthquakes and Political Power in Minoan Crete," *Annali di Geofisica* 38, nos. 5–6 (November– December 1995): 881–91, http://www.earth-prints.org/bitstream/2122/1806/1/ 38%20la%20rosa.pdf (accessed June 4, 2011). Note: Article touches on Minoan concrete, including nice photographs of it in situ (see pp. 884–85).

25. Paul J. Krumnacher, "Lime and Cement Technology: Transition from Traditional to Standardized Treatment Methods" (master's thesis, Virginia Polytechnic Institute, February 5, 2001): 4.

CHAPTER 3: THE GOLD STANDARD

1. Arthur Nussbaum, "The Significance of Roman Law in the History of International Law," *University of Pennsylvania Law Review* 110, no. 5 (March 1952): 578–687.

2. Thucydides, *The Peloponnesian War*. Many good translations are available in English.

3. Pliny the Younger (Gaius Plinius Luci), *Letters*, book 2, letters 12 and 13; book 3, letter 4. The best English translation is easily the one by Betty Radice, *The Letters of Pliny the Younger* (London: Penguin Classics, 1969). See also Marcus

Tullius Cicero, *In Verrem*, an English translation of which is available at the Society of Ancient Language website: http://www.uah.edu/student_life/organizations/ SAL/texts/latin/classical/cicero/inverrems1e.html (accessed June 5, 2011).

4. "Background to Life in the Roman Empire," Schools History, http:// www.schoolshistory.org.uk/gcseromebackround.htm (accessed June 5, 2011).

5. "Via Aurelia: The Roman Empire's Lost Highway," *Archaeology News*, http://archaeologynews.multiply.com/journal/item/773 (accessed June 22, 2011). See also *Wikipedia*, "Tabula Peutingeriana," http://en.wikipedia.org/wiki/ Tabula_Peutingeriana (accessed June 5, 2011).

6. "The Practice of Sterilization of Surgical Instruments Was Introduced by . . . ?" The Student Doctor Network, http://www.studentdoctor.net/panda bearmd/2007/11/14/everything-you-need-to-know-about-complementary-and -alternative-medicine-part-2/ (accessed June 4, 2011).

7. "Roman Military Trivia," Historum, http://www.historum.com/ancient -history/9984-roman-military-trivia.html (accessed June 4, 2011). See also "The Practice of Sterilization of Surgical Instruments."

8. John W. Humphrey, John P. Oleson, and Andrew N. Sherwood, *Greek and Roman Technology: A Source Book* (London: Routledge, 1999), p. 28.

9. Lionel Casson, *Travel in the Ancient World* (Baltimore: Johns Hopkins University Press, 1994), pp. 203–211.

10. T. R. Reid, "The World According to Rome," *National Geographic*, August 1997, p. 18.

11. Theophrastus's *On Stones* (English translation) can be found at Farlang, http://www.farlang.com/gemstones/theophrastus-on-stones/page_054 (accessed June 5, 2011).

12. Anna Marguerite McCann, "The Harbor and Fishery Remains at Cosa, Italy," *Journal of Field Archaeology* 6, no. 4 (Winter 1979): 391–411.

13. Cato the Elder (Marcus Porcius Cato), *De Agricultura* (*On Agriculture*); also known as *De Re Rustica* (*On Rural Affairs*).

14. Vitruvius (Marcus Vitruvius Pollio), *De Architectura* (*On Architecture*); also known as *The Ten Books on Architecture*.

15. Pliny the Elder (Gaius Plinius Secundus), *Naturalis Historiae* (*Natural History*).

16. Plutarch (Lucius Mestrius Plutarchus), *The Lives of the Noble Grecians and Romans*, trans. John Dryden, rev. Arthur Hugh Clough (Chicago: Encyclopaedia Britannica, 1952), pp. 276–92. Note: I updated the Dryden translation

from Greek of the little verse written about Cato's sour character to conform to more contemporary English.

17. Cato the Elder, *De Agricultura*, book 1, chap. 38, lines 1–4. I based my text on the translation by W. D. Cooper and H. B. Ash, published in the Loeb Classical Library series and largely influenced by an earlier German translation found in the Teubner edition by Goertz. The text comes from Bill Thayer's wonderful website Lacus Curtius: http://penelope.uchicago.edu/Thayer/E/Roman/Texts/Cato/De_Agricultura/A*.html (accessed August 21, 2011).

18. David Moore, "The Secrets of Roman Concrete," *Constructor*, September 2002, p. 16.

19. Vitruvius, *De Architectura*, book 5, chap. 12, paras. 2–4.

20. Vitruvius, *De Architectura*. I worked with the translation by Morris Hicky Morgan, published by Harvard University Press (1914), which is now in the public domain and can be found on Project Gutenberg's website at http://www.gutenberg.org/files/20239/20239-h/29239-h.htm (accessed August 20, 2011). As with many of the older translations (which are often based on even older versions), I may have slightly modified the English text to make its sense clearer. I invite people to compare my text with the original if they suspect that I might have strayed too far from the primary sources.

21. Derek Williams, *Romans and Barbarians—Four Views from the Empire's Edge* (New York: St. Martin's Press, 1998), p. 131.

22. *Wikipedia*, "I Quattro Libri dell'Architettura," http://en.wikipedia.org/wiki/I_quattro_libri_dell%27architettura (accessed June 5, 2011).

23. Vitruvius, *De Architectura*, book 1, chap. 1, para. 7.

24. Vitruvius, *De Architectura*, book 8, chap. 6, paras. 10–11.

25. Vitruvius *De Architectura*, book 2, chap. 4, para. 1.

26. Vitruvius *De Architectura*, book 2, chap. 5, para. 1.

27. Vitruvius *De Architectura*, book 2, chap. 6, para. 1.

28. Vitruvius *De Architectura*, book 5, chap. 12, paras. 2–4.

29. Robert L. Hohlfelder, "Beyond Coincidence? Marcus Agrippa and King Herod's Harbor," *Journal of Near Eastern Studies* 59, no. 4 (October 2000): 241–42. Note: Perhaps the best examination of the political dimensions of Herod's massive building project at Caesarea, as well as the complicated relationship the Judean king had with Rome. Analysis of the wood used for the harbor shows that it came from Central Europe. My hypothesis about its specific origin (the Danube area) and use is merely a suggestion, and I am open to alternative scenarios.

30. Ibid., pp. 249–50.

31. Ibid., pp. 248–49.

32. Ibid., p. 242–43.

33. Ibid., p. 249.

34. John P. Oleson, Robert L. Hohlfelder, Avner Raban, and R. Lindley Vann, "The Caesarea Ancient Harbor Excavation Project (C. A. H. E. P.): Preliminary Report on the 1980–1983 Seasons," *Journal of Field Archaeology* 1, no. 3 (Autumn 1984): 288–89.

35. Ibid., pp. 297–98.

36. Ibid., p. 298.

37. Ibid., pp. 297–99.

38. Ibid., pp. 284–85.

39. Ibid., p. 297.

40. Robert L. Hohlfelder, John P. Oleson, Avner Raban, and R. Lindley Vann, "Herod's Harbour at Caesarea Maritime," *Biblical Archaeologist* 47, no. 3 (Summer 1983): 137.

41. Michael Vasta, "Flavian Visual Propaganda: Building a Dynasty," *Constructing the Past* 8, no. 1 (2007): 112–14.

42. *Wikipedia*, "Caesarea Maritima," http://en.wikipedia.org/wiki/Caesarea _Maritima (accessed June 6, 2011).

43. Christopher Brandon et al., "The Roman Maritime Concrete Study (ROMACONS): The Harbour of Cheronisos in Crete and Its Italian Connection," *Méditerranée* 104 (2005): 25–29.

44. Anthony A. Barrett, *Caligula: The Corruption of Power* (New Haven, CT: Yale University Press, 1998), pp. 198–200. See also Samuel Ball Platner, *A Topographical Dictionary of Ancient Rome* (London: Oxford University Press, 1929), pp. 370–71.

45. Pliny the Elder, *Historia Naturalis*, book 36, para. 70. My quotes are based on the excellent translation by John F. Healy for Penguin Classics. Sadly, it is not the complete encyclopedia but only "a selection." Those interested in reading Pliny's entire tome can consult the version at Lacus Curtius: http:// penelope.uchicago.edu/Thayer/E/Roman/Texts/Pliny_the_Elder/home.html.

46. Ibid., p. 16, paras. 201–202.

47. Ibid., book 36, para. 53.

48. Hugh Plommer, *Vitruvius and Later Roman Building Manuals* (London: Cambridge University Press, 1973).

49. Chris Scarre, *Chronicle of the Roman Emperors* (London: Thames & Hudson, 1995), p. 52.

50. Suetonius, *The Twelve Caesars*, book 6, chap. 31.

51. Ibid.

52. Cassius Dio, *History of Rome*, book 63, chap. 29.

53. Suetonius, *The Twelve Caesars*, book 8, chap. 13.

54. M. J. Carter, "Gladiatorial Combat: The Rules of Engagement," *Classical Journal* 102, no. 2 (December–January 2006–2007): 97–114.

55. "Roman Gladiators," Ancient History Encyclopedia, http://www.ancient.eu.com/article/38/ (accessed June 6, 2011).

56. Thomas Horner-Dixon, *The Upside of Down: Catastrophe, Creativity, and the Renewal of Civilization* (Washington, DC: Island Press/Shearwater Books, 2006), pp. 31–56. Copies of his online essay on the Colosseum's construction can be downloaded at http://www.theupsideofdown.com/rome/colosseum.

57. K. M. Coleman, "Launching into History: Aquatic Displays in the Early Empire," *Journal of Roman Studies* 83 (1993): 48–49, 58–60.

58. David Moore, "How I Became Interested in Roman Concrete," Roman Concrete.com, http://www.romanconcrete.com/docs/my_interest/my_interest.htm (accessed June 6, 2011).

59. Edward Gibbon, *The History of the Decline and Fall of the Roman Empire* (New York: Fred de Fau and Co., 1906), p. 1.

60. N. S. Gill, "Hadrian's Wall," About.com, Ancient-Classical History, http://ancienthistory.about.com/cs/rome/a/aa060600a.htm (accessed June 6, 2011).

61. Leofranc Holford-Strevens, *Aulus Gellius* (London: Gerald Duckworth, 1988), pp. 72–92.

62. Adam Ziolkowski, "Was Agrippa's Pantheon the Temple of Mars in Campo?" *Papers of the British School at Rome* 62 (1994): 261–77.

63. Ibid., p. 263.

64. "Pantheon," Great Buildings, http://www.greatbuildings.com/buildings/Pantheon.html (accessed June 6, 2011).

65. Jo Marchant, "Is the Roman Pantheon a Giant Sundial?" *New Scientist* (February 4, 2009), http://www.sciencedirect.com/science/article/pii/S026240790960261X (accessed June 6, 2011).

66. Lambert Rosenbusch, "Pantheon as an Image of the Universe," European Mathematical Information Service, http://www.emis.de/journals/NNJ/conf_reps_v6n1-Rosenbusch.html (accessed June 7, 2011).

67. David Moore, "The Riddle of Ancient Roman Concrete," *Spillway* (February 1993), a newsletter of the United States Department of Interior, Bureau of Reclamation, Upper Colorado Region. A copy may be found at http://www.romanconcrete.com/docs/spillway/spillway.htm (accessed June 11, 2011).

68. Cassius Dio, *History of Rome*, book 53, sect. 27, lines 2–4.

69. "The Roman Pantheon," Socyberty, http://socyberty.com/society/the-roman-pantheon/ (accessed June 7, 2011).

CHAPTER 4: CONCRETE IN MESOAMERICA AND RENAISSANCE EUROPE

1. The subject of the amount of limestone kilned and its effect on deforestation is controversial. It must be remembered that plaster weathers away, so calculations based on remaining amounts are dicey propositions. For more information on current thinking on the subject, see Elliot M. Abrams and David J. Rue, "The Causes and Consequences of Deforestation among the Prehistoric Maya," *Human Ecology* 16, no. 4 (December 1988): 377–95; see also D. Clark Wernecke, "A Burning Question: Maya Lime Technology and the Maya Forest," *Journal of Ethnobiology* 28, no. 2 (September 2008): 200–210; also check out Edwin R. Littmann, "Ancient Mesoamerican Mortars, Plasters, and Stuccos: Palenque, Chiapas, *American Antiquity* 25, no. 2 (October 1959): 264–66, and the same author's "Ancient Mesoamerican Mortars, Plasters, and Stuccos: The Use of Bark Extracts in Lime Plasters," *American Antiquity* 25, no. 4 (April 1960): 593–97.

2. Lucia A. Ciapponi, "Fra Giocondo da Verona and His Edition of Vitruvius," *Journal of the Warburg and Courtauld Institutes* 47 (1984): 72–90.

3. Joerg Garms, "Projects for the Pont Neuf and Place Dauphine in the First Half of the Eighteenth Century," *Journal of the Society of Architectural Historians* 26, no. 2 (May 1967): 102–13.

CHAPTER 5: THE DEVELOPMENT OF MODERN CONCRETE

1. Major A. J. Francis, *The Cement Industry: 1796–1914: A History* (London: David & Charles, 1977), pp. 19–20; see also H. D. (author's initials), "Andernach Trass," *Minutes of the Proceedings of the Institution of Civil Engineers*, vol. 39, part 1, ed. James Forrest (1874–1875): 313–16.

2. R. J. M. Sutherland, Dawn Humm, and Mike Chrimes, *Historic Concrete: Background to Appraisal* (London: Thomas Telford, 2001), p. 118.

3. Ibid.

4. *Wikipedia*, "Horsepower," http://en.wikipedia.org/wiki/Horsepower (accessed June 7, 2011).

5. John Smeaton, *Reports of the Late John Smeaton*, 2 vols. (London: M. Taylor, 1832). Note: these two large volumes of John Smeaton's letters, essays, and articles collected after his death for publication lead one to suspect that he must have spent much of what little spare time he had between his major building projects and extensive experiments writing.

6. Henry Reid, *The Science and Art of the Manufacture of Portland Cement* (London: E. & F. N. Spon, 1877), p. xvii; see also *Wikipedia*, "Henry Winstanley," http://en.wikipedia.org/wiki/Henry_Winstanley (accessed June 7, 2011). See also *Smeaton's Tower*, LAL Torbay StopPress, November 2009, http://lalschools.com/uploads/media/Smeaton-Tower.pdf (accessed June 7, 2011).

7. Ibid.

8. Uriah Cummings, *American Cements* (Boston: Rogers & Manson, 1898), p. 16.

9. Reid, *The Science and Art of the Manufacture of Portland Cement*, p. xvi.

10. John Smeaton, *A Narrative of the Building and a Description of the Construction of the Eddystone Lighthouse with Stone* (London: 1791). Note: This was reprinted many times, and a poor-quality text version of the 1881 edition is available at US Archive: http://ia600208.us.archive.org/3/items/bookofbritishtop00andeuoft/bookofbritishtop00andeuoft_djvu.txt.

11. Bryan Higgins, *Experiments and Observations Made with the View of Improving the Art of Composing and Applying Calcareous Cements, and of Preparing Quicklime. Theory of These, and Specification of the Author's Cheap and Durable Cement for Building, Incrustation, or Stuccoing, and Artificial Stone* (London: T. Cadell, 1780). See also Francis, *The Cement Industry*, p. 24.

12. Francis, *The Cement Industry*, p. 26.

13. Ibid., pp. 30–31.

14. Ibid., p. 27.

15. Ibid., p. 28.

16. Ibid., p. 29.

17. Ibid., pp. 30–31.

18. Ibid., pp. 35–36, 43.

19. Ibid., p. 32.

20. Ibid., pp. 20–24.

21. Ibid., pp. 43–44.

22. Ibid., p. 43.

23. Nicky Smith, "Pre-Industrial Lime Kilns, Introduction to Heritage Assets," *English Heritage* (May 2011): 2–5. See also Francis, *The Cement Industry*, pp. 36, 71.

24. Richard Beamish, *Memoir of the Life of Sir Marc Isambard Brunel* (London: Longman, Green, Longman, and Roberts, 1862), pp. 1–2.

25. Ibid., pp. 8–9.

26. Ibid., p. 240.

27. Patrick Beaver, *A History of Tunnels* (Secaucus, NJ: Citadel Press, 1973), p. 42. See also Ted Rohrlich, "Metro Tunnel Project: Metro Rail Digs: Risky Business," *Los Angeles Times*, March 22, 1987, http://articles.latimes.com/1987-03-22/news/mn-15118_1_metro-rail-design/2 (accessed June 8, 2011).

28. Beamish, *Memoir of the Life of Sir Marc Isambard Brunel*, pp. 240–41.

29. Beaver, *History of Tunnels*, p. 42.

30. Ibid.

31. Ibid. (footnote).

32. Ibid., p. 43.

33. "A Great Bore Made Useful," in *Chambers Journal* (W. & R. Chambers, 1866): 405. Note: Beamish's biography tactfully skips over probable reasons, though the article cited here (authorship not provided) claims that Marc Brunel "became ill in consequence of the intense mental and bodily labour and excitement during this anxious period."

34. Beamish, *Memoir of the Life of Sir Marc Isambard Brunel*, pp. 260–61.

35. John Timbs, *Curiosities of London* (London: David Bogue, 1855), pp. 712–13.

36. Kendall F. Haven, *100 Greatest Science Inventions* (Greenwood, CO: Libraries Unlimited, 2006), p. 25.

37. *Wikipedia*, "Joseph Aspdin," http://en.wikipedia.org/wiki/Joseph_Aspdin (accessed August 25, 2011).

38. Francis, *The Cement Industry*, pp. 76–78.

39. Ibid., pp. 78–79.

40. A. C. Smeaton, *The Builder's Pocket Manual: Containing the Elements of Building, Surveying, & Architecture* (London: M. Taylor, 1837), p. 21.

41. Francis, *The Cement Industry*, p. 96.

42. Ibid., pp. 82–83.

43. Ibid., p. 115.

44. Ibid., pp. 83–84.

45. Ibid., p. 115.

46. Ibid., p. 113.

47. Ibid., p. 149.

48. Ibid., pp. 149–50.

49. Robert W. Lesley, *History of the Portland Cement Industry in the United States* (Chicago: International Trade Press, 1924), p. 36.

50. Francis, *The Cement Industry*, p. 124.

51. G. Haegermann, "Dokumente zur Entstehungsgeschichte des Portland-Cements," *ZKG—Zement Kalk Gips* 1, no. 1 (January 23, 1970): 10. (". . . indessen verstand der Mann sein Fach nicht.")

52. Ibid., p. 11: ("Nachdem das zwischen Herrn Ed. Fewer in Lägerdorf und mir bestehende Societätsverhältnis aufgehoben worden ist, liegt mir nicht länger eine Verantwortlichkeit für die Qualität and Güte desjenigen Cements ob, welcher von dem Herrn Ed. Fewer fortan in Lägerdorf fabriziert werden wird. Ebensowenig ist es dem letzteren fernerhin gestattet, meinen Namen für seine Fabrikmarke zu benutzen. Itzhoe, d. 9 Juli 1863.")

53. Ibid.

54. *Wikipedia*, "100 Greatest Britons," http://en.wikipedia.org/wiki/100_Greatest_Britons (accessed August 25, 2011).

55. Francis, *The Cement Industry*, p. 77.

56. Ibid., pp. 77–78.

57. Ibid., pp. 78–79.

58. Hermione Hobhouse, "The West India Docks," British History Online, http://www.british-history.ac.uk/report.aspx?compid=46495 (accessed June 9, 2011).

59. Francis, *The Cement Industry*, p. 96.

60. Sally Festing, "Great Credit upon the Ingenuity and Taste of Mr. Pulham," *Garden History* 16, no. 1 (Spring 1988): 90–102.

61. Francis, *The Cement Industry*, p. 93.

CHAPTER 6: REFINEMENTS, REINFORCEMENT, AND PROLIFERATION

1. Major A. J. Francis, *The Cement Industry: 1796–1914: A History* (London: David & Charles, 1977), p. 253.

2. Ibid., pp. 258–66.

3. David Merlin Jones, "'Rock Solid?' An Investigation into the British Cement Industry," *Civitas* (November 2010): 1–2.

4. Francis, *The Cement Industry*, pp. 231–34, 254–55.

5. Ibid., pp. 231–35.

6. Albert Wells Buel and Charles Shattuck Hill, *Reinforced Concrete* (New York: Engineering News Publishing Company, 1904), p. 205. See also Terri Meyer Boake, "Reinforced Concrete," in *Building Construction*, chap. 6, p. 3, online essay at Waterloo University's Department of Architecture website, http://www.architecture.uwaterloo.ca/faculty_projects/terri/images/course_pdf/172-ch6 .pdf (accessed June 8, 2011).

7. G. A. Wayss, *Das System Monier, Eisengerippe mit Cementumhüllung, in seiner Anwendung auf das gesammte Bauwesen* (Berlin, 1887).

8. *Wikipedia*, "François Coignet," http://127.0.0.1:4664/search?q=monier &flags=68&s=_0O5E4_ZZu0-wgHZ72GWqKBxdxg (accessed June 8, 2011).

9. Mary S. J. Gani, *Cement and Concrete* (London: Spon Press, 1997), p. 8.

10. Francis, *The Cement Industry*, p. 127.

11. Arthur S. Masten, *The History of Cohoes, From Its Earliest Settlements to the Present Time* (New York: J. Munsell, 1877): pp. 265–67.

12. Sara E. Wermiel, "California Concrete, 1876–1906: Jackson, Percy, and the Beginnings of Reinforced Concrete Construction in the United States," *Proceedings of the Third International Congress on Construction History* (May 2009): 1–2.

13. Ernest Leslie Ransome and Alexis Saurbrey, *Reinforced Concrete Buildings* (New York: McGraw-Hill, 1912), pp. 1–2.

14. Oscar Wilde, *The Picture of Dorian Gray* (London: Bibliolis Books, 2011 [ed. of 1897 original]), p. 222.

15. Rudyard Kipling, "Great Quotes," http://www.greatquotes.com/quotes/author/Rudyard/Kipling/keyword/san+francisco (accessed June 8, 2011).

16. Hinton Rowan Helper, *The Land of Gold* (Baltimore, MD: Sherwood & Co., 1855), p. 58.

17. Ransome and Saurbrey, *Reinforced Concrete Buildings*, p. 2.

18. Ibid., p. 3.

19. Wermiel, "California Concrete," pp. 2–4.

20. Ransome and Saurbrey, *Reinforced Concrete Buildings*, p. 2.

21. Raymond H. Clary, *The Making of Golden Gate Park: The Early Years: 1865–1906* (San Francisco: Don't Call It Frisco Press, 1988), pp. 100, 103, 158. See also "Sweeney Observatory," http://www.outsidelands.org/sweeney-observatory.php (accessed June 8, 2011).

22. The many accolades given to reinforced concrete's resiliency to the earthquake and fire will be covered in depth in chap. 9 of this book.

CHAPTER 7: THE WIZARD AND THE ARCHITECT

1. Michael Peterson, "Thomas Edison's Concrete Houses," *Invention & Technology Magazine* 11, no. 3 (Winter 1996).

2. Ibid.

3. Ibid.

4. "Edison Now Making Concrete Furniture," *New York Times*, December 9, 1911.

5. Ibid.

6. Ford Richardson Bryan, *Beyond the Model T: The Other Ventures of Henry Ford* (Detroit: Wayne State University Press, 1997), p. 76.

7. Ibid., pp. 77–78.

8. Roger Friedland and Harold Zellman, *The Fellowship: The Untold Story of Frank Lloyd Wright and the Taliesin Fellowship* (New York: Harper Perennial, 2006), pp. 2–11.

9. Meryle Secrest, *Frank Lloyd Wright* (New York: Alfred A. Knopf, 1992), pp. 59–61.

10. Ibid., pp. 419–20.

11. Jackie Craven, "Louis Sullivan, America's First Modern Architect," About.com, http://architecture.about.com/od/greatarchitects/p/sullivan.htm (accessed June 8, 2011).

12. Edward H. Madden, "Transcendental Influences on Louis H. Sullivan and Frank Lloyd Wright," *Transactions of the Charles S. Pierce Society* 32, no. 2 (Spring 1995): 286–321. See also Mark Mumford, "Form Follows Nature: The Origins of American Organic Architecture," *Journal of Architectural Education* 42, no. 3 (Spring 1989): 26–28.

13. Frank Lloyd Wright, *Genius and the Mobocracy* (New York: Duell, Sloan and Pearce, 1949).

14. Secrest, *Frank Lloyd Wright*, pp. 192–93, 202–209, 212–13.

15. Frank Lloyd Wright, *Frank Lloyd Wright: Essential Texts*, ed. Robert Twombly (New York: W. W. Norton, 2009), p. 86.

16. Judith A. Sebesta, "Spectacular Failure: Frank Lloyd Wright's Midway Gardens and Chicago Entertainment," *Theater Journal* 53, no. 2 (May 2001): 291–309.

17. *Wikipedia*, "Parthenon (Nashville)," http://en.wikipedia.org/wiki/Parthenon_%28Nashville%29 (accessed June 8, 2011).

18. Secrest, *Frank Lloyd Wright*, pp. 217–24.

19. William Han Kan Lee, "International Handbook of Earthquake & Engineering Seismology," *International Association of Seismology and Physics of the Earth's Interior—Committee of the International Association for Earthquake Engineering* 1, no. 1 (Orlando, FL: Academic Press, 2002): 39. See also David J. Wald, "Variable-Slip Rupture Model of the Great 1923 Kanto, Japan Earthquake: Geodetic and Body-Waveform Analysis," *Bulletin of the Seismological Society of America* 85 (1995): 163.

20. Bryce Walker, *Earthquake* (New York: Time-Life Books, 1987), p. 153.

21. Ibid., p. 152.

CHAPTER 8: THE CONCRETIZATION OF THE WORLD

1. Colbert Roberto A. Reid, "The Panama Canal Death Tolls," December 17, 2008, Silver People Foundation, http://thesilverpeoplehcritage.wordpress.com/2008/12/17/the-panama-canal-death-tolls/ (accessed June 10, 2011).

2. Raul Berreneche, "Mies in Berlin," *MoMA* 4, no. 4 (May, 2010): 2.

3. Roger Friedland and Harold Zellman, *The Fellowship: The Untold Story of Frank Lloyd Wright and the Taliesin Fellowship* (New York: Harper Perennial, 2006), pp. 405–406.

4. Meryle Secrest, *Frank Lloyd Wright* (New York: Alfred A. Knopf, 1992), p. 442.

5. Ibid., p. 443.

6. Friedland and Zellman, *The Fellowship*, pp. 310–13.

7. Ibid., p. 366; Secrest, *Frank Lloyd Wright*, pp. 486–87.

8. Friedman and Zellman, *The Fellowship*, p. 374. See also *Wikipedia*, "Hilla von Rebay," http://en.wikipedia.org/wiki/Hilla_von_Rebay (accessed June 10, 2011).

9. "The Gordon Strong Automobile Objective, Sugarloaf Mountain, Maryland, 1924–1925," in *Frank Lloyd Wright: Designs for an American Landscape, 1922–1932* (Library of Congress), http://www.loc.gov/exhibits/flw/flw02.html (accessed June 10, 2011).

10. Secrest, *Frank Lloyd Wright*, p. 551.

11. John Yeomans, *The Other Taj Mahal: What Happened to the Sydney Opera House* (London: Longmans, Green and Company, 1968), p. 63.

12. Ibid., pp. 22–23.

13. Ibid., pp. 12–13.

14. Nevill Drury, *The Art of Rosaleen Norton, with Poems by Gavin Greenlees* (London: Walter Glover, 1982).

15. *Wikipedia*, "Rosaleen Norton," http://en.wikipedia.org/wiki/Rosaleen _Norton (accessed June 10, 2011).

16. Yeomans, *The Other Taj Mahal*, p. 11.

17. Ibid., p. 34.

18. Ibid., p. 26.

19. Ibid.

20. Ibid., p. 27.

21. Robert Hughes, *The Shock of the New* (New York: Knopf Doubleday, 1991).

22. *Wikipedia*, "Robert Askin," http://en.wikipedia.org/wiki/Robert_Askin (accessed June 10, 2011).

23. Yeomans, *The Other Taj Mahal*, pp. 130–32.

24. Eric Ellis, "Utzon Speaks," *Good Weekend*, October 31, 1992. Copy can be found at http://www.ericellis.com/utzon.htm (accessed May 1, 2011).

25. Yeomans, *The Other Taj Mahal*, p. 148.

26. Ibid.

27. Ibid., p. 149.

CHAPTER 9: THE BAD NEWS

1. Myron H. Lewis and Albert H. Chandler, *Popular Hand Book for Cement and Concrete Users* (New York: Norman W. Henley Publishing Company, 1911), pp. 166–67; see also Charles W. Bates, "Some Advantages of Reinforced Concrete Construction," *Ohio Architect Engineer and Builder* 20, no. 5 (November 1912): 57–60.

2. Frederick W. Taylor and Sanford E. Thompson, *A Treatise on Concrete, Plain and Reinforced* (New York: John Wiley & Sons, 1916), p. 11. See also Charles Pallisher, *Practical Concrete Block Making* (New York: Industrial Publishing Company, 1908), p. 8.

3. "Fire and Quake Proof Building," *New York Tribune*, January 6, 1907, p. 6; see also *Cement Age* 3, no. 1 (June 1906): 111.

4. *Reinforced Concrete in Factory Construction* (New York: Atlas Portland Cement Company, 1907), pp. 47–58.

5. Joseph Freitag, *Fire Prevention and Fire Protection as Applied to Building Construction* (New York: John Wiley & Sons, 1912), pp. 209–210.

6. Emil Mörsch, *Der Eisenbetonbau—Seine Theorie und Anwendung* (Stuttgart: Konrad Wittwer, 1906), p. 1.

7. Gladys Hansen and Emmett Condon, *Denial of Disaster* (San Francisco: Cameron and Company, 1989).

8. Bobby Caina Calvin, "San Francisco Revises Death Toll for 1906 Earthquake. Tally Could Exceed 3,400," *Boston Globe*, February 27, 2005, http://www.boston.com/news/nation/articles/2005/02/27/san_francisco_revises_death_toll_for_1906_earthquake/ (accessed June 12, 2005).

9. *The San Francisco Earthquake and Fire of April 18, 1906 and Their Effects on Structures and Structural Materials, Reports by Grove Karl Gilbert, Richard Lewis Humphrey, John Stephen Sewell, and Frank Soulé* (Washington, DC: Department of the Interior, United States Geological Survey, 1907): 29–30.

10. Ibid., p. 109.

11. Ibid., p. 33.

12. Ibid., p. 125.

13. Ibid., p. 126.

14. Ibid.

15. Ibid., p. 130.

16. Ibid., p. 147.

17. Ibid., p. 150.

18. David Starr Jordan, John Casper Branner, Charles Derleth Jr., Grove Karl Gilbert, F. Omori, Stephen Taber, Harold W. Fairbanks, Mary Austin, *The California Earthquake of 1906* (San Francisco: A. M. Robertson, 1907), p. 114.

19. Ibid., p. 131.

20. Ibid., p. 133.

21. Ibid., p. 134.

22. A. L. A. Himmelwright, *The San Francisco Earthquake and Fire* (New York: Roebling Construction Company, 1906), p. 226; see also *Burnt Clay Products in Earthquake and Fire* (Los Angeles: Brick Construction Company, 1907), pp. 46–47.

23. F. W. Fitzpatrick, "The San Francisco Calamity," *Fireproof Magazine* 9, no. 1 (July 1906): 45; see also *Burnt Clay Products in Earthquake and Fire*, p. 59.

24. *The San Francisco Earthquake and Fire of April 18, 1906*, p. 109.

25. *Burnt Clay Products in Earthquake and Fire*, p. 58.

26. *The San Francisco Earthquake and Fire of April 18, 1906*, p. 33.

27. Mark A. Wilson, *Julia Morgan, Architect of Beauty* (Layton, UT: Gibbs Smith, 2007), p. 57.

28. "The Cement Market," *Cement and Engineering News*, June 1906, p. 134.

29. John R. Freeman, *Earthquake Damage and Earthquake Insurance* (McGraw-Hill, 1932).

30. Ibid., p. 321.

31. Author's interview with Robert Nason, September 21, 1996.

32. Stephen Tobriner, "An EERI Reconnaissance Report: Damage to San Francisco in the 1906 Earthquake: A Centennial Perspective," *Earthquake Spectra* 22, sect. 2 (April 2006): 22.

33. *Burnt Clay Products in Earthquake and Fire*, p. 4.

34. Fitzpatrick, "The San Francisco Calamity," p. 27.

35. Ibid., p. 29.

36. Starr Jordan et al., *The California Earthquake of 1906*, p. 134.

37. Author's interview with Robert Nason, September 21, 1996.

38. "You Build Not for a Day, But for 1,000 Years," *Cement and Engineering News* 17, no. 12 (December 1905): 40.

39. Richard L. Humphrey, "Concrete and Seawater," *Concrete-Cement Age* 7, no. 6 (December 1915): 205.

40. Richard L. Humphrey, "Address of the President," *Proceedings of the Tenth Annual Convention of the American Concrete Institute* 10 (1914; published by the Institute, 1917): 52.

41. Ibid.

42. W. G. Gregory, *Concrete and Building Work* (Hong Kong: Hong Kong University Press, 1959).

43. Ernest Leslie Ransome and Alexis Saubrey, *Reinforced Concrete Buildings* (New York: McGraw-Hill, 1912), pp. 10, 12.

44. Philip H. Perkins, *Concrete Structures: Repair, Waterproofing and Protection* (Elsevier Science, 1977), p. 33; see also Gregory, *Concrete and Building Work*, p. 37.

45. P. Kumar Mehta and Richard W. Burrows, "Building Durable Structures in the 21st Century," *Concrete International* 23, no. 3 (March 2001): 59.

46. Ibid., pp. 59–60.

47. Ibid., p. 60.

48. Ibid., p. 61.

49. "Fact Sheet: Bridges," *Infrastructure Report Card* (compiled 2006), American Society of Civil Engineers, http://www.infrastructurereportcard.org/factsheet/bridges (accessed June 13, 2011).

50. "Fact Sheet: Roads," *Infrastructure Report Card*, American Society of Civil Engineers, http://www.infrastructurereportcard.org/fact-sheet/roads (accessed June 13, 2011).

51. "Fact Sheet: Waste Water," *Infrastructure Report Card*, American Society of Civil Engineers, http://www.infrastructurereportcard.org/fact-sheet/wastewater (accessed June 13, 2011).

52. "Report Cards," American Society of Civil Engineers, http://www.infrastructurereportcard.org/report-cards (accessed June 13, 2011).

53. Deborah N. Huntzinger and Thomas D. Eatmon, "A Life-Cycle Assessment of Portland Cement Manufacturing: Comparing the Traditional Process with Alternative Technologies," *Journal of Cleaner Production* 17 (2009): 668.

CHAPTER 10: THE GOOD NEWS

1. Erica von Tassel, "Alternate Cementitious Materials," course material, Pennsylvania State College of Engineering, http://www.engr.psu.edu/ce/courses/ce584/concrete/library/materials/Altmaterials/Altmaterialsmain.htm (accessed June 13, 2011).

2. Gabriel Nelson, "EPA Agrees to Rethink Parts of New Cement Kiln Rules," *New York Times*, May 13, 2011, http://www.nytimes.com/gwire/2011/05/13/13greenwire-epa-agrees-to-rethink-parts-of-new-cement-kiln-66661.html (accessed June 13, 2011); see also "Cement Industry Challenges Pollution Cuts That Would Save Lives and Money," *Aggregate Research*, November 9, 2010, http://www.aggregateresearch.com/articles/20717/Cement-industry-challenges-pollution-cuts-that-would-save-lives—money.aspx (accessed June 13, 2011).

3. P. Kumar Mehta, "High-Performance, High-Volume Fly Ash Concrete for Sustainable Development," *International Workshop for Sustainable Development and Concrete Technology* 5, Institute for Transportation, Iowa State University, http://www.intrans.iastate.edu/pubs/sustainable/mehtasustainable.pdf (accessed June 13, 2011).

4. *Wikipedia*, "Cathodic Protection," http://en.wikipedia.org/wiki/ Cathodic_protection (accessed June 13, 2011).

5. "Galvanic Corrosion, Bimetallic Corrosion," Corrosionist, http://www .corrosionist.com/Galvanic_Corrosion.htm (accessed June 13, 2011).

6. A. Franchi and P. Crespi, "Some Recent Results of Tests on Steel Rebars," Vanadium International Technical Committee, http://www.vanitec.org/pdfs/ b92231bbe939ca130af90cbe324b524c.pdf (accessed June 14, 2010).

7. "Cathodic Protection," National Physical Laboratory (UK), http://www .npl.co.uk/upload/pdf/cathodic_protection.pdf (accessed June 13, 2011).

8. Rengaswamy Srinivasan, Periya Gopalam et al., "Design of Cathodic Protection of Rebars in Concrete Structure: An Electrochemical Engineering Approach," *Johns Hopkins APL Technical Digest* 17, no. 4 (1996): 362.

9. "FRP Rebar," Emerging Construction Technologies, Division of Construction Engineering and Management, Purdue University, http://rebar.ecn .purdue.edu/ect/links/technologies/civil/frprebar.aspx (accessed June 14, 2011).

10. Doug Gremel and Ryan Koch, "Bridge Construction," *Roads and Bridges*, January 2009, p. 31; see also R. V. Balendran et al., "Application of FRP Bars as Reinforcement in Civil Engineering Structures," *Structural Survey* 20, no. 2 (2002): 62–72.

11. Doug Gremel, "Commercialization of Glass Fiber Reinforced Polymer Rebar," SEAOH Convention, Concrete for the New Millennium, July 1999, http://www.vectorgroup.com/SEAO%20Hawaii%20Aug%2799.pdf (accessed June 14, 2011).

12. David Stein, "Aluminum Bronze Alloy for Corrosion Resistant Rebar," *IDEA Project* (Innovations Deserving Exploratory Analysis Project), Transportation Research Board, National Research Council, NCHRP-ID019 (December 1996): 1–13.

13. *Wikipedia*, "Bronze," http://en.wikipedia.org/wiki/Bronze (accessed June 14, 2011). Article discusses classic bronze; for aluminum bronze alloys, see n. 12 above).

14. "History of the Hockley Viaduct," Friends of the Hockley Viaduct,

http://www.hockleyviaduct.hampshire.org.uk/index_files/Page298.htm (accessed June 14, 2011).

15. Kathy Riggs Larsen, "Pentagon Memorial: Designed with Corrosion in Mind. Architects, Contractor Plan for Structures to Last 100 Years," *Materials Performance* (November 2008): 30–34.

EPILOGUE

1. Christina Vasquez "'Something Inside Me Said Run': Surfside Survivor Describes Escaping as Building Came Down." *Local 10 News*, WPLG, June 29, 2021. https://miami.cbslocal.com/2021/07/15/911-calls-surfside-condo-collapse. Accessed last on August 10, 2021. See also: Melissa Mahtani, Melissa Macaya, Mike Hayes, Veronica Rocha, Fernando Alfonso III, "The Latest on the Partial Building Collapse Near Miami," CNN, June 30, 2021, https://www.cnn.com/us/live-news/miami-florida-building-collapse-06-30-21/h_b6d81cbe1f1752b2c1ec202b423e6320. Last accessed August 17, 2022.

2. Sophia Ankel, "A Woman Standing on Her Balcony Was on the Phone to Her Husband as the Miami Condo Began to Collapse and Described the Unfolding Disaster before the Line Went Dead," *The Insider* (London) June 26, 2021, https://www.insider.com/woman-was-on-phone-to-husband-as-miami-block-collapsed," accessed July 11, 2021 (Note: Some early accounts have Ms. Stratton's mistakenly spelled as *Straton*. Her remains were recovered on July 12, 2021.)

3. Jessica Vallejo, "'There's People in the Rubble Yelling' 991 Calls Released from Surfside Condo Collapse," WFOR. https://miami.cbslocal.com/2021/07/15/911-calls-surfside-condo-collapse. Last accessed August 10, 2021. See also: Michelle Marchante, Martin Vassolo, "A Woman Called for Help under the Rubble of the Surfside Collapse, Fire Chief Says," *Miami Herald*, July 7, 2021. https://www.miamiherald.com/news/local/community/miami-dade/miami-beach/article252499543.html. Last accessed August 10, 2021

4. "Boy Pulled from Rubble after Surfside Collapse." *Miami Herald*. June 24, 2021. https://www.miamiherald.com/news/local/community/miami-dade/article252326838.html. Last accessed August 10, 2021.

5. Fiorella Terenze. "Prospective Buyer Tours Basement Parking Garage of Champlain Tower South," *Miami Herald*, July 8, 2021 (Video was taken on *June*

17, not July 17 as recorded in article. What is especially troubling is that stalactites appear to be forming on the ceiling. This means that enough water had penetrated the concrete to begin dissolving its calcium carbonate, the chemical which gives concrete its hardness. See also: Sarah Blaskey, Aaron Leibowitz: "Two Days before Condo Collapse, a Pool Contractor Photographed This Damage in Garage," *Miami Herald*, July 06, 2021. https://www.miamiherald.com/news/local/community/miami-dade/miami-beach/article252421658.html. Last accessed August 10, 2021.

6. Alex Harris, "Sea Rise under Scrutiny in Condo Collapse: Corrosion Likely, But No Sign of Sinkhole," *Miami Herald*, August 3, 2021, https://www.miamiherald.com/news/local /article252877743.html. Last accessed August 10, 2021.

7. Fred Grimm, "Surfside Condo Built in Era Tainted in Corruption," *South Florida Sun Sentinel*, July 2, 2021. https://www.sun-sentinel.com/opinion/commentary/fl-op-com-grimm-south-florida-building-inspectors-corruption-20210702-6u6kamfs4vcefatc6iqbidns6y-story.html Last accessed August 20, 2021.

8. Sarah Blaskey, Aaron Leibowitz, Ben Conarck, Samantha J. Gross: "Before Role in Surfside Condo That Fell, Engineer Had Hand in Another Building Mess," *Miami Herald*, July 17, 2021. See also: Brian Bandell, "Armando Codina Plans Luxury Building in Coral Gables," *South Florida Business Journal*, April 1, 2021. https://www.bizjournals.com/southflorida/news/2021/04/01/codina-partners-plans-luxury-residential.html, Last accessed: August 16, 2021

9. Author has personally seen many building sites in which *rusted rebar* was used in construction work. This gives the corrosion a head-start once the opportunity presents itself. See also: Eric J. Schindelholz, et al. *Effect of Relative Humidity on Corrosion of Steel under Acidified Artificial Seawater Particles*. Office of Scientific and Technical Information (OSTI), 2014.

10. FEMA *Summary Report on Building Performance / Hurricane Katrina, 2005*. Federal Emergency Management Authority, 2006. Page numbers not given, find at 4.1.3

11. Ibid. (At the very least, such government reports should recommend the regular inspection of these once flooded buildings.)

12. *Champlain Towers South Condominium Structural Field Survey Report – MC Job#18217*. Morabito's most quoted remarks from his report are on Page 7: "Failure to replace the waterproofing in the near future will cause the extent of the

concrete deterioration to expand exponentially." & "The failed waterproofing is causing major structural damage to the concrete structural slab below these areas." These observations would not convey the severity of the problem to a member of the condominium board who was not a structural engineer. We also have no idea—at present—whether Mr. Morabito verbally discussed the logical consequences of these problems with any member of the Champlain Towers South Condominium Board.

13. Schumacher, M. M., *The Seawater Corrosion Handbook*, U.S. Department of Energy, Office of Scientific and Technical Information (OSTI). 1979. See also: Schwartz, Dave P.E., *A Corrosive Environment*, Aquatics International, September 1, 2011. https://www.aquaticsintl.com/facilities/maintenance/a-corrosive-environment_o. Last accessed August 10, 2021 (Author also demonstrates the corrosion effects of chlorinated water on "stainless steel.")

14. Sarah Blaskey, Aaron Leibowitz, "Video Appears to Show Rubble in Garage, Water Gushing in Just Before Surfside Condo Collapse." *Miami Herald*, July 7, 2021. https://www.miamiherald.com/news/local/community/miami-dade/miami-beach/article252475248.html Last accessed, August 10, 2021.

15. Aaron Leibowitz, Mary Ellen Klas, Sarah Blaskey, "Surfside Official Was Sent Disturbing Report. He Told Board Condo Was 'in Good Shape'," *Miami Herald*, June 27, 2021, updated July 07, 2021. Note: Despite above headline, the quote attributed to the Surfside building official, Rosendo "Ross" Prieto, was "very good shape". Last accessed August 11, 2021.

16. The cost of repairing a corrosion patch varies greatly from one region to another. Another factor is competency. (The more experienced contractor may even charge less, probably because he knows how to do the job correctly, how long it will take, what tools to bring with him.) A cost-saving alternative is to have the superintendent—if the building has one—to learn how to do minor corrosion repairs. It's not rocket science and can be quickly learned. However, larger corrosion repairs should be left to veteran corrosion experts.

17. David Fischer, Terry Spencer, "Search for Bodies Concludes at Florida Condo Collapse Site," AP New Service, July 23, 2021, https://apnews.com/article/technology-florida-surfside-building-collapse-8c92c024cab1511d130898 f02cf3489a–Last accessed August 17, 2021

18. *Wikipedia*. "1974 Miami DEA Building Collapse," https://en.wikipedia.org/wiki/1974_Miami_DEA_building_collapse. See also citation 19 below.

19. Noel King, "A Structural Engineer Explains How the Florida Condo Collapse Will Be Investigated," *Morning Edition*, NPR, June 29, 2014. https://www.npr.org/sections/live-updates-miami-area-condo-collapse/2021/06/29/1011201090/questions-persist-over-what-caused-champlain-towers-south-to-collapse. John Pistorino also comments on the DEA building collapse and the pushback against the 40-year-recertification building code change.

20. Devoun Cetoute, Rob Wile: "North Miami Beach Orders 10-Story Condo Evacuated after Report Declares It Unsafe," *Miami Herald*, July 7, 2021– https://www.miamiherald.com/news/local/community/miami-dade/north-miami/article252544473.html. Last accessed: August 11, 2021. See also: Devoun Cetoute, "Crestview Towers Violated 39 Safety Codes, City Says. It Will Remain Closed to Residents," *Miami Herald*. July 16, 2021. https://www.miamiherald.com/news/local/community/miami-dade/north-miami/article252842848.html. See also: *Wikipedia*. "Crestview Towers." https://en.wikipedia.org/wiki/Crestview_Towers

21. Joey Flechas, "Safe or Unsafe? Residents at Miami Beach Condo Facing Unsafe Structure Violation," *Miami Herald*, July 15, 2021. Last accessed: July 29, 2021. Note: Photos are revealing of sorry state of Port Royale. It had already been cited for "structure deterioration" and "concrete spalling." (The terms "concrete spalling" and "concrete delamination" are just other forms of describing concrete damage caused by rebar corrosion.)

22. https://theportroyale.com Last accessed August 11, 2021

23. Joey Flechas, "Safe or Unsafe" (see 21 above), *Miami Herald*, July 15, 2021

24. Justin Gillis, "Sea Levels Rising So Fast in Some States That an Expert Thought the Data Were Wrong," *New York Times*, August 9, 2017. See also: "New Study Warns That Sea Levels Will Rise Faster Than Expected," *Monga Bay*, https://news.mongabay.com/2021/02/new-study-warns-that-sea-levels-will-rise-faster-than-expected/ 2, February 2021. Both last accessed on August 17, 2021

25. Mario Ariza, Alex Harris, "Miles of Florida Roads Face 'Major Problem' from Sea Rise. Is State Moving Fast Enough?," *Florida Climate Reporting Network*, March 19, 2021. (The article was a joint effort of *South Florida Sun Sentinel* and *Miami Herald*.)

26. Ibid.

27. *NL Times* (*Netherlands Times*) Dec. 5, 2013, https://nltimes.nl/2013/12/05/stronger-1953-storm–Last accessed August 15, 2021. See also: Mathew P.

Wadey, et al. *A comparison of the 31 January–1 February 1953 and 5–6 December 2013 Coastal Flood Events around the UK,* Frontiers of Marine Science, November 6, 2015. https://www.frontiersin.org/articles/10.3389/fmars.2015.00084/full Last accessed: August 15, 2021. Note: Some of the data obtained from the 1953 storm were muddled by overwhelmed and/or inadequate instruments. The most precise figures were obtained for the extent of the surge flooding, which could be referenced by a number of existing landmarks to the highest surge stains or water damage.

28. Billy Fleming, "The Dutch Can't Save Us from Rising Seas," *Bloomberg,* October 17, 2018. https://www.bloomberg.com/news/articles/2018-10-17/the -dutch-approach-to-rising-seas-is-not-a-universal-fix–Last accessed: August 18, 2021.

29. Ibid.

30. Stephen P. Leatherman, et al. *Vanishing Lands: Sea Level, Society and Chesapeake Bay.* Chesapeake Bay Field Office, Fish and Wildlife Service, Annapolis, Maryland. 1995. https://pubs.er.usgs.gov/publication/95388. Last accessed August 5, 2021

31. The Netherlands model might well work for a small area, like that encompassed by Washington, D.C.

32. Nathan Kensinger, "Here's How NYC Transit System Is Prepping For Sea Level Rise—And Why It May Not Be Enough," *Gothamist,* April 22, 2021. https://gothamist.com/news/heres-how-nyc-transit-system-is-prepping-for-sea -level-riseand-why-it-may-not-be-enough Last accessed: August 20, 2021

33. *Sea Level Rise Vulnerability and Consequences Assessment,* Chapter 12, Pages 242–301, February 2020, San Francisco Planning Commission. https:// sfplanning.org/sea-level-rise-action-plan , Last accessed: August 19, 2020

34. The ride has recently been resurrected as *Finding Nemo*-based attraction. Many, if not most, newer Disneyland or Disney World rides serve to promote their film franchises, many of which, are based on older theme park rides (*Pirates of Caribbean, Jungle Cruise,* etc.) Odd. Sad.

35. Taisha Walker, "Structural Engineer Says No Cause for Houston High-Rise Dwellers to Panic Following Florida Building Collapse," KRPC. *Click2Houston.com,* June 28, 2021; updated June 29, 2021. Last accessed: August 19, 2021. Personal aside: Seeing this headline after reading about Houston's history of flooding raised a number of unpleasant emotions.

36. Ibid.

37. Ibid.

38. *Harris County's Flooding History*, Harris County Flood Control District. https://www.hcfcd.org/About/Harris-Countys-Flooding-History. Last accessed: August 16, 2021

INDEX

Photos and figures indicated by *italicized* page numbers.